Series editors

Prof. Bruno Siciliano
Dipartimento di Ingegneria Elettrica
e Tecnologie dell'Informazione
Università degli Studi di Napoli
Federico II
Via Claudio 21, 80125 Napoli
Italy
E-mail: siciliano@unina.it

Prof. Oussama Khatib
Artificial Intelligence Laboratory
Department of Computer Science
Stanford University
Stanford, CA 94305-9010
USA
E-mail: khatib@cs.stanford.edu

More information about this series at http://www.springer.com/series/5208

Gentiane Venture · Jean-Paul Laumond
Bruno Watier
Editors

Biomechanics
of Anthropomorphic Systems

 Springer

Editors
Gentiane Venture
Tokyo, Japan

Bruno Watier
Toulouse, France

Jean-Paul Laumond
LAAS-CNRS
Toulouse, France

ISSN 1610-7438   ISSN 1610-742X   (electronic)
Springer Tracts in Advanced Robotics
ISBN 978-3-319-93869-1   ISBN 978-3-319-93870-7   (eBook)
https://doi.org/10.1007/978-3-319-93870-7

Library of Congress Control Number: 2018944349

Printed on acid-free paper

This Springer imprint is published by the registered company Springer Nature Switzerland AG
The registered company address is: Gewerbestrasse 11, 6330 Cham, Switzerland

Foreword

Robotics is undergoing a major transformation in scope and dimension. From a largely dominant industrial focus, robotics is rapidly expanding into human environments and vigorously engaged in its new challenges. Interacting with, assisting, serving and exploring with humans, the emerging robots will increasingly touch people and their lives.

Beyond its impact on physical robots, the body of knowledge robotics has produced is revealing a much wider range of applications reaching across diverse research areas and scientific disciplines, such as biomechanics, haptics, neurosciences, virtual simulation, animation, surgery and sensor networks among others. In return, the challenges of the newly emerging areas are proving an abundant source of stimulation and insights for the field of robotics. It is indeed at the intersection of disciplines that the most striking advances happen.

The *Springer Tracts in Advanced Robotics (STAR)* is devoted to bringing to the research community the latest advances in the robotics field on the basis of their significance and quality. Through a wide and timely dissemination of critical research developments in robotics, our objective with this series is to promote more exchanges and collaborations among the researchers in the community and contribute to further advancements in this rapidly growing field.

The volume by G. Venture, J.-P. Laumond and B. Watier provides a broad edited collection on human movement, control and robotics, which is the outcome of the workshop "Biomechanics of Anthropomorphic Systems" held at LAAS-CNRS in Toulouse in November 2016. Following a tutorial presentation of the problem, the contents are effectively organized into three main parts: biomechanics, motor control and humanoid robotics. The unique feature of the volume stands in its inherent interdisciplinary character concerned with anthropomorphic systems at large and humanoid robotics. Research challenges dealt with span from multibody optimization to dynamic modelling, from modularity to agility of human movements, and from design to control problems.

Rich of results by the leading roboticists, biomechanicists and specialists of motor control, this volume constitutes a very fine addition to STAR!

Naples, Italy   Bruno Siciliano
May 2018   STAR Editor

Contents

Foreword

Jean-Paul Laumond, Gentiane Venture and Bruno Watier

Biomechanics studies the understanding human body motions: how to describe them, analyze them and assess them. Biomechanics as an interdisciplinary field can date back to Aristotle (384–322–BC) who wrote the treatise *"About the movement of Animals"* (350 BC) and provided the first analysis of gait. Later, Herophilos (335–280 BC), Galen.

(AD 129–201), Avicenna (980–1037 AD) and Vesalius (1514–1564 AD) brought essential information about human anatomy and could explained the differences between motor and sensory nerves and agonist and antagonist muscles.

During the Renaissance, known as a prolific and fast period for sciences, numerous scientists focused their research to better understanding the mechanical behavior of humans. One can refer to the amazing works of L. da Vinci (1452–1519 AD) and G. A. Borelli (1608–1679 AD) to name only a few. Mechanical laws of motion were applied very early for better understanding anthropomorphic action as suggested in advance by Newton (1643–1727 AD): "For from hence are easily deduced the forces of machines, which are compounded of wheels, pullies, levers, cords, and weights, ascending directly or obliquely, and other mechanical powers; as also the force of the tendons to move the bones of animals." It is only in the 19th century, with the development of photography and motion picture that E. J. Marey (1830–1904 AD) and E. Muybridge (1830–1904 AD) introduced chronophotography to scientifically investigate animal and human movements. They opened the field of motion analysis by being the first scientists to correlate ground reaction forces (Figs. 1 and 2) with

G. Venture (✉)
Tokyo University of Agriculture and Technology, Tokyo, Japan
e-mail: venture@cc.tuat.ac.jp

J.-P. Laumond · B. Watier
Toulouse, France
e-mail: jpl@laas.fr

B. Watier
e-mail: bruno.watier@univtlse3.fr

© Springer International Publishing AG, part of Springer Nature 2019   1
G. Venture et al. (eds.), *Biomechanics of Anthropomorphic Systems*, Springer Tracts in Advanced Robotics 124, https://doi.org/10.1007/978-3-319-93870-7_1

Fig. 1 Left: photochronographe Right: detail of the "Dynamographe", both from Marey (*Crédit* Collège de France. Archives)

Fig. 2 "Dynamographe" from Marey embedded in the ground at "Station Physiologique" in Paris (*Crédit* Collège de France. Archives)

kinematics with an incredible accuracy (Fig. 3). Since then the basic technology hasn't changed much, indeed contemporary motion capture techniques for kinematics and kinetics data are based on the same technology (Figs. 4 and 5), only analysis techniques have changed, with the use of numerical data (Fig. 6).

Yet the mechanical modelling of such a complex structure as the human body is still an active field of research. Contrarily to most systems, the human body, though being made of links and joints, is far more complex than that of a robot. Indeed, the joints are not fixed: bones are not attached to each other strictly speaking and this very unique structure allows for many more degrees of freedom while at the same time adding some kinematics constraints. Thus, today numerous biomechanicists

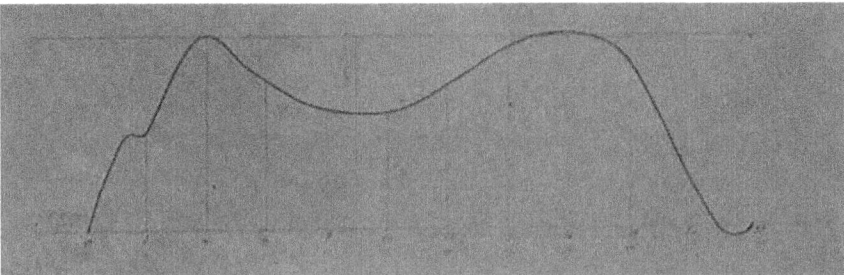

Fig. 3 Vertical force component during normal gait from Marey (*Crédit* Collège de France. Archives)

Fig. 4 "coureur de velocité 5 images à la seconde", runner captured at 5fps from Marey (*Crédit* Collège de France. Archives)

pursue their research for better understanding mechanical behavior of such complex tissues.

Nevertheless, biomechanics as an established field can be considered relatively young. The first specialized journal, the *Journal of Biomechanics* has been published since 1968, the first international research seminar took place in 1969 and the International Society of Biomechanics (ISB) was founded in 1973.

Moreover, despite of the apparent simplicity of a given movement, the organization of the underlying neuro-musculo-skeletal system remains unknown. What makes the difference when executing a task between a novice and a skilled person is yet to be found. A reason of the complexity is the redundancy of the motor system: a given action can be realized by different muscles and joint activity patterns, and the same underlying activity may give rise to several movements. Some theories, such as optimal control or motor primitives, provide tentative solutions to this conundrum, but none satisfies all empirical findings, which largely pertain to very basic movements, such as grasping or pointing. **Motor control** can also be traced back to Aristotle (384–322 BC) who was the first paying attention to the concept of motion coordination by comparing coordination with harmony. He attributed both to design of creation. For centuries, the understanding of motion coordination was ignored

444444444444r J.-P. Laumond et al.

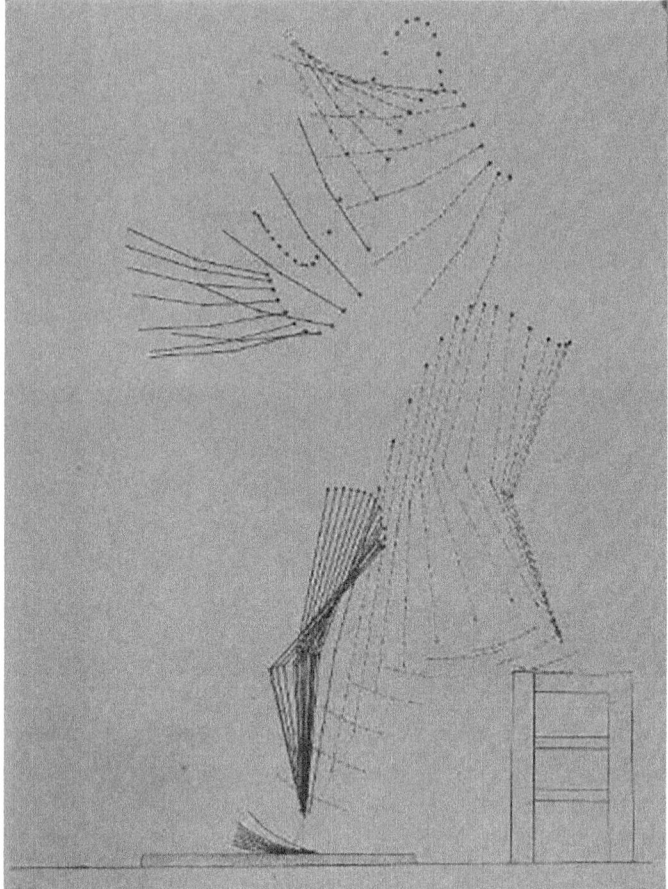

Fig. 5 "Analyse cinematique d'une chute raidie dans un saut en profondeur", kinematic analysis of a stiffen fall during a deep jump, from Marey (*Crédit* Collège de France. Archives)

until a famous experiment performed by N. Bernstein (1896–1966) on blacksmiths. After this pioneering work in the 1940's on the existence of motor synergies, numerous researchers "walking on the border" of their disciplines have tried to discover universal laws and principles underlying the human motions and they have tried to answer the open question: "how does the brain solve the redundancy of the system?". These synergies may represent the fundamental building blocks composing complex movements.

Motor control is also a recently established field. While N. Bernstein, considered as the pioneer of motor control, published his first work half a century ago, the journal of *Motor Control* exists only since 1997. The first conference "*Progress in Motor Control*" was held at about the same time and the International Society of Motor Control was established in 2001.

Fig. 6 Motion capture in the XXIth century (©LAASCNRS)

In **humanoid robotics**, researchers are facing the very same redundancy and complexity challenges as the researchers in life sciences. Indeed, a humanoid robot is a robot whose body shape and structure are made such that it resembles the human body. Attempts to make humanoid robots can be traced back to antiquity, with myths and legends such as the Golem. The first mechanical systems were then developed in the middle age and later during the Renaissance with L. da Vinci's automaton. Humanoid robotics took a new turn with WABOT-1 (1972), built by Waseda University in Japan, which was the first humanoid robot per se in history. It was able to walk, recognize objects and Japanese language, to manipulate objects, and to synthesize voice. Remarkable evolutions of humanoid robots followed then with the development of ASIMO or HRP-2 to name only a few. In 2017, the Gepetto team at LAAS-CNRS in France also presented a powerful humanoid robot designed for industrial applications called Pyrène, the first of the TALOS series developed by PAL-Robotics Company in Spain.

Naturally, the researches focusing on such complex anthropomorphic systems are very recent. The first International Conference on Humanoid Robots was first held in Boston (USA) in 2000 and Kajita published one of the first dedicated book on 2003. Humanoid robotics is facing two important challenges, the first one is on the design of the robots. Far from imitating human except a few such as the robots from M. Inaba at the Univ. of Tokyo, most of these humanoid robots are made of gears and motors, pullies and cables that actuate the joints of the robot. The sensing also is extremely different. Secondly, controlling these systems, even for a task as simple as walking, is not trivial because of the redundancy and the underactuation. Optimal control can be used but the function to optimize is often chosen randomly somehow.

The objective of the present book is to gather the work of biomechanicians, specialists of motor control and roboticists to promote an interdisciplinary research on anthropomorphic systems at large and on humanoid robotics. This is quite a challenge to gather such knowledge in interdisciplinary projects.

Still, there is a strong interest around human motion analysis and its transfer to robotic platforms. Better identifying the structuration of human movement is a necessary and a promising undertaking. Indeed, our capacity to simulate the motion or generate it on robotics platform demonstrates our understanding of the human motion. However, it is obvious to note, despite the accelerating knowledge in these areas, the development of humanoid robots in natural dynamic environment remains an open challenge. It reveals simultaneously our misunderstanding of the human motion generation and the lack of communications between biomechanicists, specialists of motor control and roboticists.

It is a major challenge facing the human-robots interaction.

This book keeps traces of a unique workshop[1] that was held at LAAS-CNRS in Toulouse (France) in November 2016, where roboticists, biomechanicists and specialists of motor control discussed the different aspects of their fields. In addition to that, a selection of chapters has been added to reflect the importance of the intertwinement of their research fields. We warmly thank the authors for sharing their own experiences while being accessible for a broad audience.

The multidisciplinary perspective on this book as previously presented above is reflected in the book by its table of content. The chapters are gathered within three mains parts addressing respectively Biomechanics (F. Valero Cuevas et al., B. Watier et al., R. Dumas., G. Venture et al.), Motor Control (Berret et al., M. Latash et al., T. Flash et al., I. Frakhatdinov) and Humanoid Robotics (R. Featherston et al., F. Nori et al., M. Benallegue et al., O. Stasse et al.).

Editing a book with a multidisciplinary perspective is not an easy task. We thank all the authors for their effort in making their own research field accessible to others. We also thank also the reviewers and the editorial committee (R. Dumas, I. Farkhatdinov, T. Flash, M. Latash, and F. Nori) for their help in editing the book.

Toulouse, January 18th 2018.

---

[1]*The workshop took place in the framework of the Anthropomorphic Motion Factory launched by the European project ERC-ADG 340050 Actanthrope (2014–2018) devoted to exploring the computational foundations of anthropomorphic action.*

Should Anthropomorphic Systems be "Redundant"?

Ali Marjaninejad and Francisco J. Valero-Cuevas

Abstract We explore the conceptual design and implementation of muscle redundancy and kinematic redundancy for anthropomorphic robots from three perspectives: (i) The control of tendon-driven systems, (ii) How the number of muscles define functional capabilities, and (iii) How too few synergies can be detrimental to functional versatility. Historically, roboticists prefer either rotational actuators located at each joint (i.e., rotational degree-of-freedom, DOF), or few linear actuators (i.e., two dedicated muscles per joint) for tendon-driven robots. In contrast, biological limbs have evolved to include too many muscles (Valero-Cuevas in Fundamentals of neuromechanics. Springer, Berlin (2015) [1]), which are thought to unnecessarily complicate their anatomy and control. The question, then, is why has evolution converged on these apparently under-determined (or redundant) solutions? If we really have extra muscles, then which muscle would you give up? By taking a formal mathematical approach to the control of tendons—which is the actual problem that confronts the nervous system—we have proposed a resolution to this apparent paradox by proposing that vertebrates may have, in fact, barely enough muscles to meet the numerous physical constraints for ecological functions (as opposed to simple laboratory tasks) (Valero-Cuevas in Fundamentals of neuromechanics. Springer, Berlin (2015) [1]; Loeb in Overcomplete musculature or underspecified tasks? Mot Control 4(1):81–83 (2000) [2]). This approach can be called Feasibility Theory, which describes how the anatomy of the system, and the constraints defining the task define the set of feasible actions the system can produce. The role of the (neural or engineered) controller is then, to find ways to use the mechanical capabilities of the combined controller-plant system to the fullest (Valero-Cuevas in Fundamentals of neuromechanics. Springer, Berlin (2015) [1]. Similarly, the effective mechanical

A. Marjaninejad
Department of Biomedical Engineering, University of Southern California,
Los Angeles, CA, USA
e-mail: marjanin@usc.edu

F. J. Valero-Cuevas (✉)
Departments of Biomedical Engineering, Electrical Engineering (Systems),
Computer Science, and Mechanical & Aerospace Engineering, and Division
of Biokinesiology & Physical Therapy, University of Southern California,
Los Angeles, CA, USA
e-mail: valero@usc.edu

© Springer International Publishing AG, part of Springer Nature 2019
G. Venture et al. (eds.), *Biomechanics of Anthropomorphic Systems*, Springer Tracts
in Advanced Robotics 124, https://doi.org/10.1007/978-3-319-93870-7_2

design of a robotic limb, at a minimum, requires controllability (i.e., enough control degrees of freedom, or muscles) to produce arbitrary forces and movements (i.e., changes of state; Ogata in Modern control engineering. Prentice hall, India (2002) [3]). Force and movement capabilities have distinct governing equations and are, in fact, in competition with one another (e.g., a see-saw demonstrates, as per the Law of Conservation of Energy, how producing higher forces is associated with lower velocities and vice versa). Therefore, we explored the potential evolutionary pressures that may have shaped vertebrate limbs by evaluating how the number of muscles affects the competing demands to produce endpoint forces and velocities. A related concept that cuts across biological and robotic systems is the idea that the kinematics and kinetics of a wide variety of actions exhibit a low-dimensional structure that can be approximated with a few principal components (sometimes called descriptive synergies; Brock and Valero-Cuevas in Transferring synergies from neuroscience to robotics comment on hand synergies: integration of robotics and neuroscience for understanding the control of biological and artificial hands by M. Santello et al. Phys Life Rev 17:27 (2016) [4]; Rieffel et al. in Automated discovery and optimization of large irregular tensegrity structures. Comput Struct 87(5):368–379 (2009) [5]). This has been taken to mean that a few degrees of freedom suffice to produce versatile behavior in the real world. However, the fine behavioral details that distinguish different actions are, by definition, not captured by the commonalities among them. Thus, versatility in the real world likely depends on recognizing and executing fine distinctions among actions; which implies that more degrees of freedom of control are critical for true functional versatility. These three independent arguments support the perspective that creating anthropomorphic systems requires apparently redundant structures, because only then can they truly execute a wide variety of real-world tasks. In addition, we also present an open-access MATLAB toolbox that allows users from different backgrounds to explore these concepts in detail. We believe this new perspective will improve the conceptualization, understanding, and design of anthropomorphic systems.

1   Mechanical and Neural Foundations of Feasibility Theory

In this section, we provide fundamental concepts required to study neuromechanical systems, which will set fundamentals for the following sections of this chapter. We begin with limb kinematics, providing a common conceptual language to muscle mechanics. Next, we introduce motor control and feasible movements of tendon driven limbs to show how tendon-driven systems are in fact over-determined. Due to the inherent properties of biological muscles, the nervous system is likely not as redundant as when considering the force control problem in isolation. Moreover, we have described how different task constraints (i.e., the mechanical definition of the task) naturally limit feasible actions. Thus, each additional muscle adds an additional control DOF—and therefore the ability to meet more simultaneous functional constraints and produce a wider variety of tasks—which is the origin of versatility. Many of the materials and concepts of this section are summarized or first introduced in [1].

1.1 Limb Kinematics and Limb Mechanics

We first start with limb kinematics, which characterize the motions and positions of rigid bodies, regardless of the forces which produce them. In this chapter, in order to simplify the governing equations, we consider limbs as rigid bodies. Subsequently, we introduce equations for limb mechanics which involve limb kinematics as well as the forces and torques that interact with the limb.

1.1.1 Limb Kinematics

We first define a limb as a set of connected links and hinges. The endpoint of a multi-joint limb is defined by the homogeneous transformation matrix $T_{base}^{endpoint}$. $T_{base}^{endpoint}$ can also be written as the multiplication of transformation matrices of all DOFs:

$$T_{base}^{endpoint} = T_0^N = T_0^1 T_1^2 \cdots T_{N-2}^{N-1} T_{N-1}^N \tag{1}$$

where each transformation matrix is defined as:

$$T_i^j = \left\{ \begin{matrix} R_i^j & \mathbf{p}_{i,j} \\ 0\,0\,0 & 1 \end{matrix} \right\} \tag{2}$$

In the above equation, R_i^j represents the rotation matrix and $\mathbf{p}_{i,j}$ represents the displacement for each DOF [1]. A schematic representation of the system for Eq. 1 is plotted in Fig. 1. Furthermore, the forward kinematic model (also known as the geometric model), G(q), for a planar system (in two-dimensional space) is defined as follows (given that a rigid body on the plane has three degrees of freedom defining its location and orientation):

$$G(q) = \begin{pmatrix} \textit{displacement in the direction of } i_0 \\ \textit{displacement in the direction of } j_0 \\ \textit{rotation about the } k_0 \textit{ axis} \end{pmatrix} = \begin{pmatrix} x \\ y \\ \alpha \end{pmatrix} \tag{3}$$

For non-planar limbs, displacement in the direction of k_0, rotation about the i_0, and rotation about the j_0, will also need to be included in G(q) [1]. The endpoint velocities are obtained by differentiating G(q) with respect to time:

$$\dot{G}(\mathbf{q}) = \frac{dG(\mathbf{q})}{dt} = \frac{\partial G(\mathbf{q})}{\partial \mathbf{q}} \frac{d\mathbf{q}}{dt} = \frac{\partial G(\mathbf{q})}{\partial \mathbf{q}} \dot{\mathbf{q}} = \begin{pmatrix} \dot{x} \\ \dot{y} \\ \dot{\alpha} \end{pmatrix} \tag{4}$$

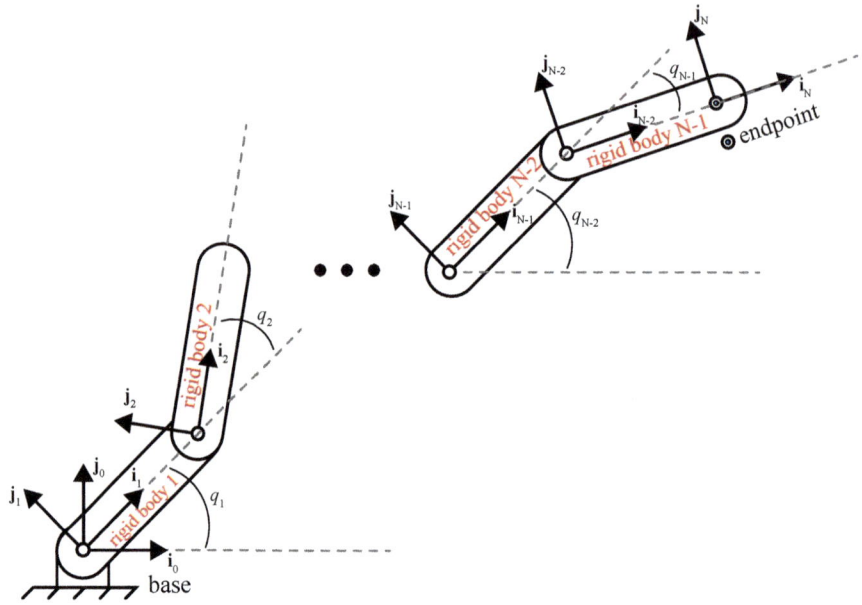

Fig. 1 A schematic representation of a simple multi-link system (Reproduced, with permission, from [1])

We call $\frac{\partial G(\mathbf{q})}{\partial \mathbf{q}}$ as the Jacobian matrix. For a limb with N degrees of freedom, it is defined as:

$$\frac{\partial G(\mathbf{q})}{\partial \mathbf{q}} = J(\mathbf{q}) = \begin{bmatrix} \frac{\partial G_x(\mathbf{q})}{\partial q_1} & \frac{\partial G_x(\mathbf{q})}{\partial q_2} & \cdots & \frac{\partial G_x(\mathbf{q})}{\partial q_N} \\ \frac{\partial G_y(\mathbf{q})}{\partial q_1} & \frac{\partial G_y(\mathbf{q})}{\partial q_2} & \cdots & \frac{\partial G_y(\mathbf{q})}{\partial q_N} \\ \frac{\partial G_\alpha(\mathbf{q})}{\partial q_1} & \frac{\partial G_\alpha(\mathbf{q})}{\partial q_2} & \cdots & \frac{\partial G_\alpha(\mathbf{q})}{\partial q_N} \end{bmatrix} \tag{5}$$

1.1.2 Limb Mechanics

As mentioned earlier, limb mechanics involve limb kinematics as well as the forces and torques the limb can produce. In this section, we will explain how to relate joint torques to endpoint forces using limb kinematics. Let's begin by defining the internal and external work for the two DOF systems shown on Fig. 2 as:

$$External\ work = \mathbf{f} \cdot \Delta \mathbf{x} \tag{6}$$
$$Internal\ work = \tau \cdot \Delta \mathbf{q} \tag{7}$$

where $\mathbf{f}$ and $\Delta \mathbf{x}$ are scalar values representing endpoint force and endpoint velocity, while τ and $\Delta \mathbf{q}$ are torque and joint angle rotation vectors. Please note that,

Fig. 2 An illustration of the geometric relationship between the endpoint displacement (Δ**x**) and the rotation of the first joint (Δq_1 in a 2-DOF limb) (Reproduced, with permission, from [1])

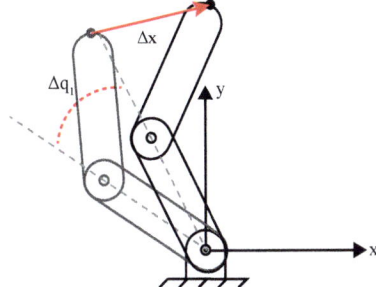

depending on the kinematic DOFs of the system, the endpoint can also produce a torque ($\tau_{endpoint}$), whose external work would be the torque times the rotation of the endpoint. For the sake of simplicity, here we present the derivation only considering endpoint forces. Following with Eqs. 7 and 8, from the conservation of energy law, we have:

$$\mathbf{f}.\Delta\mathbf{x} = \tau.\Delta\mathbf{q} \tag{8}$$

Changing the dot product in Eq. 8 to its equivalent inner product while substituting Δ for both **x** and **q** with derivatives, we have:

$$\mathbf{f}^T\dot{\mathbf{x}} = \tau^T\dot{\mathbf{q}} \tag{9}$$

Using Eqs. 4 and 5, we rewrite Eq. 8 as:

$$\mathbf{f}^T J(\mathbf{q})\dot{\mathbf{q}} = \tau^T\dot{\mathbf{q}} \tag{10}$$

Eliminating $\dot{\mathbf{q}}$ from both sides of the equation leads to Eqs. 11 and 12, which define the relationship between the joint torques and endpoint forces.

$$\mathbf{f}^T J(\mathbf{q}) = \tau^T \tag{11}$$
$$\mathbf{f}^T = J(\mathbf{q})^{-1}\tau^T \tag{12}$$

Extensions of this concept for 3-dimensional space with detailed examples are provided in [1].

1.1.3 Tendon-Driven Limb Mechanics

Most robotic limbs are driven by either rotational or linear actuators that drive each kinematic DOF [1]. In the robotics literature, the so-called torque-driven formulation assumes symmetric actuators. That is, equal torque capabilities in both clockwise and counterclockwise directions.

In the so-called tendon-driven systems, actuators are connected to the limbs using strings, cables, or tendons. It is clear that these actuators can only pull (and not push) on the tendons, thus they can only drive the DOF in one direction. Therefore, each DOF requires, on average, more than one actuator and symmetry of actuation is not guaranteed. In addition, tendon-driven systems are flexible because the routing of their tendon can allow one actuator to drive more than one DOF—and therefore impose correlations in actuation across DOFs [6, 7]. Moreover, the moment arms (i.e., minimal perpendicular distance between the tendon path and the center of rotation of the DOF) can be arbitrarily set within and across tendons and DOFs.

This flexibility of actuation that can be built into the morphology of the design, compared to torque-driven systems, introduces unique flexibility and challenges to their construction and control. To some, this means that tendon-driven systems are unnecessarily difficult to build and control. However, we and others also argue that they have much to offer [7–13]. Using tendons to apply torque to the DOFs, as opposed to having actuators directly apply torque, makes tendon-driven systems capable of remote actuation. This means that the designer can place the actuator far from the joint itself. Although making the system harder to control, flexible tendon routing can provide much more versatility and preferentially larger feasible end-point forces and velocities in directions of interest.

1.1.4 Tendon Actuation

To explore how tendons create torque in tendon driven systems, see Fig. 3. This simplified model illustrates a planar, one-joint limb using one muscle. The torque at the joint of this model is equivalent to the cross-product of the force and moment arm r:

$$\tau = r \times f_m = \|r\| \, \|f_m\| \, sin(\alpha) \tag{13}$$

where $\times$ represents the cross-product, f_m represents the muscle force, and α represents the angle between the force and the moment arm.

As mentioned earlier, in tendon-driven systems, one actuator can exert torque in multiple DOFs. Here, we are going to study an example of such while introducing the moment arm matrix. We begin by illustrating the relationship between the torque vector (here, a vector of length two, representing the two DOFs) and muscle force vector for the two-joint planar limb shown in Fig. 4.

$$\begin{pmatrix} \tau_1 \\ \tau_2 \\ \vdots \\ \tau_M \end{pmatrix}_{M \times 1} = R(\mathbf{q})_{M \times N} \begin{pmatrix} f_1 \\ f_2 \\ \vdots \\ f_N \end{pmatrix}_{N \times 1} \tag{14}$$

where $R(\mathbf{q})$ represents the moment arm matrix, which maps the muscle forces to the joint torques. In example illustrated in Fig. 4, the moment arm matrix $R(\mathbf{q})$ is defined as:

$$R(\mathbf{q}) = \begin{bmatrix} -r_1 & -r_1 & r_2 & r_2 \\ -r_3 & r_4 & -r_3 & r_4 \end{bmatrix} \tag{15}$$

By convention, each entry in the moment arm matrix on the ith row and jth column will be the coefficient which transforms the force induced by the ith muscle to the torque exerted at the jth joint (as seen in Eq. 9). The positive value of an element in this matrix means that counterclockwise (positive) torque will be applied when tension is applied to the tendon through muscle contraction (applying concentric force).

In order to relate muscle excursions to joint movements in the model shown above, we produce a set of equations. Again, by convention, we consider counterclockwise rotations as positive rotations. Following the conventions we have mentioned so far, a positive joint rotation with a positive moment arm induces a shortening in the length of its muscle and tendons and vice versa. Therefore, the set of equations relating joint angles and muscle excursions for the example case provided in Fig. 4 will be as follows:

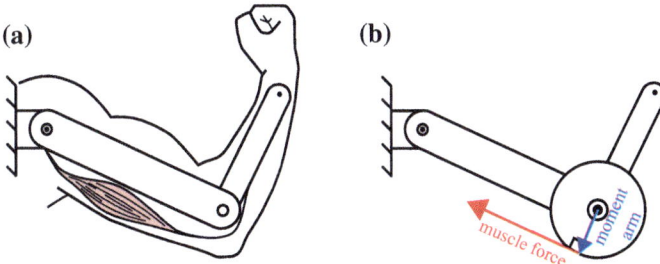

Fig. 3 Schematic representation of a planar, two joint limb with only one muscle (**a**) and its equivalent simplified model (**b**)

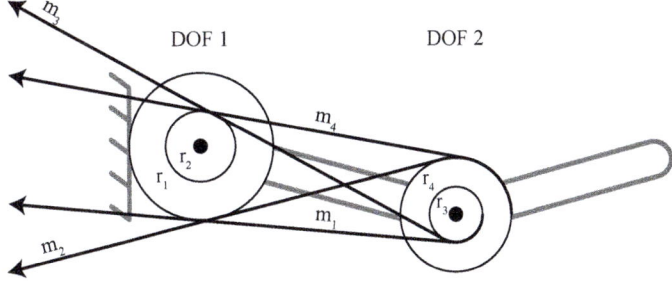

Fig. 4 A sample two-joint limb with four muscles (Reproduced, with permission, from [1])

$$
\begin{pmatrix} \partial S_1 \\ \partial S_2 \\ \partial S_3 \\ \partial S_4 \end{pmatrix}_{4\times 1} = \begin{bmatrix} r_1 & r_3 \\ r_1 & r_4 \\ -r_2 & r_3 \\ -r_2 & r_4 \end{bmatrix}_{4\times 2} \begin{pmatrix} \partial q_1 \\ \partial q_2 \end{pmatrix}_{2\times 1} \tag{16}
$$

Using Eq. 9, we can rewrite Eq. 10 in the general case with M joints and N muscles as:

$$
\begin{pmatrix} \partial S_1 \\ \partial S_2 \\ \vdots \\ \partial S_N \end{pmatrix}_{N\times 1} = (-R_{M\times N})^T \begin{pmatrix} \partial q_1 \\ \partial q_2 \\ \vdots \\ \partial q_N \end{pmatrix}_{M\times 1} \tag{17}
$$

Taking a closer look at Eqs. 14 and 17, we see a very important distinction. Equation 14 is under-determined, meaning that there is more than one solution in the force space to achieve the desired set of torques. This is one of the main reasons that neuromuscular systems are thought to be redundant. However, looking at Eq. 17, we notice that this equation is over-determined, meaning there is, at maximum, only one set of values for the joint angles fulfilling this set of equations [1, 6, 14]. In other words, you cannot contract and shorten a specific muscle without having changes in the lengths of other muscles connecting to or passing through the same joint. It illustrates that our nervous system must consider a complex variety of constraints when pulling a tendon. Failure in fulfilling the requirements of this complex control problem, especially failure in relaxing muscles that are being lengthened as a result of joint rotations, might disrupt movement or injure muscles or tendons [6, 15]. Note that this over-determined case only arises when treating limbs as tendon-driven systems. This is one of the important, yet mostly overlooked aspects of robotics today.

1.2 Motor Control of Tendon Driven Limbs and Feasibility Theory

In this section, we describe the tendon-driven system that the nervous system faces. First, we present a conceptual framework where one can think of neural commands as being a high-dimensional activation vector that is mapped into lower-dimensional "spaces", that capture its transformation into endpoint forces. Next, we describe Feasibility Theory, which defines how different constraints can limit the feasible activations in different spaces (activation, muscle force, torque, and end-point force spaces). If we are looking for a versatile system to deal with day to day activities, then a larger number of DOFs are required as constraints are added.

1.2.1 Motor Control of Tendon Driven Limbs

Although the nervous system activates muscles through the recruitment of motor neurons and modulation of their firing rates, we can, without loss of generality, simplify the problem by assigning a value between 0 and 1 to the activation level of each muscle, where 0 represents complete inactivation and 1 represents maximal activation. The activation vector α is described as Eq. 18 for an N muscle system:

$$\alpha = \begin{pmatrix} \alpha_1 \\ \alpha_2 \\ \vdots \\ \alpha_N \end{pmatrix}, \quad 0 \leq \alpha_i \leq 1 \; for \; i = 1, \; \ldots, \; N \tag{18}$$

where α_i is the activation value for the ith muscle. Now the set of the generated muscle forces at a particular moment and at a particular activation level can be defined as:

$$\mathbf{f_m} = F_0(\mathbf{l_m}, \mathbf{v_m})\alpha \tag{19}$$

where F_0 is the diagonal matrix. Each element on the main diagonal of this matrix will represent the maximum force that the corresponding muscle can exert. These diagonal values depend on many factors such as muscle architecture, pennation angle, physiological cross-sectional area, as well as the fiber length (l_m) and velocity (v_m) of the muscle at every time point. Now, we rewrite Eq. 14 by substituting the force vector from Eq. 19, which leads to the following equation:

$$\tau_{M \times 1} = R(\mathbf{q}_{M \times N}) \, F_0(\mathbf{l_m}, \mathbf{v_m})_{N \times N} \, \alpha_{N \times 1} \tag{20}$$

Equation 20 shows that the control of joint torques in tendon-driven limbs is an under-determined set of equations. However, adding a cost function which needs to be minimized (e.g. the total sum of the activation values), will force this set of equations to have fewer possible solutions (or even just one).

Considering the fact that maximum muscle force values (the diagonal values on the F_0 matrix) are also functions of muscle length and velocity, we see that discovering the activation values which result in a desired set of joint torques (Eq. 20) is difficult. In fact, it will apply constraints to the solutions that the nervous system can produce. We discuss the effects of these constraints in greater detail in the following subsection, which explains how our nervous system faces a much more complex control problem than initially hypothesized in the literature [16].

Reconsidering the set of tendon excursions, we can rewrite Eq. 17 in terms of the muscle length vector, $\mathbf{l_m}$, and muscle velocity vector, $\mathbf{v_m}$ (Eqs. 21 and 22).

$$(\delta\mathbf{l}_m)_{N \times 1} = (-R_{M \times N})^T \delta\mathbf{q}_{M \times 1} \tag{21}$$

$$(\mathbf{v}_m)_{N \times 1} = (-R_{M \times N})^T \dot{\mathbf{q}}_{M \times 1} \tag{22}$$

These last two sets of equations illustrate how control of tendon excursions is an over-determined problem (i.e. there is, at most, only one set of solutions for it). Therefore, our nervous system faces a biomechanical limitation [1].

1.2.2   Feasibility Theory: Defining Feasible Sets of Actions in Tendon-Driven Limbs

In this section, we first explore how a neural activation vector is mapped into the torque space and the effect of different biomechanical parameters on this mapping. Next, we demonstrate how functional constraints can limit the feasible action spaces, therefore making it harder for the controller to find a solution within these spaces [16].

Let's assume we have three muscles, representing their activation values as a_1 to a_3. A rectangular cuboid can be used as a visual representation of these activations. This cuboid is shown in Fig. 5a. The corresponding muscle force vector is determined by Eq. 19 and represented in Fig. 5b. Now, let's say we have a two DOF joint. This assumption means that our feasible muscle force set (Fig. 5b), which is a three-dimensional cuboid, will be mapped into the two-dimensional feasible torque space. The amount of torque generated at each joint is determined by Eq. 20. The feasible torque set for this example is shown in Fig. 5c. We utilize the appropriate Jacobean matrix, mapping this feasible torque set to the feasible end-point force set using Eq. 12. The feasible force set for this example is shown in Fig. 5d. Looking at the different parts of Fig. 5 we can see the contributions of having extra muscles in each of these feasible action spaces.

For the example case in Fig. 5, let's now assume that there is a functional constraint. This may include a certain constraint on the magnitude of the force in one axis in the end-point force space, or a certain torque in the torque space, etc. Since the torque or the end-point force spaces are two-dimensional in this example, these constraints can only be points or lines in these spaces. However, when we track them back into the muscle activation space (or muscle force space), they can be points, lines, or planes (or hyper-planes in more than three dimensional spaces) since all the transformations are linear. Say we have defined a constraint whose representation in the muscle activation space is a plane (i.e. the constraint plane). The new feasible activation set now is the intersection of the feasible activation set, without any constraints, with this constraint plane. This new feasible activation set is shown in Fig. 6a. If more constraints are added, the feasible muscle activation set will lose even more dimensions and might become a line, a point, or an empty set (Fig. 6b–d). This shows that while creating a specific amount of end-point force or torque in a joint can be an under-determined problem, functional constraints and feasible action spaces (as well as the mapping functions between these spaces) can limit the abilities of our neuromechanical system to a great extent [1, 16]. Therefore, the control problem which the nervous system faces is a very complicated one.

This suggests that having extra muscles is not redundant, but is a necessary requirement for versatility. Extra sets of muscles will increase the DOF of the system while

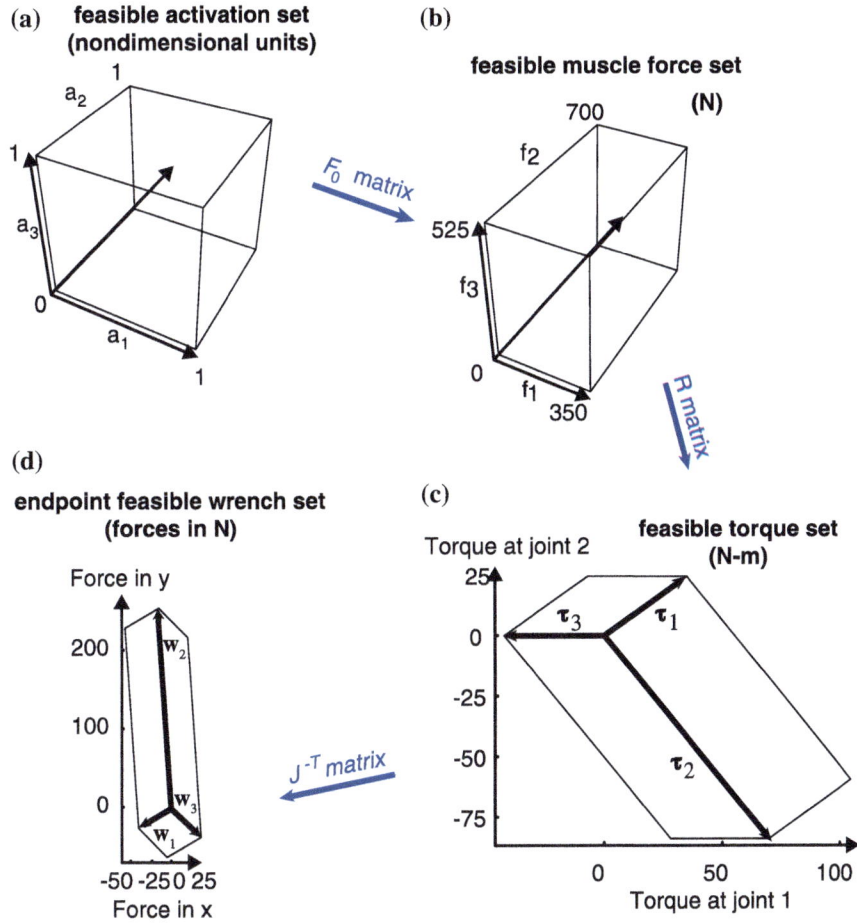

Fig. 5 The representation of the feasible actions in different spaces for the toy example discussed in the text. **a** the feasible set of activations. **b** the feasible set of muscle forces. **c** the feasible set of joint torques. **d** the feasible set of end-point forces (Reprinted, with permission, from [1])

enabling the nervous system to find solutions for different sets of problems we face daily.

2 An Evolutionary Fitness Approach to the Relationship Between the Number of Muscles and Versatility

Throughout the years, many have wondered why the anatomy of vertebrates has evolved to include a seemingly redundant number of muscles. Here we show how

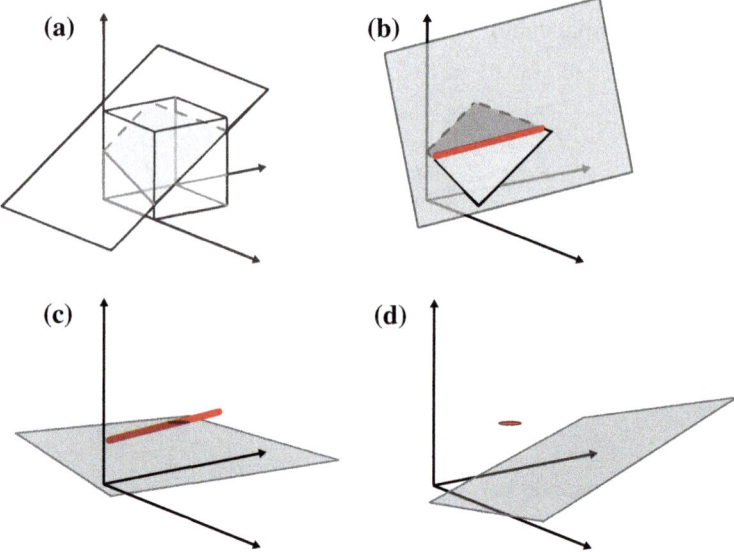

Fig. 6 The effects of adding functional constraints on the feasible actions spaces. More and more constraints are applied as we move from (**a**) to (**d**) (Reproduced, with permission, from [1])

extra muscles can enhance mechanical versatility using an evolutionary approach while clarifying why muscle redundancy is not a comprehensive belief. We begin our study by observing how our biomechanical tendon-driven model performs a set of specified kinetic tasks while changing the number of muscles. We study the optimal number of muscles with three main fitness functions. Namely, the Effectiveness, Agility, and Phenotypical cost. During all tasks, the goal was to apply maximum force in a specific direction. We found the optimal activation values as a function of task restraints using a linear optimization algorithm.

2.1   Background

Muscle redundancy has been discussed extensively ever since the earliest neuromuscular studies. Evolution from the earliest ape species, *Nakalipithecus nakayamai*, to modern *Homo Sapiens*, exemplifies the growth and development of muscles, enabling our species to perform multiple tasks. There are certain questions that resurface each time someone tries to explore this field. Why do we a so many muscles despite the limited degrees of freedom in our limbs or fingers? Why do we have a specific number of muscles? What are the costs and benefits of having this set of muscles?

In this section, we explore answers to these questions with an evolutionary fitness approach. We study how extra sets of muscles affect Effectiveness, Agility, and Phenotypical cost (we will describe each later in this section). In addition to this, we explore how decreasing the number of muscles affects performance in different tasks. In this study, we have used a 2-DOF arm model with three different muscle sets.

2.2 Method

We first explain our model and assumptions. Next, we describe the simulated tasks used in this study. Finally, we introduce our fitness functions for each of the elements mentioned earlier as well as the Overall Fitness, which is the weighted linear combination of all the individual fitness values. Note that we use the term "fitness" in the general sense where it is not necessarily tied to a single specific cost function. Rather, fitness in the biological sense indicates the ability to meet current multi-dimensional requirements and perform well in a given environment. For the sake of simplicity, we define fitness as the ability to meet a compound cost function—but other cost functions may also be suitable depending on the functional goals at hand.

2.2.1 Model

We begin with a simplified model of the human arm with a two DOF planar structure. One can also generalize this model to other body parts with similar structure e.g. fingers. We select four postures based on common tasks performed by the arm, as shown in Fig. 7. Each posture was held static while force was maximized in four directions: upward, downward, frontward, and backward.

Three muscle sets were designed to compare the effects of varying the number of muscles. These muscle sets, shown in Fig. 8, were designed with the intent of recreating a model with a realistic set of arm muscles (although this model is constrained to two-dimensional space) as well as models with fewer or more muscles than a real arm. Across muscle sets, the number of muscles was decreased while keeping the same original routing configuration of the previous muscle set. This was done to compare the effects of decreasing or increasing the number of muscles only. The moment arms were estimated with reference to [1].

We selected 3, 7, and 14 as the number of muscles which are the minimum number of muscles for a 2-DOF system, the real number of muscles in a human arm and two times of the number of muscles in a human arm, respectively. In the Monte Carlo analysis, the moment arm values were varied by ±20% over 100 simulations to test the sensitivity of the results to these values.

2.2.2 The Model Assumption

In this study, all muscle lengths and maximum muscle forces were assumed to be equal enabling us to address only a specific set of questions. We study how the number of muscles affects a specific set of outputs in the absence of other variables.

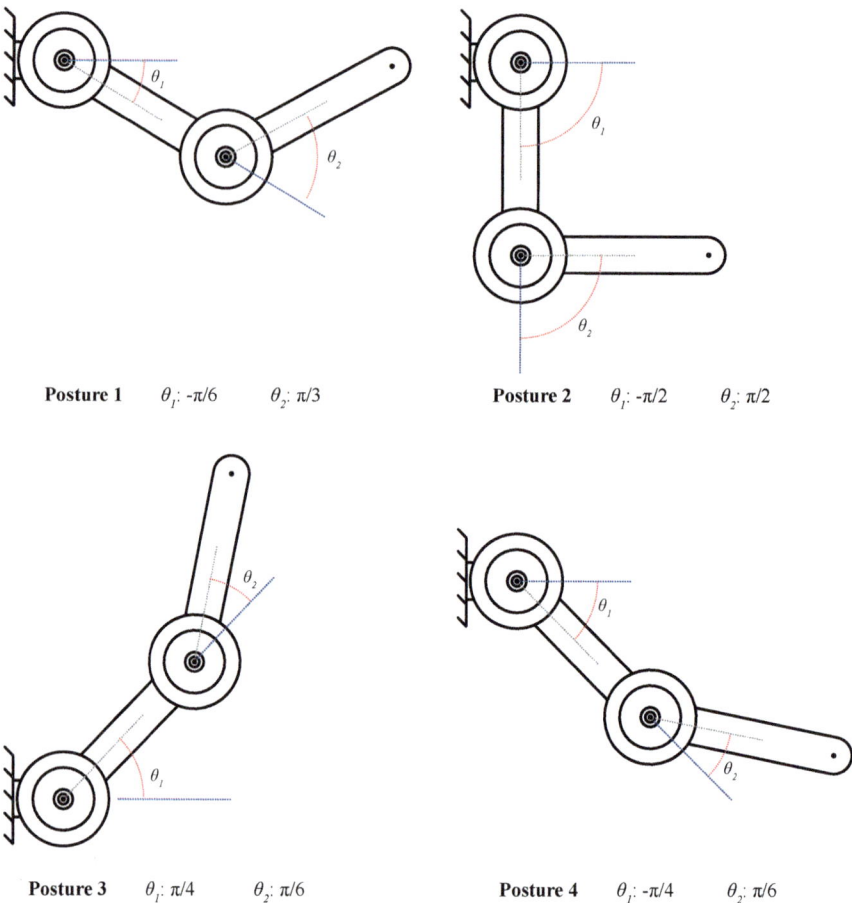

Posture 1 θ_1: -π/6 θ_2: π/3   Posture 2 θ_1: -π/2 θ_2: π/2

Posture 3 θ_1: π/4 θ_2: π/6   Posture 4 θ_1: -π/4 θ_2: π/6

Fig. 7 Four different postures used for simulations. These postures were inspired by day to day activities

Also, muscle forces were assumed to be independent from muscle lengths or muscle velocities.

2.2.3   Cost Formulation

Here we describe the performance metrics studied. First, we define the Average Energy as the summed square of muscle activation values, divided by the number of muscles. Although muscle activation is a neural action, it will be proportional to physical muscle activity (and therefore requires energy) since we assumed that all muscles have equal maximum force values. Average Energy shows how hard muscles

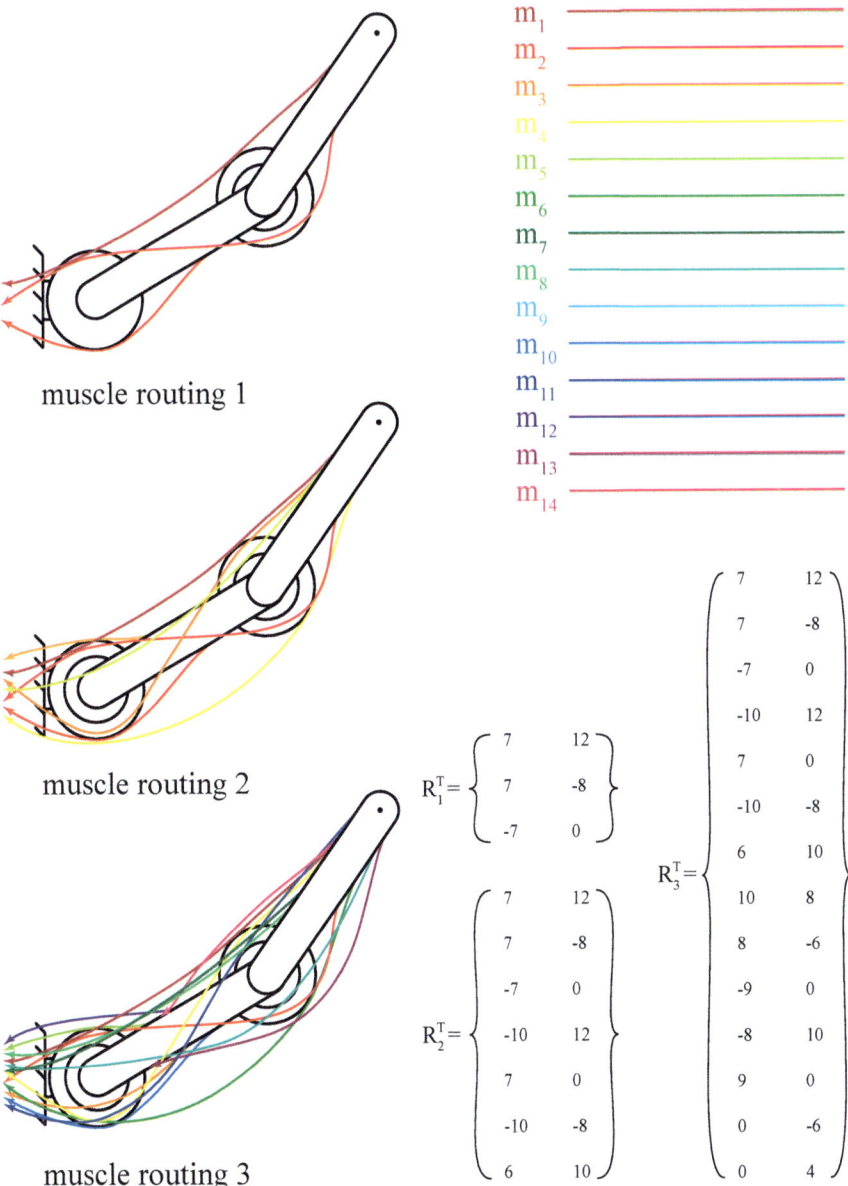

m_1
m_2
m_3
m_4
m_5
m_6
m_7
m_8
m_9
m_{10}
m_{11}
m_{12}
m_{13}
m_{14}

muscle routing 1

muscle routing 2

muscle routing 3

$$R_1^T = \left\{ \begin{array}{rr} 7 & 12 \\ 7 & -8 \\ -7 & 0 \end{array} \right\}$$

$$R_2^T = \left\{ \begin{array}{rr} 7 & 12 \\ 7 & -8 \\ -7 & 0 \\ -10 & 12 \\ 7 & 0 \\ -10 & -8 \\ 6 & 10 \end{array} \right\}$$

$$R_3^T = \left\{ \begin{array}{rr} 7 & 12 \\ 7 & -8 \\ -7 & 0 \\ -10 & 12 \\ 7 & 0 \\ -10 & -8 \\ 6 & 10 \\ 10 & 8 \\ 8 & -6 \\ -9 & 0 \\ -8 & 10 \\ 9 & 0 \\ 0 & -6 \\ 0 & 4 \end{array} \right\}$$

Fig. 8 Muscle routings and moment arm matrices used in the simulations

pull on average to perform the task. The Average Energy will be as defined in the quadratic Eq. 23.

$$Average\ Energy = \frac{1}{N} \sum_{i=1}^{N} \alpha_i^2 \qquad (23)$$

where α_i is the activation of the ith muscle (between 0 and 1) and N is the number of muscles [17, 18].

Although Average Energy shows the total amount of energy used to perform a task, it is not the best way to calculate Effectiveness since the maximum force output of a task is also very important. Therefore, we define Effectiveness as the maximum output force divided by the Average Energy. Note that Effectiveness is distinct from efficiency as we are using it to reflect overall ability after normalizing for energy consumption. Phenotypical cost (see below) already considers metabolism.

$$Effectiveness = \frac{maximal\ force}{Average\ Energy} \qquad (24)$$

We know that there is a limit on how fast a muscle can contract (the maximum speed for muscle excursion). We define "Agility" for each joint as the maximum rate of change in the joint (regardless of the direction), assuming that the muscle excursions for all muscles have the same upper limit. This is similar to the concept of manipulability, which considers the transformation of joint angular velocities into endpoint velocities as per the Jacobian of the limb [19]. Therefore, Agility for each joint is defined as:

$$Agility = \frac{dq}{ds} = max_{ij}\left(abs\left(\frac{1}{r_{ij}^T}\right)\right) \qquad (25)$$

where i is the muscle index, r_i is the corresponding moment arm value, max stands for maximum, and abs stands for absolute value. Please note that to maximize the $\frac{dq}{ds}$, r_{ij}^T cannot be equal to zero, since this would mean that the muscle is not connected to the intended joint.

Lastly, the "Phenotypical" cost is related to the number of muscles due to the nature of muscle packaging [20]. In particular, the Phenotypical cost can consider both the cross-sectional area and volume of muscles and muscle groups. Therefore, we explored this value using the square and the cube of the number of muscles (Overall Fitness A and Overall Fitness B, respectively). The quadratic version preferentially considers metabolic and phenotypical costs associated with muscle stress and physiological cross-sectional area [21]; whereas the cubic version attempts to further penalize the complexity of vascularizing, repairing, maintaining, packaging and controlling more muscles [20].

To compute the fitness of alternative embodiments for a multi-muscle limb, we compute Overall Fitness as shown below. It is the weighted sum of the above-mentioned elements:

$$Overall\ Fitness = w_1 \times Effectiveness_N + w_2 \left(\frac{Agility_{N,1} + Agility_{N,2}}{2} \right)$$
$$-w_3 \times Phenotypical\ cost \qquad (26)$$

where $Effectiveness_N$ and $Phenotypical\ cost_N$, are the normalized (between 0 to 1) Effectiveness and Phenotypical Cost, respectively. $Agility_{N,k}$ is the normalized Agility in the kth joint while w_1, w_2, and w_3 are weights for $Effectiveness_N$, average $Agility_{N,k}$ (averaged over k) and the $Phenotypical\ Cost_N$ respectively.

We used linear programming [22] to find the solutions in the activation space for the constrained problem of maximizing the force in only the specified direction in each task.

2.3 Results

Figures 9 and 10 respectively demonstrate Average Energy and Effectiveness as a function of number of muscles for each posture. In addition, Agility was plotted as a function of number of muscles for each joint (Fig. 11). We then performed Monte

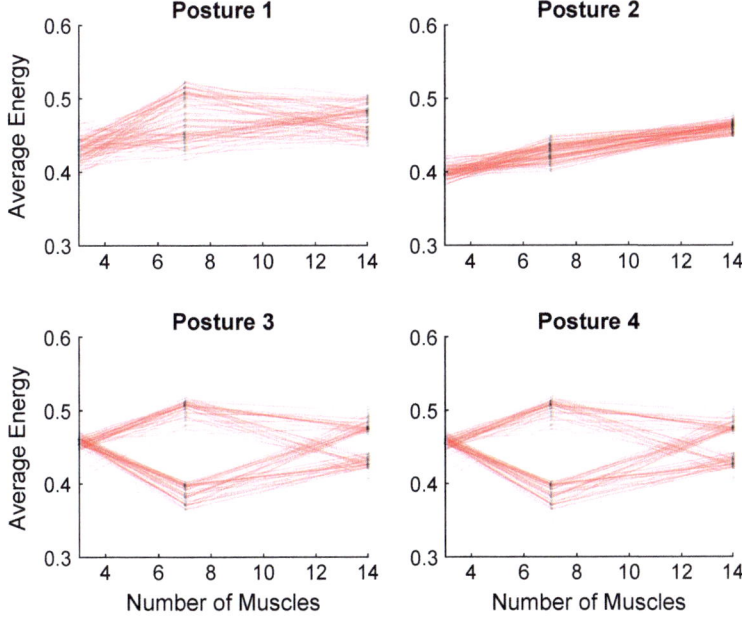

Fig. 9 Average energy plots (Monte Carlo analysis)

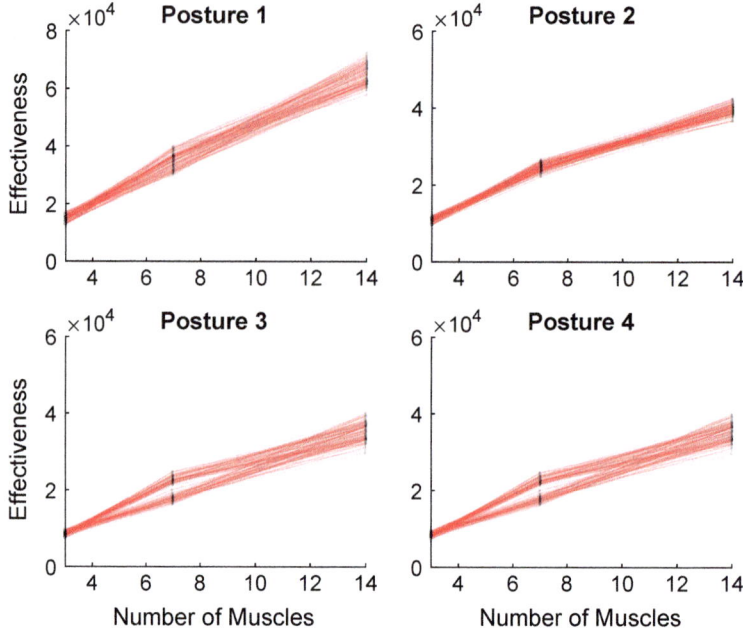

Fig. 10 Effectiveness plots (Monte Carlo analysis)

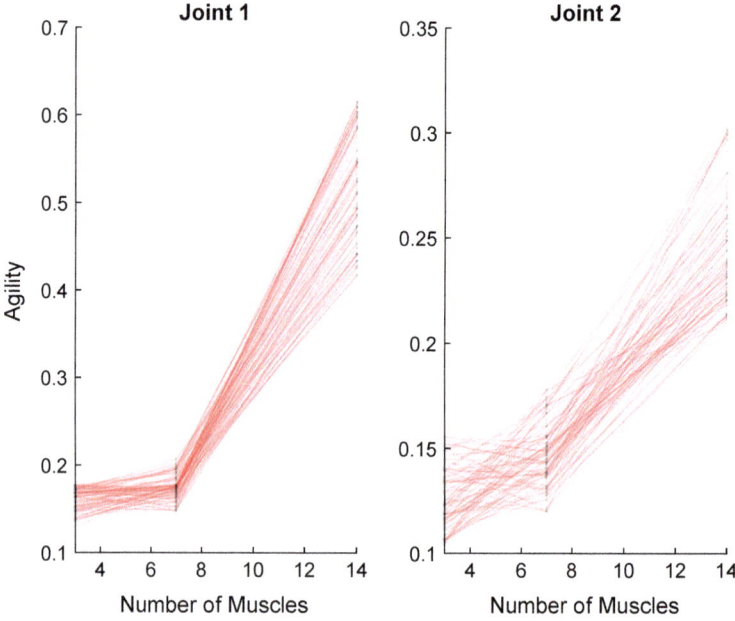

Fig. 11 Agility plots for each of the two joints (Monte Carlo analysis)

Carlo analysis to determine the sensitivity of the results to the assumed values for the moment arm matrix. The results follow the same patterns regardless of the variation in values for the moment arm matrix, demonstrating that this analysis is generalizable to a large variation in moment arm values.

Our results show that increasing the number of muscles increases Effectiveness and Agility. However, it is clear that more muscles also have more Phenotypical cost. This is why we believe that there is a "sweet spot" for an optimal number of muscles, based on how much weight each of these goals have in an anthropomorphic system. We believe that, these weights are set during evolution to find the optimal number of muscles to be as versatile as possible, while maintaining a reasonable Phenotypical cost. That is where the Overall Fitness, introduced in the methods section (Eq. 26), will be useful. The Overall Fitness plays a pivotal role by combining the weighted effect of each element and providing a single measure that needs to be optimized. Again, while adding extra muscles will increase Effectiveness (Fig. 10) and Agility (Fig. 11), it will also increase the Phenotypical cost. Therefore, for each system, the number of muscles for which this Overall Fitness is minimized is the optimal muscle number for that system.

By changing the weights in the Overall Fitness function, we can easily find the set of weights where 7 muscles (the real number of muscles in a human arm) are the

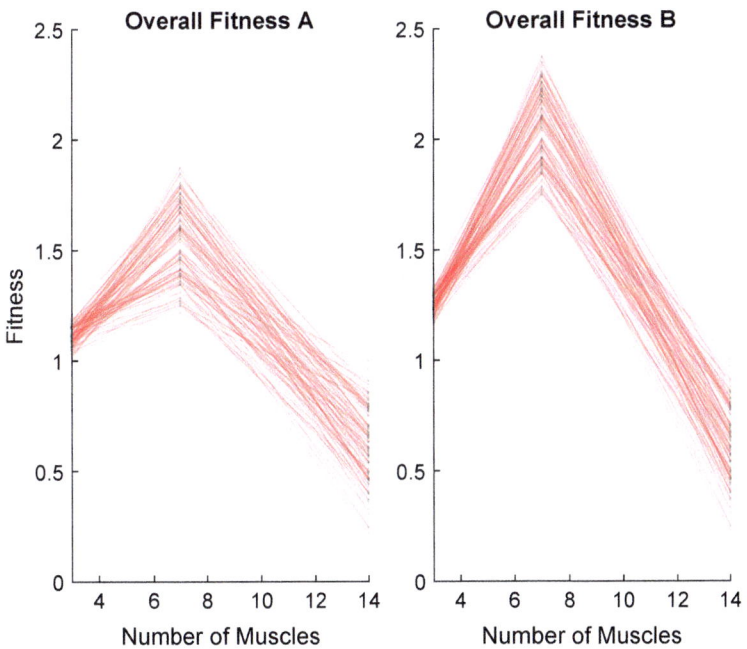

Fig. 12 Overall Fitness **a** and **b** as a function of the number of muscles (Monte Carlo analysis). Note that Phenotypical costs are quadratic (**a**) or cubic (**b**) functions that dominate for larger numbers of muscles

optimal choice. As described before, these weights can be interpreted as the relative importance of each goal (Effectiveness, Agility, and low Phenotypical cost) to vertebrates, from an evolutionary point of view. By setting w_1, w_2, and w_3 to 4, 1, and 4 respectively, we have Fig. 12, which shows that the optimal number of muscles is 7.

Please note that there are many combinations of weights that can lead to a specific number of muscles. Similarly, any choice of cost function in the literature can be a matter of choice and preference [23]. We chose three fitness functions (Effectiveness, Agility and Phenotypical cost) to reflect the multi-dimensional nature of functional fitness. This, in fact, is best addressed as multi-objective optimization that allows espousing any one to the exclusion of others. This confronts us with the fact that any cost function is, in essence, a reflection of the multiple fitness criteria that may have been achieved, are being pursued, or are even changing in the environment. Thus, a change in environment, goals and life habits, over time, may naturally change the number and/or routing of muscles in a given anthropomorphic system. In addition, although the general patterns between simulations and real systems match, it is important to keep in mind the simplifications that were made, when comparing results from simulations to physiological recordings.

3 Does a Minority of the Variance Contain a Majority of Information?

In this section, we explore the risks of assuming that a low-dimensional approximation suffices to capture the versatility of anthropomorphic systems. We have shown that, although a few Principal Components (PCs) can explain most of the variance in a specific movement or a set of gestures, the remaining variance can in fact contain critical details. This highlights that the reduction of DOF will come at the cost of versatility. Therefore, the fact that a few PCs capture the most of the variance does not mean that anthropomorphic systems should be low-dimensional.

The problem of face recognition serves as a useful analogy. Human faces all share common features, e.g. we generally have two eyes, two ears; and the general placement of the mouth, nose, eyes and ears follows a specific pattern. However, we can recognize a particular face from among many only due to its small differences compared to the others. Similarly, when talking about hand gestures, postures and functions, the details can become very important to a specific task.

3.1 Background

It is known that different sets of motor actions share many commonalities. For example, a linear combination of a small set of basis vectors in a set of movements can explain large amounts of variance for each movement pattern in the set [24]. It is also true for static postures, which means a linear combination of a few basis vectors can explain large amounts of variance in a multi-DOF system; like a human hand [25]. Unfortunately, this is often over-interpreted as a sign that we have more than enough DOF, or as a sign of redundancy in anthropomorphic systems.

We have used principal component analysis (PCA) to extract the principal components (PCs) in a set of hand gestures and shown that although the first few PCs will explain a large amount of variance, the details that make differentiating between these different movements or postures possible are present in the higher PCs. That is, although higher PCs explain less variance and are generally smaller in amplitude, they are the most important in making postures different from one another. Therefore, these extra DOF are the main contributors to versatility in anthropomorphic systems.

In this section, we further demonstrate this concept with a special focus on hand gestures. We simulate five different hand gestures, comparing and contrasting their representations in the joint angle space (19 joints). We also apply PCA to the joint angle data of these three gestures and demonstrate the effects of utilizing only the first two PCs as compared to all PCs involved.

3.2 Method

Grasping gestures of the hand are historically categorized into two main sets: "precision grasp" and "power grasp" [26]. In the former, the thumb and one or more of the remaining fingers will contact the object or apply force in opposition to each other. In the latter, the object will be grasped such that the palm of the hand comes into contact with it [27].

To model our distinct hand gestures, we used MuJoCo, a physics engine which provides accurate simulations for applications including robotics and biomechanics [28]. Five different hand gestures were modeled; two power grasps, two precision grasps, and a non-practical posture which we refer to as the Claw gesture. These five gestures are represented in Fig. 13.

In both power grasps, the fingers opposing the thumb follow similar flexion/extension patterns in their joints. The main difference between power grasps 1 and 2 are the finger abduction values. Index, middle, ring, and pinky fingers have more space between them in power grasp 2 as compared to the power grasp 1. In precision grasp 1, only the index finger opposes the thumb while the other fingers are flexed. In precision grasp 2, the middle finger also opposes the thumb while the other fingers are less flexed. In the Claw gesture, index and ring group together with middle and pinky fingers respectively, and are opposing the thumb.

We extract 19 different joint angles for these three gestures from MuJoCo. These angles are namely Wrist PRO, Wrist UDEV, Wrist FLEX, Thumb ABD, Thumb MCP, Thumb PIP, Thumb DIP, Index ABD, Index MCP, Index PIP, Index DIP, Middle MCP, Middle PIP, Middle DIP, Ring ABD, Ring MCP, Ring PIP, Ring DIP, Pinky ABD, Pinky MCP, Pinky PIP, and Pinky DIP. We then apply PCA (similar to [25]) and compare joint angles of all five gestures before and after applying dimensionality reduction. In the reduced dimension case, we map only the first two PCs back to the joint angle space. We also calculate the Pearson's correlation coefficient for each pair of gestures in the joint angle space for before and after dimensionality reduction. The Pearson's correlation coefficient of two different gestures in the joint angle space is defined as follows:

$$Corr_{i,j} = \frac{\sum_{k=1}^{19}(angle_{i,k} - \overline{angle_i})(angle_{j,k} - \overline{angle_j})}{\sqrt{\sum_{k=1}^{19}(angle_{i,k} - \overline{angle_i})^2}\sqrt{\sum_{k=1}^{19}(angle_{j,k} - \overline{angle_j})^2}} \qquad (27)$$

where $Corr_{i,j}$ stands for the Pearson's correlation coefficient between the ith and the jth gestures and $angle_{i,j}$ represents the angle in the kth joint of the ith posture. Moreover, $\overline{angle_x}$ represents the sample average of xth gesture and is defined as:

$$\overline{angle_i} = \frac{1}{19}\sum_{k=1}^{19} angle_{i,k} \qquad (28)$$

In addition, to show the correlations for all five pairs (with and without dimensionality reduction), we created a five by five matrix in which the color of the element on the ith row and jth column represents the Pearson's correlation coefficient between the ith and the jth gesture in the joint angle space. This correlation matrix is defined as follows:

$$C = \begin{bmatrix} Corr_{1,1} & \cdots & Corr_{1,j} \\ \vdots & \ddots & \vdots \\ Corr_{i,1} & \cdots & Corr_{i,j} \end{bmatrix} \qquad (29)$$

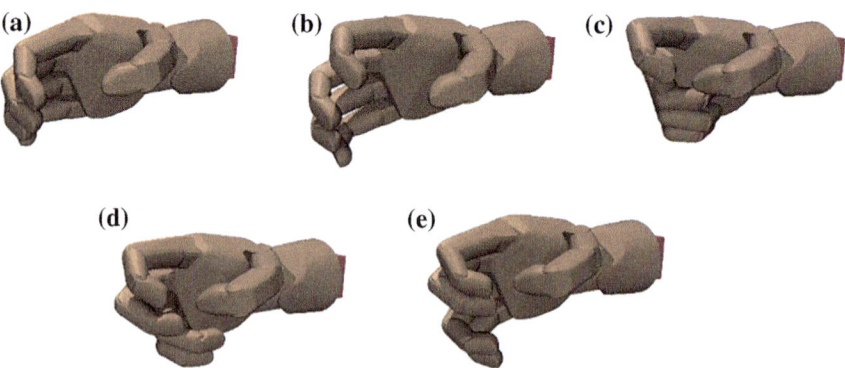

Fig. 13 3D model of the five different hand gestures studied in this section. **a** Power grip 1. **b** Power grip 2. **c** Precision grip 1. **d** Precision grip 2. **e** The Claw gesture

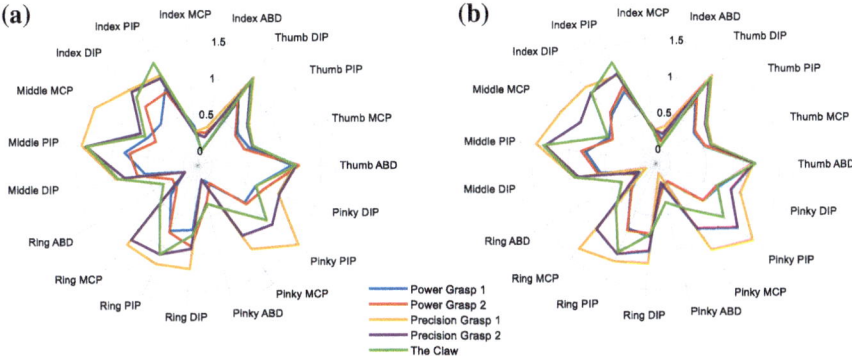

Fig. 14 Spider plot representations of the 19 joint angles **a** Without dimensionality reduction. **b** With only the first two PCs

3.3 Results

The joint angle representation of the hand gestures shown in Fig. 13 are illustrated in Fig. 14a. We apply PCA (as explained in the method section) to extract the most common components for the five gestures. 91.40% of the variance is explained by the first two PCs. The resulting joint angle space representation for the first two PCs is shown in Fig. 14b.

As can be seen in Fig. 14, with only considering the first two PCs, power grasp 1 and power grasp 2 have grouped together. The same pattern is observed with precision grasp 1 and precision grasp 2. This was within our expectations since by saving the first two PCs, we are ignoring the smaller differences and paying attention to the commonalities.

Figure 15 shows the correlation values for the two cases studied on Fig. 14 colored boxes in the matrix shown in Fig. 15a represent the correlation coefficient between the joint angle representation vectors of different gestures. Figure 15b represents the same measure for the case with each gesture only being represented by their first two PCs. Comparing Fig. 15a and b, we can make three very important observations.

First, in Fig. 15b, gestures are clustered into three main groups; namely, Power grasp, Precision grasp, and the Claw. These clusters are represented as red squares in Fig. 15b. This shows that by keeping only the first two PCs, the intra-group correlation values have increased i.e. intra-group dissimilarities have decreased. In other words, all group members have lost their distinctions in the reduced dimension space to some extent.

Second, the correlation values between gestures in power grasp and gestures in precision grasp groups are converged to the almost same values for any pair of gestures from these two groups. This is illustrated as the dark and light blue lines on the intersections of the power grasp and precision grasp clusters in Fig. 15b. This is important because in the reduced dimension space, dissimilarity values between

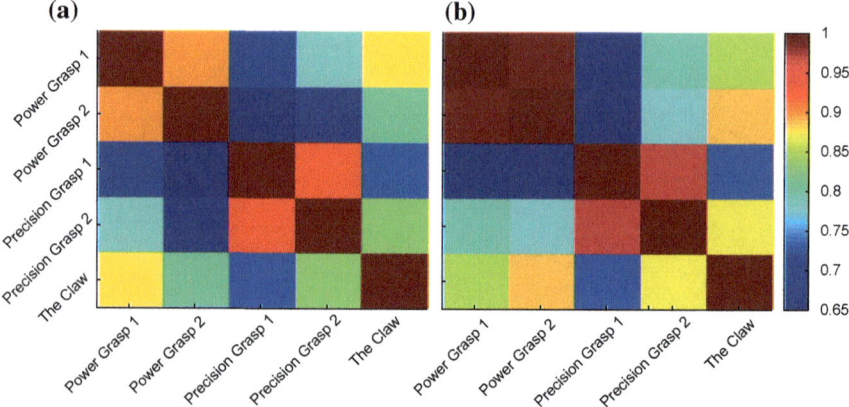

Fig. 15 The correlation coefficients for each of the gesture pairs. **a** Without dimensionality reduction. **b** With only the first two PCs

any of the power grasp gestures with any of the precision grasp gestures are almost the same. This means that distinguishing different gestures from each other is much more difficult in the reduced dimension space.

Third, the Claw gesture is much more closely correlated to other gestures in the reduced dimensional space. This is mostly observed between the correlation values of the Claw and power grasp 2, and also between the Claw and precision grasp 2.

This, again, makes an accurate distinction between the different gestures more challenging. This was observed even though the Claw gesture is a non-practical gesture, which is unlikely to be used in day to day activities.

4   Conclusions and Future Work

In this chapter, we took three different approaches to address the question: "Should anthropomorphic systems be redundant?"

In the first section, we presented the classical approach to muscle redundancy for joint torque and endpoint force production in tendon-driven systems. The notion of muscle redundancy holds that there are many ways in which tendon-driven anthropomorphic systems can activate muscles to generate the desired net torques at each joint [17, 18]. However, we underscore that tendon-driven systems are, in fact, *overdetermined* from the perspective of *tendon excursions* [1, 6, 14]. That is, the lengths and velocities of all muscles crossing a joint, or set of joints, are determined by the rotations at those joints. From a mechanical perspective, this means that a given limb movement defines a unique set of tendon excursions and velocities. Muscles that shorten during the movement can, in principle, go slack (but then they do not contribute to torque production). However, muscles that lengthen during the movement must do so as specified by the joint rotations. This poses a practical problem in the case where motors are not backdrivable or muscles have stretch reflexes: any muscle that does not lengthen appropriately will disrupt the movement. Therefore, the controller (be it neural or engineered) seeking to produce smooth and accurate movement in a tendon-driven system is not necessarily confronted with a redundant system, as is typically assumed. Rather, it must excite muscles to produce the necessary time history of joint torques *while* allowing muscles to lengthen in the precise way needed. This perspective is not new. Sherrington emphasized the importance of inhibition as a central requirement for the production of movement over 100 years ago [29, 30]. We therefore propose that it is critical for researchers today to pursue a neo-Sherringtonian research direction to understand the robotic and neural control of movement.

The fact that moving smoothly and accurately is neither a redundant, simple, nor a forgiving control problem for tendon-driven systems poses several critical research directions. For example, why does such behavior take years to perfect during typical development in humans (and still not fully available in robot), and why is it so susceptible to developmental and neurological conditions? We propose that the over-determined nature of muscle excursions makes the problem of producing smooth and accurate movements unforgiving to even small errors in development and neurological conditions, which requires further study [1, 6, 14].

In addition, there is emerging evidence that cardinal features of healthy force and movement variability (which are often considered to have cortical origins) can arise naturally as a consequence of the neural control of afferented muscles (i.e., where regulating reflex gains is critical) [31–33]. This opens new research directions to begin to explain, from a purely spinal and peripheral perspective, the clinical presentation of at least some types of tremor in neurological conditions.

The second approach described some aspects of Feasibility Theory, which helps us understand how the anatomy of the plant, and the mechanical constraints that define the task, determine the dimensionality and structure of its feasible activation set (i.e., the family of all feasible commands that can accomplish the task). Such feasible activation sets are well-structured, low-dimensional subspaces embedded in the high-dimensional space of muscle activations. Thus, future research should focus on how the controllers of anthropomorphic systems can explore, identify, exploit, and remember those feasible activation sets. After all, the most any neural or engineered controller can do is explore and exploit the capabilities of the tendon-driven system as a whole [1].

Another important aspect of Feasibility Theory is that the number of independently-controllable muscles also determines the number of independent task constraints that tendon-driven systems can satisfy [1, 16]. Thus, adding and having more (appropriately placed) muscles may, in fact, be the critical enabler of ecological (i.e., real-world) function. That is, more muscles enable performing more complex tasks—where complexity is taken to mean the need to meet more task constraints simultaneously. Thus, it is important to investigate how ecological tasks necessitate having more muscles than are apparently necessary when studying "simpler" experimental tasks [2]. This implies that failure to control all muscles independently (as is common in, say, stroke) will reduce functional capabilities because independent muscle control is necessary to meet the multiple requirements of ecological tasks. Thus, "redundant" systems with many muscles (or control DOFs in general) are functionally desirable.

Importantly, these findings also motivate further research on the advantages and disadvantages of muscle synergies (where several muscles are activated in a correlated manner, effectively reducing the number of independent control DOFs). Given that implementing muscle synergies can be an effective way to control robots [34]—and assess limb movement [35]—exploring the relationship between the number of independent control DOFs and functional versatility requires further study [4, 11, 24, 36].

The second section also approached the classical problem of muscle redundancy from the perspective of multi-objective optimization. That is, tackling simultaneous and independent functional goals. Here, we have explored three: maximal end-point force, maximal joint angular velocities, and the Phenotypical cost of having additional muscles. We showed how an anthropomorphic system can adapt to consider all three goals to arrive at the most desirable number of muscles (yet sub-optimal with respect to individual cost functions). Moreover, this desirable number of muscles is a function of the relative weighing across goals. We find that more muscles allow the limb to be better at multiple goals. Moreover, this study underscored how a change in environment, goals, and life habits may, over time, naturally change the number and/or routing of muscles in a given anthropomorphic system, and vice versa.

The third and last section highlights our final approach to the question: "Should anthropomorphic systems be redundant?" We first showed how, in agreement with [25], more than 90% of the variance in different hand gestures studied here can be explained with only two principal components. However, disregarding the remaining PCs will naturally make it more difficult to distinguish and/or implement each gesture. This example highlights a little-appreciated—in our opinion—consequence in dimensionality reduction: that it will make it very difficult to distinguish similar hand gestures with different functional roles as per well-known grasp taxonomies [37]. For example, a power grasp with the thumb abducted serves to oppose the fingertips. If the thumb is slightly adducted, it can press against the side of the fingers to roll an object with high precision (see Figure 3 in [37]). Thus, although the higher PCs explain ever-decreasing percentages of variance, they nevertheless have important functional consequences. This is in agreement with recent findings in the field of soft robotics. Such studies show how small amounts of passive deformation, provided by non-stiff materials, can significantly increase the functional capabilities, robustness, and versatility of under-actuated hands [4, 38, 39]. Therefore, finding that some PCs that explain relatively little variance does not necessarily mean that the system has unnecessary DOFs or is functionally redundant.

MATLAB toolbox

Most of the simulations demonstrated in this chapter were performed using a custom Neuromechanics toolbox written in MATLAB. This toolbox is available at https://github.com/marjanin/Neuromechanics-Toolbox and at https://ValeroLab.org.

Acknowledgements This study was supported by the National Institute of Arthritis and Muscu-loskeletal and Skin Diseases of the National Institute of Health (NIH) under award numbers R01 AR-050520 and R01 AR-052345 to FVC, and by the Department of Defense under award number MR150091 to F.V-C., and University of Southern California Graduate School's Provost Fellowship to AM. The content is solely the responsibility of the authors and does not necessarily represent the official views of the funding agencies. We acknowledge Dr. Christopher Laine and Amir Ahmadi for their helpful comments on this manuscript. We thank the University of Southern California for facilities provided during the course BME/BKN 504 and Antara Dandekar and Monica Guerrero for their help with the conceptualization of Section II.

References

1. Valero-Cuevas, F.J.: Fundamentals of Neuromechanics. Springer, Berlin (2015)
2. Loeb, G.E.: Overcomplete musculature or underspecified tasks? Mot. Control **4**(1), 81–83 (2000)
3. Ogata, K., Yang, Y.: Modern Control Engineering, vol. 4. Prentice hall, India (2002)
4. Brock, O., Valero-Cuevas, F.: Transferring synergies from neuroscience to robotics comment on hand synergies: integration of robotics and neuroscience for understanding the control of biological and artificial hands by M. Santello et al. Phys. Life Rev. **17**, 27 (2016)
5. Rieffel, J., Valero-Cuevas, F., Lipson, H.: Automated discovery and optimization of large irregular tensegrity structures. Comput. Struct. **87**(5), 368–379 (2009)
6. Hagen, D.A., Valero-Cuevas, F.J.: Similar movements are associated with drastically different muscle contraction velocities. J. Biomech. **59**, 90 (2017)
7. Odhner, L.U., Jentoft, L.P., Claffee, M.R., Corson, N., Tenzer, Y., Ma, R.R., Buehler, M., Kohout, R., Howe, R.D., Dollar, A.M.: A compliant, underactuated hand for robust manipulation. Int. J. Robot. Res. **33**(5), 736–752 (2014)
8. Lee, Y.-T., Choi, H.-R., Chung, W.-K., Youm, Y.: Stiffness control of a coupled tendon-driven robot hand. IEEE Control Syst. **14**(5), 10–19 (1994)
9. Kobayashi, H., Hyodo, K., Ogane, D.: On tendon-driven robotic mechanisms with redundant tendons. Int. J. Robot. Res. **17**(5), 561–571 (1998)
10. Fu, J.L., Pollard, N.S.: On the importance of asymmetries in grasp quality metrics for tendon driven hands. In: 2006 IEEE/RSJ International Conference on Intelligent Robots and Systems, pp. 1068–1075. IEEE, New York (2006)
11. Inouye, J.M., Kutch, J.J., Valero-Cuevas, F.J.: A novel synthesis of computational approaches enables optimization of grasp quality of tendon-driven hands. IEEE Trans. Rob. **28**(4), 958–966 (2012)
12. Inouye, J.M., Valero-Cuevas, F.J.: Anthropomorphic tendon-driven robotic hands can exceed human grasping capabilities following optimization. Int. J. Robot. Res. **33**(5), 694–705 (2014)
13. Mardula, K.L., Balasubramanian, R., Allan, C.H.: Implanted passive engineering mechanism improves hand function after tendon transfer surgery: a cadaver-based study. Hand **10**(1), 116–122 (2015)
14. Valero-Cuevas, F.J., Cohn, B., Yngvason, H., Lawrence, E.L.: Exploring the high-dimensional structure of muscle redundancy via subject-specific and generic musculoskeletal models. J. Biomech. **48**(11), 2887–2896 (2015)
15. Proske, U., Morgan, D.: Muscle damage from eccentric exercise: mechanism, mechanical signs, adaptation and clinical applications. J. Physiol. **537**(2), 333–345 (2001)
16. Inouye, J.M., Valero-Cuevas, F.J.: Muscle synergies heavily influence the neural control of arm endpoint stiffness and energy consumption. PLoS Comput. Biol. **12**(2), e1004737 (2016)
17. Crowninshield, R.D., Brand, R.A.: A physiologically based criterion of muscle force prediction in locomotion. J. Biomech. **14**(11), 793–801 (1981)

18. Chao, E., An, K.-N.: Graphical interpretation of the solution to the redundant problem in biomechanics. J. Biomech. Eng. **100**(3), 159–167 (1978)
19. Yoshikawa, T.: Foundations of Robotics: Analysis and Control. MIT press, Cambridge, MA (1990)
20. Leijnse, J.: A generic morphological model of the anatomic variability in the m. flexor digitorum profundus, m. flexor pollicis longus and mm. lumbricales complex. Cells Tissues Organs **160**(1), 62–74 (1997)
21. Zajac, F.E.: Muscle and tendon properties models scaling and application to biomechanics and motor. Crit. Rev. Biomed. Eng. **17**(4), 359–411 (1989)
22. Chvatal, V.: Linear Programming. Macmillan, New York (1983)
23. Prilutsky, B.I.: Muscle coordination: the discussion continues. Mot. Control **4**(1), 97–116 (2000)
24. Kutch, J.J., Valero-Cuevas, F.J.: Challenges and new approaches to proving the existence of muscle synergies of neural origin. PLoS Comput. Biol. **8**(5), e1002434 (2012)
25. Santello, M., Flanders, M., Soechting, J.F.: Postural hand synergies for tool use. J. Neurosci. **18**(23), 10105–10115 (1998)
26. Napier, J.R.: The prehensile movements of the human hand. Bone Joint J. **38**(4), 902–913 (1956)
27. Johansson, R.S., Cole, K.J.: Sensory-motor coordination during grasping and manipulative actions. Curr. Opin. Neurobiol. **2**(6), 815–823 (1992)
28. Todorov, E., Erez, T., Tassa, Y.: Mujoco: a physics engine for model-based control. In: 2012 IEEE/RSJ International Conference on Intelligent Robots and Systems (IROS), pp. 5026–5033. IEEE, New York (2012)
29. Sherrington, C.S.: Reflex inhibition as a factor in the co-ordination of movements and postures. Exp. Physiol. **6**(3), 251–310 (1913)
30. Sherrington, C.S.: Inhibition as a coordinative factor https://www.Nobelprize.org (1932)
31. Laine, C.M., Nagamori, A., Valero-Cuevas, F.J.: The dynamics of voluntary force production in afferented muscle influence involuntary tremor. Front. Comput. Neurosci. **10**, 86 (2016)
32. Jalaleddini, K., Nagamori, A., Laine, C.M., Golkar, M.A., Kearney, R.E., Valero-Cuevas, F.J.: Physiological tremor increases when skeletal muscle is shortened: implications for fusimotor control. J. Physiol. **595**, 7331 (2017)
33. Nagamori, A., Laine, C.M., Valero-Cuevas, F.J.: Cardinal features of involuntary force variability can arise from the closed-loop control of viscoelastic afferented muscles. PLoS Comp Biol (2017, in press)
34. Santello, M., Bianchi, M., Gabiccini, M., Ricciardi, E., Salvietti, G., Prattichizzo, D., Ernst, M., Moscatelli, A., Jörntell, H., Kappers, A.M., et al.: Hand synergies: integration of robotics and neuroscience for understanding the control of biological and artificial hands. Phys. Life Rev. **17**, 1–23 (2016)
35. Ting, L.H., McKay, J.L.: Neuromechanics of muscle synergies for posture and movement. Curr. Opin. Neurobiol. **17**(6), 622–628 (2007)
36. Valero-Cuevas, F.J., Santello, M.: On neuromechanical approaches for the study of biological and robotic grasp and manipulation. J. Neuroeng. Rehabil. **14**(1), 101 (2017)
37. Feix, T., Romero, J., Schmiedmayer, H.-B., Dollar, A.M., Kragic, D.: The grasp taxonomy of human grasp types. IEEE Trans. Human-Mach. Syst. **46**(1), 66–77 (2016)
38. Deimel, R., Brock, O.: A novel type of compliant and underactuated robotic hand for dexterous grasping. Int. J. Robot. Res. **35**(1–3), 161–185 (2016)
39. Catalano, M.G., Grioli, G., Farnioli, E., Serio, A., Piazza, C., Bicchi, A.: Adaptive synergies for the design and control of the Pisa/IIT SoftHand. Int. J. Robot. Res. **33**(5), 768–782 (2014)

From Biomechanics to Robotics

Galo Maldonado, Philippe Souères and Bruno Watier

Abstract How does the central nervous system select and coordinate different degrees of freedom to execute a given movement? The difficulty is to choose specific motor commands among an infinite number of possible ones. If some invariants of movement can be identified, there exists however considerable variability showing that motor control favors rather an envelope of possible movements than a strong stereotypy. However, the central nervous system is able to find extremely fast solutions to the problem of muscular and kinematic redundancy by producing stable and precise movements. To date, no computational model has made it possible to develop a movement generation algorithm with a performance comparable to that of humans in terms of speed, accuracy, robustness and adaptability. One of the reasons is certainly that correct criteria for the synthesis of movement as a function of the task have not yet been identified and used for motion generation. In this chapter we propose to study highly dynamic human movements taking into account their variability, making the choice to consider performance biomechanical variables (tasks) for generating motions and involving whole-body articulations.

1 Introduction

Humanoid robotics is growing towards agile, robust, and powerful robots able to interact with environments in which humans might be present [48]. To control these robots, motion generation algorithms are being improved to increase the robot stability, robustness, and efficiency [4, 36]. Though a large part of control frameworks are purely computational, some of them take inspiration from humans. This is not an

G. Maldonado (✉) · B. Watier
LAAS-CNRS, Université de Toulouse, CNRS, UPS, Toulouse, France
e-mail: gmaldona@laas.fr

B. Watier
e-mail: bruno.watier@univ-tlse3.fr

P. Souères
LAAS-CNRS, Université de Toulouse, CNRS, Toulouse, France
e-mail: soueres@laas.fr

© Springer International Publishing AG, part of Springer Nature 2019
G. Venture et al. (eds.), *Biomechanics of Anthropomorphic Systems*, Springer Tracts in Advanced Robotics 124, https://doi.org/10.1007/978-3-319-93870-7_3

arbitrary choice. Understanding human motion could provide the information needed to improve the design and control of humanoid robots. Modeling human motion has lead to applications in the movie and gaming industries for realistic animation of characters. In the robotic industry, collaborative robots have emerged over the past years with the objective of assisting humans in their daily life activities. Examples include assistance robots for elderly people, exoskeletons to extend human physical capabilities, brain-machine interfaces connected to robotic arms, virtual reality training systems for surgery, or controlled prostheses.

Humans and robots share some important characteristics such as under-actuation and redundancy. Under-actuation means that the number of actuated degrees of freedom "DoF" of the system is lower than the total number of available DoF. The under-actuated part are 6 DoF that define the body pose: position and orientation. In robotics these 6 parameters describe the pose of a reference frame usually called root frame. Redundancy expresses the fact that the system has more DoF than necessary to achieve a given task. The difficulty is then to select a motor command among an infinite of possible ones. In humans, redundancy appears at the level of neurons, muscles, and joints. In robots, it requires the design of motion generation algorithms to select the best solution for controlling a movement. In biomechanics, redundancy can be studied by computing the kinematics and dynamics of human motion which allows to explain motor control strategies in terms of performance variables (tasks).

In this chapter, a set of tools and methods in biomechanics and robotics and a case study of human-inspired motion generation is presented. The chapter is organized as follows. Section 2 provides a brief summary of the methods used in biomechanics to study human motion. A summary of the robotic framework used to generate whole-body motion is then given in Sect. 3. Section 4 presents a case study of a human dynamic motion which was analyzed through biomechanical methods and generated using the robotics framework. Lastly, a summary of the method used to generate human-inspired motion from biomechanics to robotics is given in Sect. 5.

2 Biomechanics Background

Biomechanics aims at applying the physical laws of mechanics to study human motion in order to provide a better understanding of the mechanical strategies used by the central nervous system (CNS) to coordinate motions. This information can be used to generate human-inspired motions with humanoid robots or anthropomorphic systems such as animation avatars. This section provides a brief summary of the mathematical and computational background needed to study the biomechanics of the human motion.

2.1 Motion Capture

Motion capture "MoCap" is a technology that allows motion to be recorded online. In biomechanics, it is used to record the movements of humans and animals for quantitative analyses. The motion is usually tracked using cameras that record the 2D positions of markers placed on the body. When more than one camera is available, the 3D marker coordinates can be computed by the MoCap system. Two types of markers can be used depending on the motion capture system: active and passive markers.

To reconstruct the motion, active systems use infra-red active markers with light-emitting diodes stimulated in a predefined sequence. On the other hand, passive systems use markers covered by a reflective tape which are lighted by infra-red cameras. Coordinate data are calculated by the system with respect to a laboratory fixed reference frame. Data can then be processed to obtain kinematic variables describing segment or joint movements. A motion capture system might also synchronize other information such as reaction forces through force sensors or muscular activity through electromyography (EMG) sensors. In order to communicate these data, an experimental protocol must be carefully described and followed. The recording protocol should take into account recommendations for reporting biomechanics data based on international standards. For example, kinematic data are collected and reported based on recommendations from the International Society of Biomechanics (ISB) [52, 53] and surface muscular activity is recorded and communicated based on the recommendations from the European project of Surface ElectroMyoGraphy for the Non-Invasive Assessment of Muscles (SENIAM).

2.2 Human Data Processing

Recorded motion data have to be reconstructed and processed. Data reconstruction consists in labeling markers and interpolating their position to reconstruct their trajectories. Next, data are filtered to remove noise from the recorded and reconstructed signals. For filtering the reconstructed data, a low-pass Butterworth digital filter applied in a zero-phase is commonly used in biomechanics. In order to filter force and marker trajectories, the same cut-off frequency is chosen to avoid inconsistencies with inverse dynamics computations (inverse dynamics of human motion will be presented in Sect. 2.5) [31, 37]. To select the cut-off frequency for filtering a signal, two types of analyses are commonly performed: power spectral analysis and/or residual analysis [51]. Power spectral analysis is a technique in which the power of the recorded signals can be studied in the frequency domain. Based on this analysis, a decision can be made to define which frequencies are to be accepted or rejected from the signals. Residual analysis [51] is used to evaluate noise by comparing the difference between the unfiltered signals and the signals filtered at different cut-off frequencies. Let N_f be the number of sampled frames in a signal, the residual ε at a given cut-off frequency f_c is calculated as follows:

$$\varepsilon(f_c) = \frac{1}{N_f} \sum_{i=1}^{N_f} \left(s_i - \hat{s}_i(f_c)\right)^2, \tag{1}$$

where s_i is the raw signal, $\hat{s}_i$ is the filtered signal with cut-off frequency f_c. Equation (1) is calculated recursively for a set of given cut-off frequencies. The set of cut-off frequencies is arbitrary but should be chosen so that the frequencies of the signal can be represented. Residuals are later displayed as a function of these cut-off frequencies to evaluate the contained information. Criteria related to the choice of the cutoff frequency based on residual analysis can be found in [51].

After data have been filtered, joint centers of rotation can be estimated from markers' trajectories. Three methods can be used to this end: virtual models (e.g. in OpenSim software [9]), regression tables (e.g. [12]) and/or functional methods (e.g. [15]). In general, the center of rotation of ball-modeled joints such as the shoulder and the hip are computed by using functional methods. The SCoRe method provides a good estimation [15] and can be computed as:

$$\min_{c1,c2} \sum_{i=1}^{N_f} \left\| R_{1_i} c_1 + t_{1_i} - (R_{2_i} c_2 + t_{2_i}) \right\|^2 , \tag{2}$$

where $c_1 \in \mathbb{R}^3$ and $c_2 \in \mathbb{R}^3$ are the centers of rotation in the body segment coordinate system, (R_{1_i}, t_{1_i}) and (R_{2_i}, t_{2_i}) are respectively the transformation matrices of body 1 and body 2 from segment (local) to world (global) coordinates. The solution to this minimization problem provides two trajectories c_1 and c_2 for the same joint and thus the mean value of both can be used as an estimate. Functional methods can also be used to determine the optimal axes of joint rotations (e.g. for calculating the knee axes [16]). Other joints center can be computed based on regression tables or based on a virtual model (which will be introduced in Sect. 2.3.1).

2.3   Scaling of Human Anthropometry

The human body is composed of hundreds of muscles and bones. In order to calculate the kinematics and dynamics, a physical model representing the skeletal or musculosketal system is needed. The model is simplified by the assumption that the human body can be described by a collection of rigid bodies (segments) representing bones or a combination of them (for example when modeling the torso or the foot segments). In spite of this simplification, this model is usually called skeletal model and the same convention will be used in this chapter. In the case of muscle studies, a more complex model of the system is needed: musculoskeletal models.

The scaling of anthropometry is used to estimate the properties of the human body segments such as segment lengths, inertia matrices, and center of mass positions. To estimate human anthropometry, cadaveric studies [12], mathematical modeling [24], scanning and imaging techniques [14, 54], kinematic measurements [23], or motion capture based identification with kinetic measurements [49], can be used. Commonly, experimental marker data from static trials are used to scale the anthropometry of the recorded participant using the regression equations provided by cadaveric studies

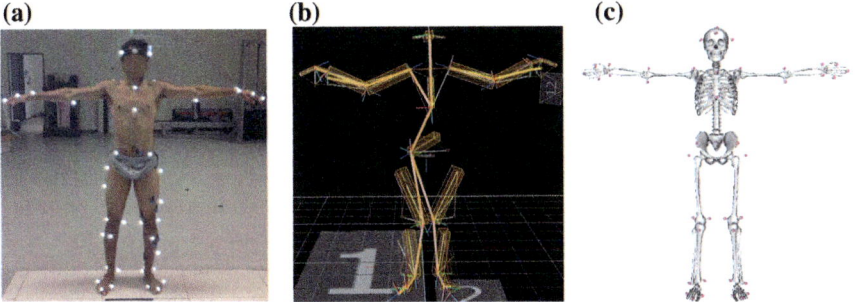

Fig. 1 In (**a**), a human with passive markers on his body. In (**b**), the model created for motion reconstruction with the Vicon Nexus software. In (**c**), a virtual skeleton based on OpenSim models which can be used for scaling human anthropometry

or by fitting these data with a virtual skeletal model (Fig. 1). Further details about the virtual skeletal model are given in Sect. 2.3.1 while the data fitting technique is presented later in Sect. 2.4.

2.3.1   Human Skeletal Model

A skeletal model is an effective tool for visualizing and analyzing human motion. When building a virtual 3D-model, the following information has to be considered:

- **The model** contains the description of the kinematic chain including joint types and joint ranges of motion.
- **Segment data** provide specifications of the physical characteristics of the body segments such as masses, inertia matrices, and positions of the center of mass of each segment.
- **Virtual markers** contain the positions of the markers placed on the model according to the experimental protocol. Markers are normally placed in accordance with the International Society of Biomechanics (ISB) standards with a minimum of 3 virtual markers per segment for a 3D analysis [52, 53].
- **The visual elements** are the 3D meshes that will be displayed, which can be created using a 3D software. These data are useful for visualization purposes but do not interfere with mathematical calculations.

2.4   Inverse Kinematics of Human Motion

Inverse kinematics of human motion consists in converting experimental marker positions into joint angles by minimizing the error coming from the reconstructed motion and from the soft tissue artifact "STA". STA corresponds to the relative motion of soft tissues with regards to the underlying bones. Two classes of methods are commonly used in the literature to calculate 3D angles: segment optimization [7] and global optimization [34]. In segment optimization, a reference frame is positioned and oriented on each modeled body segment according to the ISB recommendations. Angles are then calculated following Euler angle sequences by computing the optimal bone pose from a marker cluster. Let N_f be the total number of frames recorded in a given motion and let $N_{m,s}$ be the number of markers in segment s (at least 3 markers are needed to reconstruct the 3D coordinates of a segment). At a given frame, let $x_{s,prev}$ represent the 3D position of marker s in the previous frame and $x_{s,next}$ the 3D position of marker s in the next frame. Segment optimization methods are based on the following least-square problem:

$$\min_{d,R} \sum_{i=1}^{N_{m,s}} \left\| R x_{i,prev} + d - x_{i,next} \right\|^2, \tag{3}$$

where R and d are the rotation matrix and the translation vector respectively, which map coordinates $x_{i,prev}$ to $x_{i,next}$ so that $(R, d) : x_{i,prev} \mapsto x_{i,next}$. Global optimization methods, usually referred to as "Inverse Kinematics", can also be used to calculate joint angles and constitute a promising methodology [13]. Global optimization makes it possible to add physically realistic joint constraints while taking into account the whole kinematic chain structure, and joint ranges of motion. Inverse kinematics solves the following problem:

$$\min_{q} \sum_{i=1}^{N_m} w_i \left\| x_i^{exp} - x_i(\mathbf{q}) \right\|^2 \tag{4}$$
$$\text{s.t.} \quad \underline{q} \leq q \leq \overline{q},$$

where N_m is the total number of markers, q are the generalized coordinates, $\underline{q}$ and $\overline{q}$ are minimum and maximum ranges of motion of the joint coordinates q, x_i^{exp} is the experimental marker position of the ith marker, $x(q)$ is the corresponding virtual model marker position, and w_i is the marker weight, which specifies how strongly marker error should be minimized.

2.5 Inverse Dynamics of Human Motion

Inverse dynamics aims at determining internal forces and joint torques that generate a given motion. To this end, the body model described before (skeletal model), the movement kinematics and the measured external forces produced if there is contact with the environment are used. Three formalisms are available to compute inverse dynamics: Hamiltonian, Euler-Lagrange and Newton-Euler. In particular, Newton-Euler formulation expresses dynamic equations for each link and performs calculations recursively by propagating reaction forces and applying Newton's third law of motion (principle of action and reaction). Euler's first and second equations of motion form the so called Newton-Euler equations. The first equation states that the sum of external forces equals the variation of linear momentum:

$$\sum f_{ext} = \frac{d}{dt}(m\boldsymbol{v}) = m\boldsymbol{a}, \tag{5}$$

where m is the mass of the body and $\boldsymbol{a}$ its the center of mass acceleration. The second equation states that the sum of external torques equals the variation of angular momentum at the center of mass (CoM):

$$\sum \tau_{ext} = \frac{d}{dt}(I_G\boldsymbol{\omega}) = I_G\dot{\boldsymbol{\omega}} + \boldsymbol{\omega} \times I_G\boldsymbol{\omega}, \tag{6}$$

where I_G is the inertia, $\boldsymbol{\omega}$ and $\dot{\boldsymbol{\omega}}$ are respectively the body inertia, angular velocity and angular acceleration expressed at the CoM. Equations (5) and (6) are commonly propagated at each joint using a bottom-up approach. This method can be computed numerically through the application of the recursive formulation of Newton-Euler equations [32]. Four inverse dynamics methods have been proposed in the literature based on vectors and Euler angles, wrenches and quaternions, homogeneous matrices, or generalized coordinates and forces [11].

3 Whole-Body Motion Generation of Anthropomorphic Systems

Whole-body motion generation of anthropomorphic systems, such as humanoid robots, requires to model the system dynamics, in order to create stable and feasible movements, and to solve redundancy. In this section, the task function approach [38, 43] and the formalism of poly-articulated systems are recalled. The inverse kinematics and inverse dynamics formulation for solving the equations of motion in robotics are briefly introduced. Finally, a hierarchical task controller which aims at solving strict hierarchy problems is also reviewed.

3.1 Dynamic Model

The dynamic model of humanoid robots can be written by using the Euler-Lagrange equation. The robotic system is modeled as an under-actuated kinematic-tree chain composed of rigid bodies with a free-floating base (also called root frame) subject to external contact forces as follows:

$$M(q)\ddot{q} + b(q, \dot{q}) = S^T \tau_{int} + \sum_{k=1}^{K} J_k^T(q)\lambda_k, \qquad (7)$$

where $M(q)$ is the mass matrix, $b(q, \dot{q})$ contains gravitational, centrifugal and Coriolis forces, $S = \begin{bmatrix} 0_{n \times 6} & I_{n \times n} \end{bmatrix}$ is a matrix that selects the joint torques τ_{int} of the actuated part of Eq. (7), $J_k(q)$ is the Jacobian matrix of the kth external contact and $\lambda_k = \begin{bmatrix} f_{ext_k} & \tau_{ext_k} \end{bmatrix}^T$ is the vector of the external forces and torques induced by the kth contact.

3.2   Task Formalism for Motion Generation

Let n be the number of DoF of the system, Q the configuration space formed by $n - 6$ joints plus 6 parameters of pose (position and orientation) of the system root frame and $q(t) in Q$ the configuration vector at time t. Dependencies on time will be dropped for notation convenience when necessary. Let m be the dimension of the task space also called operational space [30]. A task function $e(q) \in \mathbb{R}^m$ comes down to an output error function whose regulation to zero corresponds to the execution of the task. For instance, a pointing task can be defined by the task function $e(q) = hand(q) - hand_{target}$, which describes the gap between the current hand position $hand(q)$ when the body is at configuration q and the expected hand position $hand_{target}$. In order to compute how the task function varies with respect to the body configuration, roboticists use the so called task Jacobian $J_e = \frac{\partial e_i}{\partial q_j}$. The derivative of the task function $e(q)$ with respect to time gives the first order kinematics relation:

$$\dot{e}(q, \dot{q}) = J_e(q)\dot{q}. \qquad (8)$$

The execution of the task can be regulated with a control law by specifying a reference behavior of the task $\dot{e}^*$, for example with a proportional derivative (PD) control law. The gains of the proportional derivative task can be tuned to obtain different reference behaviors such as exponential decays or adaptive gains of the task. Exponential decay control laws of the form $\dot{e}^* = -\lambda e$ are commonly used in robotics to make the task function converge quickly to a desired value. If reference behaviors can be extracted from humans in terms of performance variables, they can be used instead to designing human-inspired decay dynamics for the task function. For example, the minimum jerk criterion observed in human motions [20], has been used in the control of reaching tasks [27].

The task function approach can be extended to higher order derivatives for studying dynamic behaviors. By differentiating Eq. (8), the second order kinematics relation is obtained:

$$\ddot{e}(q, \dot{q}, \ddot{q}) = J_e(q)\ddot{q} + \dot{J}_e(q, \dot{q})\dot{q}, \qquad (9)$$

where $\dot{J}_e \dot{q}$ can be considered as a drift of the task.

The same formalism can be used to express other types of tasks. For example, it can be used to formulate tasks for controlling the behavior of the center of mass "CoM". Let n_s represent the number of segments of the system, $m_{s_i} \in \mathbb{R}$ the mass of the ith segment, and $c_i \in \mathbb{R}^3$ its CoM position. The CoM of the system $c \in \mathbb{R}^3$ can be computed as follows:

$$c(q) = \frac{\sum_{i=1}^{n_s} m_{s_i} c_i}{\sum_{i=1}^{n_s} m_{s_i}}. \tag{10}$$

The CoM task can then be expressed as $\mathbf{e}(\mathbf{q}) = c(q) - c^*$, where c^* is the expected CoM position.

Dynamic tasks can also be expressed in terms of momenta in order to consider inertial effects. Momenta tasks can be computed based on the centroidal dynamics: the dynamics computed at the CoM of the system [40]. A momenta task can be formulated by using the centroidal momentum matrix $A_G(q)$ which is the product of the inertia matrix of the system and the Jacobian matrix of the system ($A_G(q) = I_{sys} J_{sys}$) [39], which maps the joint velocities $\dot{q}$ to the centroidal momenta h_G as follows:

$$h_G = A_G(q)\dot{q}. \tag{11}$$

This relation has the form of Eq. (8). Moreover Eq. (11) can be differentiated with respect to time as:

$$\dot{h}_G = A_G(q)\ddot{q} + \dot{A}_G(q, \dot{q})\dot{q}, \tag{12}$$

which matches the pattern of Eq. (9).

A task can be expressed directly in the configuration space to control the posture of the robot. Instead of the task Jacobian matrix, a selection matrix is used to select the joints that will be controlled.

3.3 Inverse Kinematics Control

The inverse kinematics problem consists in finding the joint kinematics that allows the robot for accomplishing a reference kinematic task behavior. In the next subsections we show how to control the execution of tasks expressed in terms of velocities (first order kinematics) and accelerations (second order kinematics).

3.3.1 First Order Kinematics

Solving the first order kinematics comes to determine suitable joint velocities to generate the desired task velocity $e(q, \dot{q})$ using the relationship obtained in Eq. (8). In order to control the task performance, a reference task behavior e^* is provided as input and the control problem comes to solve the following unconstrained minimization problem:

$$\min_{\dot{q}^*} \left\| \dot{e}^* - J_e(q)\dot{q}^* \right\|_2^2. \tag{13}$$

The solution to this problem provides the control law of the system as follows:

$$\dot{q}^* = J_e^{\#}(q)\dot{e}^* + P_{J_e}\dot{q}_2, \tag{14}$$

where $\{.\}^{\#}$ represents the generalized inverse, P_{J_e} is the projector onto the null space of $J_e(q)$ (e.g. $P_{J_e}J_e = 0$ and $P_{J_e}P_{J_e} = P_{J_e}$) and $\dot{q}_2$ is a secondary control input that can be used to exploit the systems redundancy with respect to the task.

3.3.2 Second Order Kinematics

The second order kinematics problem deals with the relationship between task acceleration and joints acceleration provided by Eq. (9). The control problem, which consists in finding the suitable joint accelerations that generates the task reference behavior, can be expressed by the following minimization problem:

$$\min_{\ddot{q}^*} \left\| \ddot{e}^* - J_e(q)\ddot{q}^* + \dot{J}_e(q)\dot{q} \right\|_2^2. \tag{15}$$

Using the same notation as before, the control law expressed in terms of the joint accelerations is then written as:

$$\ddot{q}^* = J_e^{\#}(q)\left(\ddot{e}^* + \dot{J}_e(q)\dot{q}\right) + P_{J_e}\ddot{q}_2. \tag{16}$$

3.4 Inverse Dynamics Control

The inverse dynamics problem aims at determining the suitable joint torques to generate a reference task acceleration behavior $\ddot{e}^*$. By multiplying the actuated part of Eq. (7) by JM^{-1} and replacing Eq. (9) in Eq. (7) we obtain the following relation:

$$\ddot{e} + J_e(q)M^{-1}b - \dot{J}_e(q,\dot{q})\dot{q} = J_e(q)M^{-1}\tau. \tag{17}$$

The inverse dynamics control law can be written as:

$$\tau^* = \left(J_e(q)M^{-1}\right)^{\#}\left(\ddot{e}^* + J_e(q)M^{-1}b - \dot{J}_e(q,\dot{q})\dot{q}\right) + P_{J_e M^{-1}}\tau_2, \tag{18}$$

where $(P_{J_e M^{-1}})$ is the projector onto the null space of $J_e(q)M^{-1}$ and τ_2 is an arbitrary vector that can be used to control other tasks. Equation (18) can be extended to include rigid contact constraints [42]. The control of humanoid robots interacting with the environment has to take into account external forces. Thus, the control problem consists in finding a control law that achieves a desired task behavior while respecting the dynamic model of the system and additional constraints that ensure the feasibility of the motion. This problem can be solved by setting a minimization problem under equality and inequality constraints as in [41]:

$$\min_{\ddot{q},\tau,\lambda} \left\| \ddot{e}^* - \ddot{e}(q,\dot{q},\ddot{q}) \right\|_N^2$$

$$s.t. \quad M(q)\ddot{q} + b(q,\dot{q}) - g(q) - \sum_{k=1}^{K} J_k^T(q)\lambda_k = S^T \tau_{int} \tag{19}$$

$$J_k \ddot{q} + \dot{J}_k \dot{q} = 0$$

$$\lambda_k^{\perp} \geq 0$$

where J_k is the contact Jacobian associated to the kth contact point. The first equality of Eq. (19) ensures the respect of the dynamic model of the system. The inequality constraint $\lambda_k^{\perp} \geq 0$ guarantees that the contact forces are correctly oriented and that there is no interpenetration (rigid contact). In the same manner, other inequality constrains can be added such as joint limits ($\overline{q} \geq q \geq \underline{q}$), torque limits ($\overline{\tau} \geq \tau \geq \underline{\tau}$) or other tasks (as given in Sect. 3.2).

3.4.1 Hierarchical Control

Solutions to problems in the form of Eq. (19) can be formulated based on null space projections or optimization methods. Prioritization schemes are based on projections onto the null space of higher order priority tasks in the form of Eqs. (14) and (16). Unfortunately, these methods do not allow to cope with inequality constraints. Instead, numerical optimization techniques can be used to solve problems of the form of Eq. (19). A typical method is to use Hierarchical Quadratic Programming (HQP) [18, 29]. In order to compute HQP, basis multiplication was applied according to [17]. Computational details about this approach are given in the sequel.

Basis Multiplication

Linear equality constraints of the form $Ax = b$ such as task functions, can be solved using hierarchical control. Let us consider the case of L linear constraints (e.g. tasks) $(A_1, b_1) \cdots (A_l, l_l) \cdots (A_L, b_L)$ that have to be satisfied at best and let us consider that constraints are conflicting with each other. A strict hierarchy of constraints can be used to solve this problem [46]. The constrain with the highest priority (A_1, b_1) can be solved at best in a least-square sense with the pseudo-inverse. Then the second constraint (A_2, b_2) is solved in the null space of the first constraint. The generic solution to solve the p levels of the hierarchy can be written as:

$$x_l^* = \sum_{l=1}^{L} (A_l P_{l-1})^+ (b_l - A_l x_{l-1}^*) + \tilde{P}_L \tilde{x}_{L+1}, \tag{20}$$

with $P_0 = \mathbb{1}$, $x_0 = 0$ and $\tilde{P}_L = P_{L-1} P_L$ is the projector in the null space of $(A_l P_{l-1})$, x^* denotes the solution for the hierarchy of constraints composed of L linear constraints, P is a projector onto the null space of A ($AP = 0$ and $PP = P$) and $\tilde{x}_{L+1}$ is any vector of the configuration space that can be used to accomplish another objective. In order to fasten the numerical resolution of Eq. 20 a basis multiplication approach has been proposed [17, 18]. Given a basis Z_1 of the null space of $A1$ ($A_1 Z_1 = 0$), the projector in the null space of A_1 can be written as $P_1 = Z_1 Z_1^T$. Equation 20 can be rewritten as:

$$x_l^* = \sum_{l=1}^{L} Z_{l-1}(A_l Z_{l-1})^+ (b_l - A_l x_{l-1}^*) + \tilde{Z}_L \tilde{x}_{L+1}, \qquad (21)$$

which is more efficient to compute than Eq. (20) due to the size of the matrices. Note that Eqs. (14), (16) and (18) can be mapped to Eqs. (20) and (21).

4 Case Study

Future generations of robots will include agile, robust, efficient and powerful robots that perform more dynamic tasks. The utilization of such kind of robots will require understanding how dynamic motions have to be generated. One of the choices is, without doubt, to take inspiration from humans. As previously mentioned, the framework described in this chapter can be applied to a wide range of motions without loss of generality. This section presents a practical case of application to a highly dynamic and complex Parkour motion — called Parkour precision technique. First Parkour is introduced and the state of the art related to the Parkour precision technique is presented. Then, a skeletal model that is used for analyzing and generating the motion is described. Afterwards, the biomechanics methodology used for analyzing the Parkour precision technique is given. Finally, this analysis is used to parameterize a hierarchical controller in order to simulate similar Parkour motion with the skeletal model.

4.1 Biomechanics of Parkour Landing

Parkour is a discipline where movements are highly dynamic and complex. It requires practitioners — called traceurs — to adapt their motion to the environment in order to overcome obstacles quickly and efficiently. Motion strategies are derived from a military method developed in France after World War I, which was inspired by natural movements observed in skilled indigenous African tribes [25]. This method includes a combination of motions with variations of jumping, landing, climbing and vaulting strategies. Common to most Parkour techniques are the jumping and landing strategies. Mastering jumping and landing techniques allows practitioners for

(a)   **(b)**

(c)   **(d)**

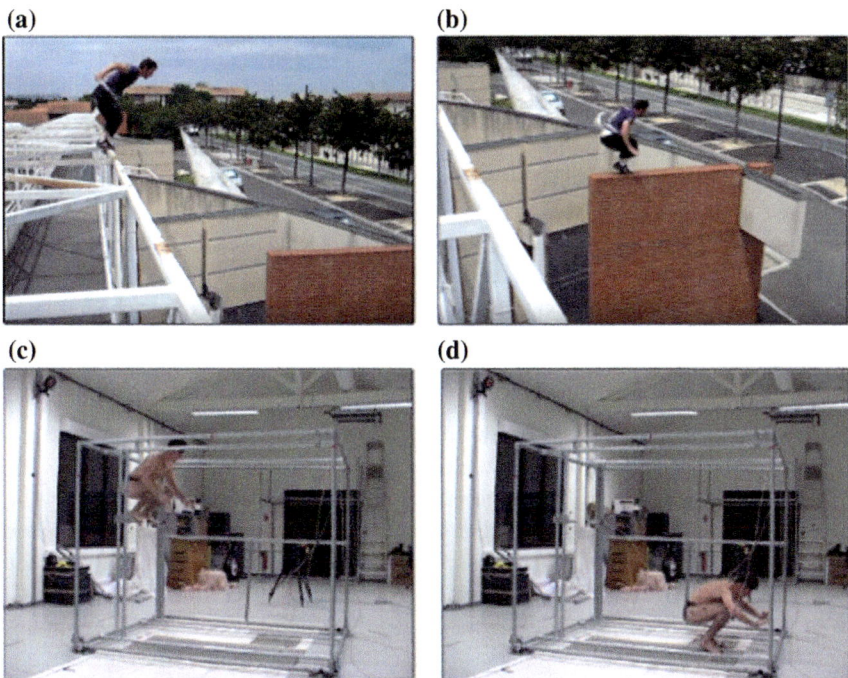

Fig. 2 Parkour precision jump and landing techniques performed in a urban space in (**a**) and (**b**), and inside a motion capture laboratory in (**c**) and (**d**)

executing efficient and safe motions. In our study, we analyzed the Parkour precision jumping and landing strategies (Fig. 2).

4.1.1 Parkour Precision Jump and Landing Techniques

Precision landing is commonly used for relatively low dropping heights. For executing this motion, trainers instruct practitioners to land and stay with precision on their forefoot avoiding heel contact with the ground, bend their lower limbs without any varus-valgus motion of the knees, and use their arms to counterbalance the movement and stabilize themselves. In what follows, we explain the selection of the performance criteria used to study the Parkour motion, which allowed us for identifying the tasks to generate the movement using the robotic framework of hierarchical task control.

The considered Parkour motion can be decomposed into three phases: take-off, flight and landing (in Sect. 4.1.5 we will detail how motion phases were divided). The take-off phase of the Parkour technique is similar to the standing long jump technique used in athletics. It requires practitioners to perform an horizontal jump without a preparatory running. Previous studies on long jump technique have suggested that during take-off motion, the principal performance factor is the velocity of the center of mass "CoM" [50] which defines the ballistic trajectory during the flight phase. Another performance criterion that has been proposed, is the angular momentum generated with the upper limbs. It was shown that using arms to generate angular momentum around the medial-lateral axis contributes to reduce the torques requested to the lower-limb joints [6]. This means that a person can counteract gravity effects by using the angular momentum generated with the upper limbs, whereas lower limbs will be more involved in the production of the force to project the CoM. This strategy contributes also to avoid excessive forward rotation [3] positioning the body in preparation to the landing phase [2]. According to the momenta conservation principle, the angular momentum remains constant during the flight phase. As a consequence the taking-off controller that generates the flight motion must be parameterized consistently. On the other hand, during the flight phase the linear momentum is only affected by the gravity. The landing phase of the Parkour technique induces practitioners to decrease peak ground reaction forces "GRFs" [47] and to stabilize better [35]. Decreasing vertical GRF and controlling stability through the antero-posterior and medial-lateral components of the GRFs is equivalent to control the derivative of linear momentum. Note that the derivative of momenta is equal to the net external forces/torques according to Euler's laws of motion. To avoid falling down after a highly dynamic landing, angular momentum might also contribute to stability by reducing the angular accelerations at the CoM. In fact, during our pre-tests, we observed that practitioners swang using their upper limbs in a three dimensional fashion.

4.1.2 Whole-Body Model

The choice of the physical model depends on the type of analysis. We are interested in a whole-body 3D analysis able to reproduce highly dynamic Parkour movements. To this end, we decided to use a whole-body 3D model that represents the main elementary movements of a Parkour jumper including 42 DoF (Fig. 1b). Note that a 3D model is necessary, because as shown by previous studies [26], a sagittal model is not sufficient to study the upper body motion during standing long jumps. The characteristics of the model are presented below:

- The lower limb, pelvis and upper limb anthropometry are based on the running model of Hamner et al. [22]. Mass properties of the torso and head segments (the head and neck segments were modeled as one segment) are estimated from the regression equations of Dumas et al. [10, 12]. Hands anthropomorphic data are based on regression equations of De Leva [8].
- Each lower extremity has seven DoF. The hip is modeled as a ball-and-socket joint, the knee is modeled as a revolute joint, the ankle is modeled as 2 revolute joints (flexion-extension and inversion-eversion) and the toes with one revolute joint at the metatarsals.

- The pelvis joint is modeled as a free flyer joint to permit the model to translate and rotate in the 3D space. This 6D joint is attached to the free-floating base (root frame) of the under-actuated systems as described in Sect. 3.1. The lumbar motion is modeled as a ball-and-socket joint [1] and the neck joint is also modeled as a ball-and-socket joint.
- Each arm includes 8 DoF. The shoulder is modeled as a ball-and-socket joint, the elbow and forearm rotations are modeled with revolute joints to represent flexion-extension and pronation-supination [28], the wrist flexion-extension and radial-ulnar deviations are modeled with revolute joints, and the hand fingers are modeled with one revolute joint for all fingers.
- The model includes a whole-body marker set with 48 markers placed in anatomical landmarks as suggested by Wu et al. [52, 53]. The visual elements are also based on the running model of Hamner et al. [22].

4.1.3   Experimental Protocol

Five healthy trained male traceurs (age: 22.2 ± 4.8 y, height: 1.73 ± 0.04 m, mass: 66.6 ± 5.1 kg) volunteered for the study. The traceurs' experience in Parkour practice was 5.4 ± 2.1 years. The subject exclusion criterion was based on history of lower extremity injuries or diseases that might affect jump and landing biomechanics. The experiments were conducted in accordance with the standards of the Declaration of Helsinki (rev. 2013) and approved by a local ethics committee. Participants performed a warming up session followed by a familiarization period during which the protocol instructions were provided to them, and during which the participant familiarized with the lab environment. The landing protocol was designed to include a jump height of 75 % of the height of the participant and a landmark placed at a horizontal distance equal to the square of the jump height [(See Eq. (22)]. Participants were asked to land on the target specified by the landmark and to stabilize using the Parkour precision technique.

4.1.4   Data Acquisition

A total of 8 successful repetitions per participant were recorded. Whole-body 3D kinematic data were collected using 14 infra-red cameras sampling at 400 Hz (Vicon, Oxford Metrics, Oxford, UK) and recording 48 reflective markers placed on the participant's body. Two force plates (AMTI, Watertown, MA, USA) embedded into the floor in order to record landing GRFs and two rigid handle bar sensors (SENSIX, Poitiers, Vienne, France) with a diameter of 63 mm placed on a Parkour tubular structure to record take-off GRFs, were used sampling at 2000 Hz. Force data were used to define the onsets used to divide the Parkour motion into phases. Markers were located on the participants body based on Wu and Dumas recommendations [10, 12, 52, 53] as follows: the first and fifth metatarsal, second toe tip, calcaneus, lateral and internal malleolus, anterior tibial tuberosity, lateral and medial epicondyles of knee, greater trochanter, posterior superior iliac spine and anterior superior iliac spine, procesuss xiphoideus, incisura jugularis, seventh cervicale, tenth thoracic vertebra, acromioclaviculare, medial and lateral epicondyle, ulnar and radial styloid, second and fifth metacarpal heads, second fingertip, sellion, occiput, right and left temporal (Fig. 1).

4.1.5   Data Analysis

Kinematics and kinetics were processed with the same cut-off frequency [31] using a low-pass Butterworth digital filter of 4th order applied in a zero-phase. A cut-off frequency of 35 Hz was used after a residual analysis [51]. All computations were performed using a custom made program with a whole-body model and a physics engine. Inverse kinematics computations were solved by minimizing the squared distance between recorded and virtual markers [(Eq. (4)] and using Euler xyz body-fixed rotation angles [45] with body frames defined according to ISB recommendations [52, 53].

The motion was divided into three phases: take-off, flight, and landing (Fig. 2). The take-off phase was defined from the minimum vertical position of the CoM until the last foot contact. The flight phase was defined between the end of the take-off phase until the initial contact "IC" with the ground, that we identified as the instant when the vertical ground reaction force reached 50 N. The landing phase was defined from IC until the CoM reached its minimum vertical position. Each phase was normalized by its time duration from 0 to 100% (Fig. 2). The linear momentum was normalized by the participant weight, and the angular momentum was normalized by the participant weight and height. Mean and standard deviation were calculated for each participant and for the whole group. Group means at key frames were used to parameterize the algorithm for motion generation.

4.2   Motion Generation

In order to generate the motion, we utilized the skeletal model introduced in Sect. 4.1.2 for human motion analysis. The robotics framework of hierarchical task control introduced in Sect. 3 was considered. Tasks were set and parameterized based on the biomechanical study made in Sect. 4.1. The motion was generated by using the method of null space projections and second order kinematics introduced in Sect. 3. The rigid body dynamics computations were done by using the Pinocchio library developed by the Gepetto team of LAAS-CNRS [4].

4.2.1   Description of the Tasks

As for motion analysis, the motion generation is divided into three phases: take-off, flight and landing. In order to apply the hierarchical control framework, tasks are stacked in a hierarchical manner for each motion phase. The selection of tasks and their hierarchy are based on the previous biomechanical study using mainly linear and angular momentum tasks. Additional tasks for foot placement, CoM and posture at key points (beginning/end of each motion phase) were also considered to better specify the motion. The desired values for parameterizing the controller were deduced from the biomechanical study. The motion was controlled using joint accelerations according to Eq. (12).

Preparation Phase

A preparation phase was added before the take-off phase for parameterizing the simulation with the initial conditions. Note that this phase is not strictly necessary as the controller can be parameterized directly with the initial conditions of the take-off phase which can be obtained from the motion analysis. This motion is generated through the control of a hierarchy of tasks organized as follows:

- At the highest priority we set a 3D foot placement task. This task was used to control the 3D position of each forefoot so that contact with the handle-bars can be maintained during this phase. The orientation components of these tasks were not constrained.
- At the second level of the hierarchy, we set a task specifying a desired 3D position of the CoM which was deduced from motion analysis.
- The remaining DoF were used to control the whole-body posture which was also extracted from motion analysis.

Take-off Phase

The take-off phase follows the preparation phase. The motion generated during this phase is organized in terms of tasks as follows:

- At the highest priority level, the 3D foot placement task of the preparation phase is kept.
- To generate the ballistic trajectory of the CoM, a linear momentum task is added at the second level of the hierarchy. The antero-posterior and vertical components provide the modulation of the CoM velocity for generating a desired ballistic trajectory. The medial-lateral component is regulated to zero to avoid undesired deviations of the CoM trajectory during the flight phase.
- In order to control the angular momentum at the CoM, a third task is stacked. This task imposes zero momentum around the vertical and antero-posterior (A-P) axes, and a desired angular momentum around the medial-lateral (M-L) axis. It allows the body model to reach a desired posture before landing and avoid somersaults. This task should also alleviate torques at the lower limbs as suggested in the literature (see Sect. 4.1).

Flight Phase

When the velocity and take-off angle of the CoM trajectory are suitable to reach a desired horizontal distance, the top level task of the foot placement is removed from the stack of tasks and the flight phase begins. The horizontal distance d_{flight} that the CoM will travel is calculated during the take-off phase according to the ballistic equations as:

$$d_{flight} = \frac{v \cos \theta}{g} \left(v \sin \theta + \sqrt{(v \sin \theta)^2 + 2gh_0} \right), \tag{22}$$

where v is the initial speed of the CoM (before the flight phase begins), θ is the take-off angle, g is the gravity acceleration and h_0 is the initial height of the CoM. During the flight phase, the momentum is conserved. A second level task is added to impose a desired posture before contacting the ground.

Landing Phase

The landing phase starts at the end of the flight phase and is set as follows:

• At the highest level of the hierarchy, the vertical feet position is regulated to keep the current contact position with the ground. A desired flexion of the toes and ankle joints is also imposed.
• At the second level, a 3D linear momentum task is added to decrease the velocity of the CoM to zero.
• At the third level, a task is added to regulate the angular momentum to zero in order to avoid tipping motions of the model.
• At the forth level of the hierarchy, a task is added to keep the CoM inside the vertical projection of the support polygon (medial-lateral and antero-posterior axis) in order to provide a static equilibrium state until the end of the motion.

4.3 Temporal Sequence

The chronological sequence of tasks stacked during each motion phase with their hierarchical order is depicted in the Table 1.

Table 1 Hierarchy of tasks used for generating motion in each phase

Hierarchy	Preparation	Take-off	Flight	Landing
1	Feet		Momenta	Feet
2	CoM	Linear Mom.	Posture	Linear Mom.
3	Posture	Angular Mom.		Angular Mom.
4				CoM

4.4 Tasks Behavior

Exponential decays of the task function were used to specify the task behavior. Weighting matrices multiplying the control input ($\ddot{q}^*$) were tuned for the take-off and landing phases in order to apply higher gains to lower joints when controlling the linear momentum, and higher gains to upper joints when controlling the angular momentum, in accordance to our biomechanical study. As the trunk segment has a significant mass compared to the upper limbs, the gain for the lumbar flexion was lowered to avoid undesirable behaviors of momenta.

4.5 Results

The experimental results and the generated motions present strong similarities, specially when comparing the motion of the lower body. Figure 3 shows snapshots of the human motion reconstructed by means of the inverse kinematics method of biomechanics, and snapshots of the motion generated using the hierarchical task control. The profiles of the linear and angular momenta of the experimental and generated motions are shown in Figs. 4 and in 5. The momenta profiles of humans represent the mean behavior of the Parkour experts with the corresponding standard deviation. Table 2 shows the difference between the ranges of motion (RoM) of the human group and the simulation model for the principal segments in the motion during each phase. In the sequel we refer to "the humans" to describe the average human motion and "the model" to refer to the generated one.

4.5.1 Take-off

- At the beginning of the motion, linear momentum values were different between the humans and the model. Throughout the motion, the linear momentum was similar in the medial-lateral component while the antero-posterior and vertical component of the linear momentum behaved differently. More antero-posterior linear momentum was generated by the humans at the end of the take-off phase, and more vertical linear momentum was generated by the model at the end of the motion phase.
- Although at the beginning of the motion angular momentum values were different, angular momentum behaved similarly in the humans and the model in the sequel. The angular momentum components around the antero-posterior and vertical axes were almost zero at the end of the motion phase, while the angular momentum component around the medial-lateral axis was not zero at the end of the motion phase.
- Figure 3 shows that upper-limbs were coordinated differently with the time evolution and that the trunk and hips were more flexed in humans.

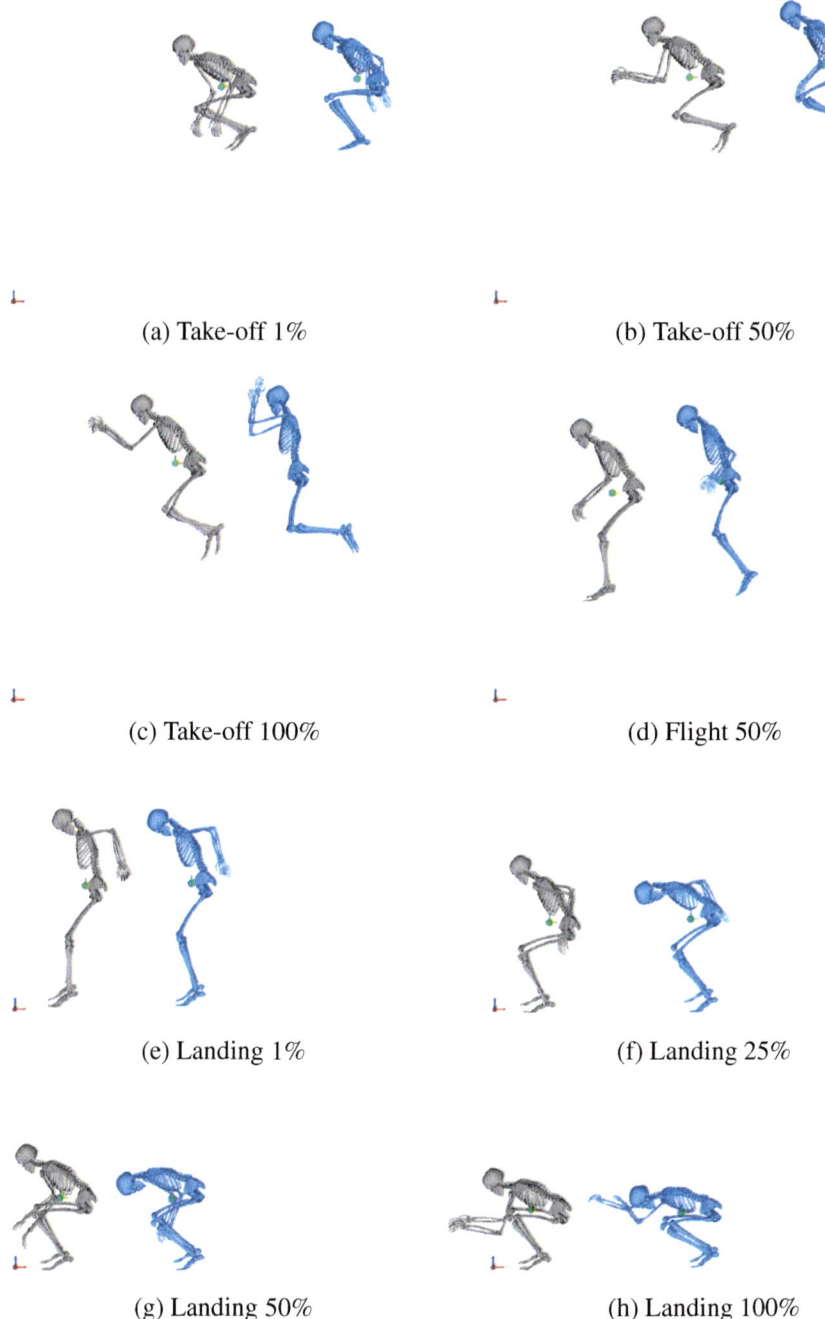

(a) Take-off 1% (b) Take-off 50%

(c) Take-off 100% (d) Flight 50%

(e) Landing 1% (f) Landing 25%

(g) Landing 50% (h) Landing 100%

Fig. 3 Snapshots of the take-off phase (**a–c**), flight phase (**d**), and landing phase (**e–h**) at different percentages of the motion phases. The skeleton on the left, represents the result of the inverse kinematics from motion analysis of a Parkour practitioner, whereas the skeleton on the right is the motion generated through hierarchical control

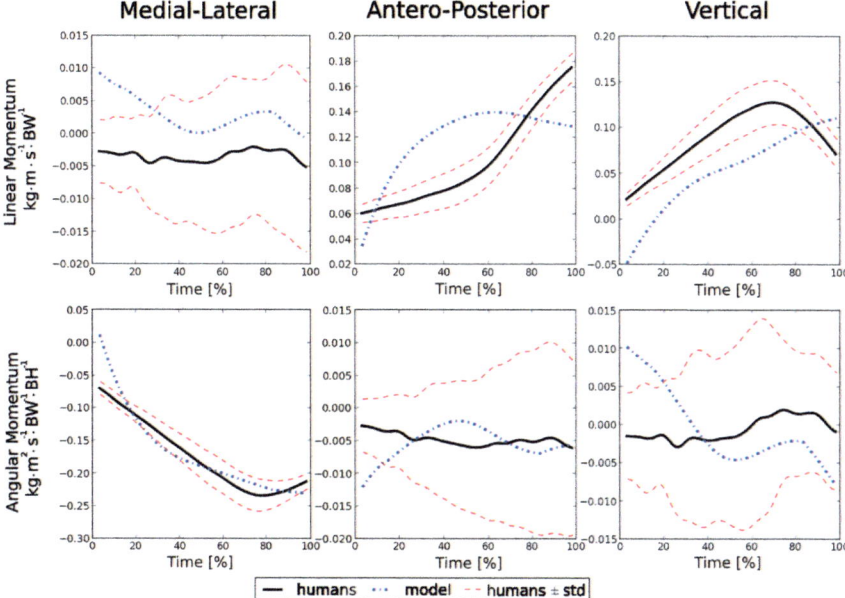

Fig. 4 Momenta profiles of the humans ($\pm$SD), and of the simulation model during the take-off phase. The first row shows the medial-lateral, antero-posterior and vertical components of the linear momentum normalized by the body weight. The second row shows the angular momentum normalized by the body weight and height, about the the medial-lateral, antero-posterior and vertical axis

- RoM of all coordinates were similar (Table 2). RoM of thigh abduction-adduction and forearm flexion-extension were slightly higher with the model, while upper arm abduction-adduction and rotation appear to be higher with the humans.

4.5.2   Flight

Momenta were not compared during this motion (see Sect. 4.1.1). Figure 3c and 3d show that the motion looks different at the beginning and at 50% of the flight. RoM were also different, specially in the case of the upper limb movement and the shank flexion-extension (Table 2).

4.5.3   Landing

- Linear momentum profiles of the humans and the model were globally similar.
- At the beginning of the landing phase, the model generated less angular momentum around the medial-lateral axis, while higher angular momentum was observed around the antero-posterior and vertical axis.

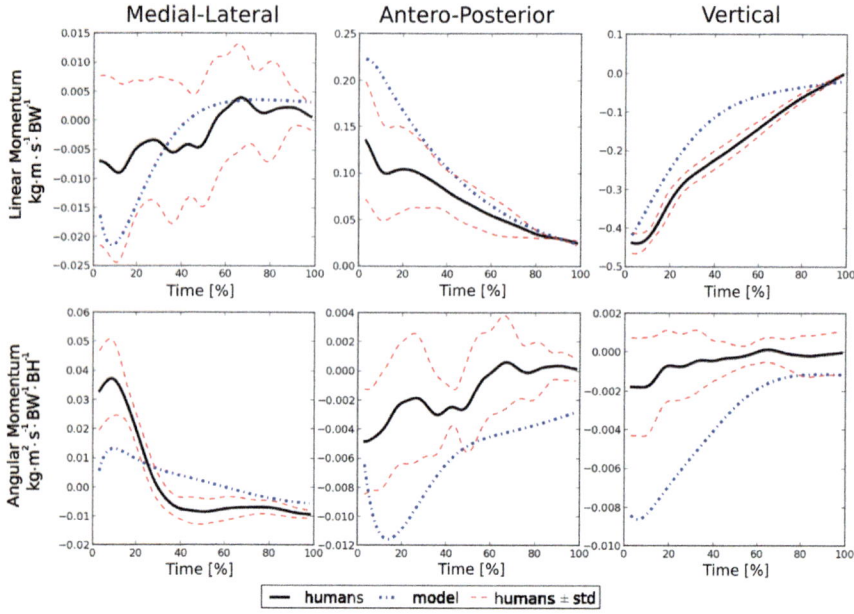

Fig. 5 Momenta profiles of the humans (± SD), and of the simulation model during the landing phase. In the first row: medial-lateral, antero-posterior and vertical components of the linear momentum normalized by the body weight are displayed. In the second row: angular momentum normalized by the body weight and height around the the medial-lateral, antero-posterior and vertical axis are displayed

Table 2 Difference in the ranges of motion (RoM), in degrees, of the analyzed human motions and the generated motion. The table shows the most relevant coordinates during the take-off, flight and landing phases of the Parkour technique. Negative values mean that the RoM of the generated motion is higher than the RoM of the human experts

	Take-off RoM (deg)		Flight RoM (deg)		Landing RoM (deg)	
	L	R	L	R	L	R
Neck-head flexion-extension	19		22		15	
Trunk flexion-extension	−20		−10		−5	
Upper arm flexion-extension	−1	−1	65	66	−44	−30
Upper arm abduction-adduction	17	33	−2	5	10	9
Upper arm rotation	17	24	44	68	47	54
Forearm flexion-extension	−17	−24	−9	−4	−15	3
Thigh flexion-extension	−2	−2	16	22	2	0
Thigh adduction-abduction	−5	−9.6	1	−4	2	2
Thigh rotation	−6	0	1	9	8	6
Shank flexion	12	12	30	44	20	23

- From Fig. 3, we can observe that the humans and the model landed with a similar posture. At 25 and 50% of the landing phase, the trunk of the model is more flexed than the the the trunk of humans.
- RoM of the upper arm coordinates appear to differ slightly while the RoM of the other coordinates look similar.

4.6   Discussion

The analyzed and the generated motion were compared in terms of kinematics and momenta. The results showed that the kinematics of the humans group were similar to the kinematics of the simulated model, specially for the lower limbs. Time evolution of momenta was sometimes slightly different during the take-off phase, while it was comparable during the landing phase. Whole-body coordination was congruent between the humans and the model, although the upper limbs strategy did not evolve similarly with time. In the next subsections we analyze the results per phase.

4.6.1   Take-off

During the take-off phase, we observed more antero-posterior linear momentum in the motions generated with the model. For generating the desired ballistic profile, the model compensated for the lack of antero-posterior linear momentum by increasing the vertical linear momentum. This strategy allowed the model to reach the initial CoM speed for landing at the requested distance. This increase of angular momentum of the model with respect to the humans one might explain why the model jumped with the trunk more extended than the human (Fig. 3c), and why the RoM of the trunk was higher. The time behavior of the vertical and antero-posterior linear momentum was different in the model. This might be due to the task reference behavior which imposes an exponential decay of the error between the actual and desired value (as explained in Sect. 3).

The profiles of angular momentum were similar for the humans and the model. The arms, which contribute to angular momentum, behaved differently with time evolution as shown in Fig. 3. The reason might be that the upper limbs contributed also to increase the vertical linear momentum (which behaved differently in the model). Note also that the RoM of the forearm flexion-extension motion was higher in the simulation model. In spite of this, the model jumped with an angular momentum that allowed it to prepare properly the body posture for landing (Fig. 3e). Controlling the posture before landing allows for a better control of stability and impact damping (See Sect. 4.1).

On angular momentum control during the take-off phase

We carried on simulations by decreasing and increasing the desired angular momentum during the take-off phase. The results showed that when the angular momentum

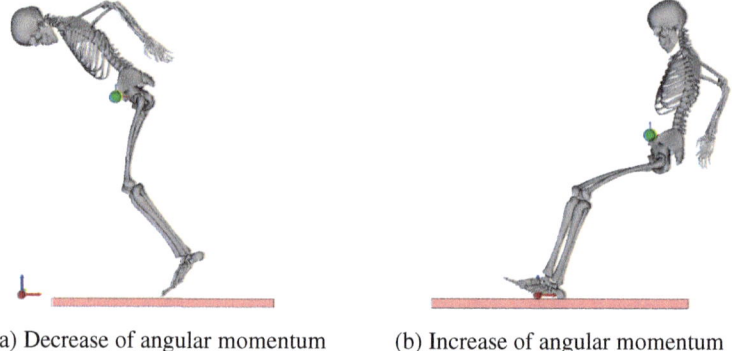

(a) Decrease of angular momentum   (b) Increase of angular momentum

Fig. 6 Effects on posture at landing when modifying the desired angular momentum during the take-off phase. The motion was generated by using the same hierarchical controller. Only the desired values of the angular momentum were modified during the take-off phase

is decreased, the model lands with a posture that makes it fall forwards (Fig. 6a). Conversely, if the angular momentum is increased, the model lands with a posture that makes it fall backwards (Fig. 6b). These results highlight the importance of controlling the angular momentum when jumping.

4.6.2   Flight

The kinematics of the humans and the model were different during the flight phase. This might be due to the fact that the flight motion of the model was generated without constraining the trunk which has the highest mass. Thus, segments with small masses might have contributed less importantly to keep momenta constant during flight. In fact, the RoM of the trunk was higher in the model whereas the RoM of the upper arm coordinates was higher in the humans. Note that the final posture (before landing) does not reflect the higher excursion of the upper arm coordinates in the humans. Instants before landing, arms are swung backwards. Later on, when contacting the ground, arms have already been swung forwards in preparation to the landing phase. In the model, this swing strategy of the arms was not considered.

4.6.3   Landing

Linear and angular momentum were similar in the humans and the model. Nevertheless, upper-body coordination during the landing phase looks different (Fig. 3). The model landed with the trunk more flexed than humans. Thus, upper-limbs might compensate for this strategy by generating a counterbalancing angular momentum.

In fact, the RoM of the upper arm flexion-extension was higher in the model. It turns out that the considered decay rate and hierarchy of tasks allowed the model to replicate the evolution of the momenta in humans.

5 Discussion

In this chapter, we presented an interdisciplinary methodology for generating human inspired motions with anthropomorphic systems such as robots. The approach involves robust biomechanics methods to analyze the human movements and the robotics framework to generate the motion. We showed that our approach is suitable to reproduce highly dynamic and complex motions comparable to those of humans. In this final section we summarize our methodology, we provide a discussion about the interest on using the proposed methodology, we show the limitations of the proposed approach, we give some perspectives for the future work, and we conclude with remarks about the interpretation of our results.

5.1 Summary of the Method "From Biomechanics to Robotics"

Important aspects of this approach can be summarized as follows:

- A physical model suitable for analyzing and/or generating the motion of interest is selected/created.
- An experimental protocol for motion analysis is designed and the human movements are recorded by using motion-capture techniques.
- The recorded motion is analyzed using robust biomechanics techniques and key performance variables of the movement are identified. Performance variables provide the information needed to create a set of tasks.
- By understanding how the identified performance variables favors the motion generation in humans, the set of tasks is organized in a hierarchical manner according to their importance in the execution of the motion. A stack of tasks is created.
- The robot controller is further parameterized using information from control strategies observed in humans by weighting the controller input. This information can be obtained from biomechanical studies.
- The motion is generated through the task function approach in robotics using the identified tasks and the physical model for motion generation.
- The human motion and the motion artificially generated are compared using the same physical model created for analyzing the motion.

5.2   Interest of the Proposed Approach

There are interesting aspects of this approach that deserve further discussion. First, our methodology differs from conventional task-space approaches, because it allows for generating whole-body motion based on the biomechanics of human movement. Quantifying and understanding the mechanics of human motion offers the possibility to parameterize robotic algorithms based on human expertise to execute movements. This provides a way to generate more robust and efficient motions. Furthermore, generating anthropomorphic motions that look more human-like is of great interest for human-robot interaction. Second, comparing the analyzed human motion with the motion generated based on the robotics task function approach provides an interesting tool for validating hypotheses about the organization of human movement based on task hierarchies.

5.3   Limitations and Future Work

Thought the biomechanics methodology provides an efficient tool for analyzing human movement, it is known to have some limitations. Some source of errors are the simplifications made when modeling the human body (e.g. modeling segments as rigid bodies, number of DoF of the model, or simplifications for modeling complex human articulations), the estimation of anthropomorphic parameters and the data processing (e.g. motion reconstruction, filtering of the noise, soft tissue artifacts). The feasibility of the motion generated can be assessed in terms of joint torque and joint power by adding inequality constraints [(Eq. (19)]. To this end, the hierarchical controller can be designed using quadratic programming QP as suggested in Sect. 3.4.1. A deeper analysis between the human and the model strategies to compare other performance variables such as reactions forces, torques, power and energy dissipation is also desirable. Furthermore, the behavior of the tasks (decay rate of the task function) can also be modified to generate smoother trajectories (see for example [27]). A vision task [5] could also be added to reflect the importance of vision in humans when performing this type of motions. Other approaches can be used to generate highly dynamic and complex motions by considering the whole trajectory along a finite time horizon, e.g. optimal control. In that case hybrid cost functions, usually described as weighted-sums of elementary criteria, are considered. The difficulty is then to identify the set of weights that lead to the best replication of the observed human movement.

Finally, though our results suggest that the proposed approach is suitable to generate motion similar to that of humans, we are not attempting to propose that humans compute inverse kinematics or inverse dynamics problems. This is still an open question, which is outside the scope of this chapter. Nevertheless, we could point out that strategies of task control have been observed in human motion. It was shown that the central nervous system finds stable solutions of motor tasks in accordance with the uncontrolled manifold theory "UCM" [44] and the motor abundance principle [21]. It is also suggested that hierarchies in human motion appear at different levels such as neurons, muscles, joints and tasks [21, 33]. Lastly, the UCM theory is linked to the equilibrium point [19] which is a physiological approach to understand human motions.

Acknowledgements Part of this work is supported by the European Research Council for the project Actanthrope (ERC-ADG347 340050) and the Flag-Era European project RoboCom++.

References

1. Anderson, F.C., Pandy, M.G.: A dynamic optimization solution for vertical jumping in three dimensions. Comput. Methods Biomech. Biomed. Eng. **2**(3), 201–231 (1999)
2. Ashby, B.M., Delp, S.L.: Optimal control simulations reveal mechanisms by which arm movement improves standing long jump performance. J. Biomech. **39**(9), 1726–1734 (2006)
3. Ashby, B.M., Heegaard, J.H.: Role of arm motion in the standing long jump. J. Biomech. **35**(12), 1631–1637 (2002)
4. Carpentier, J., Valenza, F., Mansard, N. et al.: Pinocchio: fast forward and inverse dynamics for poly-articulated systems https://stack-of-tasks.github.io/pinocchio (2015)
5. Chaumette, F., Hutchinson, S.: Visual servo control, Part I: basic approaches. IEEE Robot. Autom. Mag. **13**(4), 82–90 (2006)
6. Cheng, K.B., Wang, C.H., Chen, H.C., Wu, C.D., Chiu, H.T.: The mechanisms that enable arm motion to enhance vertical jump performance-A simulation study. J. Biomech. **41**(9), 1847–1854 (2008)
7. Chèze, L., Fregly, B.J., Dimnet, J.: A solidification procedure to facilitate kinematic analyses based on video system data. J. Biomech. **28**(7), 879–884 (1995)
8. De Leva, P.: Adjustments to Zatsiorsky-Seluyanov's segment inertia parameters. J. Biomech. **29**(9), 1223–1230 (1996)
9. Delp, S.L., Anderson, F.C., Arnold, A.S., Loan, P., Habib, A., John, C.T., Guendelman, E., Thelen, D.G.: Opensim: open-source software to create and analyze dynamic simulations of movement. IEEE Trans. Biomed. Eng. **54**(11), 1940–1950 (2007)
10. Dumas, R., Chèze, L., Verriest, J.-P.: Corrigendum to adjustments to McConville et al. and Young et al. body segment inertial parameters [J. Biomech. 40 (2007) 543553]. J. Biomech. **40**(7):1651–1652 (2007)
11. Dumas, R.: Influence of the 3D inverse dynamic method on the joint forces and moments during gait. J. Biomech. Eng. **129**(5), 786 (2007)
12. Dumas, R., Chèze, L., Verriest, J.P.: Adjustments to Mcconville et al. and Young et al. body segment inertial parameters. J. Biomech. **40**(3), 543–553 (2007)
13. Duprey, S., Chèze, L., Dumas, R.: Influence of joint constraints on lower limb kinematics estimation from skin markers using global optimization. J. Biomech. **43**(14), 2858–2862 (2010)
14. Durkin, J.L., Dowlingm, J.J., Andrews, D.M.: The measurement of body segment inertial parameters using dual energy X-ray absorptiometry. J. Biomech. **35**(12), 1575–1580 (2002)
15. Ehrig, R.M., Taylor, W.R., Duda, G.N., Heller, M.O.: A survey of formal methods for determining the centre of rotation of ball joints. J. Biomech. **39**(15), 2798–809 (2005)
16. Ehrig, R.M., Taylor, W.R., Duda, G.N., Heller, M.O.: A survey of formal methods for determining functional joint axes. J. Biomech. **40**(10), 2150–7 (2007)
17. Escande, A., Mansard, N., Wieber, P.B.: Fast resolution of hierarchized inverse kinematics with inequality constraints. In: 2010 IEEE International Conference on Robotics and Automation, pp. 3733–3738 (May 2010)

18. Escande, A., Mansard, N., Wieber, P.-B.: Hierarchical quadratic programming: fast online humanoid-robot motion generation. Int. J. Robot. Res. **33**(7), 1006–1028 (2014)
19. Feldman, A.G.: Functional tuning of the nervous system with control of movement or maintenance of a steady posture. II. Controllable parameters of the muscle. Biophysics **11**, 565–578 (1966)
20. Flash, T.: The coordination of arm movements: mathematical model'. J. Neurosci. **5**(7), 1688–1703 (1985)
21. Gera, G., Freitas, S., Latash, M., Monahan, K., Schoner, G., Scholz, J.: Motor abundance contributes to resolving multiple kinematic task constraints. Mot. Control **14**(1), 83–115 (2010)
22. Hamner, S.R., Seth, A., Delp, S.L.: Muscle contributions to propulsion and support during running. J. Biomech. **43**(14), 2709–2716 (2010)
23. Hatze, H.: A new method for the simultaneous measurement of the moment of inertia, the damping coefficient and the location of the centre of mass of a body segment in situ. Eur. J. Appl. Physiol. **34**, 217–226 (1975)
24. Havana, E.P.: A mathematical model of the human body (Report AMRL-TR-64-102). Technical report, Aerospace Medical Research Laboratory, Ohio (1964)
25. Hébert, G.: L'éducation physique, virile et morale par la Méthode Naturelle. Tome I: Exposé doctrinal et principes directeurs de travail. Vuibert (1936)
26. Hickox, L.J., Ashby, B.M., Alderink, G.J.: Exploration of the validity of the two-dimensional sagittal plane assumption in modeling the standing long jump. J. Biomech. **49**(7), 1085–1093 (2016)
27. Hoff, B., Arbib, M.: A model of the effects of speed, accuracy, and perturbation on visually guided reaching. Exp. Brain Res. **22**, 285–306 (1992)
28. Holzbaur, K.R.S., Murray, W.M., Delp, S.L.: A model of the upper extremity for simulating musculoskeletal surgery and analyzing neuromuscular control. Ann. Biomed. Eng. **33**(6), 829–840 (2005)
29. Kanoun, O., Lamiraux, F., Wieber, P.B., Kanehiro, F., Yoshida, E., Laumond, J.P.: Prioritizing linear equality and inequality systems: application to local motion planning for redundant robots. In: 2009 IEEE International Conference on Robotics and Automation, pp. 2939–2944 (2009)
30. Khatib, O.: A unified approach for motion and force control of robot manipulators: the operational space formulation. IEEE J. Robot. Autom. **3**(1), 43–53 (1987)
31. Kristianslund, E., Krosshaug, T., Van den Bogert, A.J.: Effect of low pass filtering on joint moments from inverse dynamics: implications for injury prevention. J. Biomech. **45**(4), 666–671 (2012)
32. Kuo, A.D.: A least-squares estimation approach to improving the precision of inverse dynamics computations. J. Biomech. Eng. **120**(1), 148–59 (1998)
33. Latash, M.L., Gorniak, S., Zatsiorsky, V.M.: Hierarchies of synergies in human movements. Kinesiology **40**(1), 29–38 (2008)
34. Lu, T.-W., O'Connor, J.J.: Bone position estimation from skin marker co-ordinates using global optimisation with joint constraints. J. Biomech. **32**(2), 129–134 (1999)
35. Maldonado, G., Bitard, H., Watier, B., Souères, P.: Evidence of dynamic postural control performance in parkour landing. Comput. Methods Biomech. Biomed. Eng. **18**(1), 1994–1995 (2015)
36. Mansard, N., Stasse, O., Evrard, P., Kheddar, A.: A versatile generalized inverted kinematics implementation for collaborative working humanoid robots: the Stack of Tasks. In: ICAR'09: International Conference on Advanced Robotics, pp. 1–6, Munich, Germany (June 2009)
37. Mccaw, S.T., Gardner, J.K., Stafford, L.N., Torry, M.R.: Filtering ground reaction force data affects the calculation and interpretation of joint kinetics and energetics during drop landings. J. Appl. Biomech. **29**(6), 804–809 (2013)
38. Nakamura, Y.: Advanced Robotics: Redundancy and Optimization. Addison-Wesley Pub. Co, Reading, MA (1991)
39. Orin, D.E., Goswami, A.: Centroidal momentum matrix of a humanoid robot: structure and properties. In: 2008 IEEE/RSJ International Conference on Intelligent Robots and Systems, pp. 653–659 (September 2008)

40. Orin, D.E., Goswami, A., Lee, S.-H.: Centroidal dynamics of a humanoid robot. Auton. Robot. **35**(2), 161–176 (2013)
41. Saab, L., Mansard, N., Keith, F., Fourquet, J.Y. , Soueres, P.: Generation of dynamic motion for anthropomorphic systems under prioritized equality and inequality constraints. In: 2011 IEEE International Conference on Robotics and Automation, pp. 1091–1096 (2011)
42. Saab, L., Ramos, O.E., Keith, F., Mansard, N., Soures, P., Fourquet, J.Y.: Dynamic whole-body motion generation under rigid contacts and other unilateral constraints. IEEE Trans. Rob. **29**(2), 346–362 (2013)
43. Samson, C., Espiau, B., Le Borgne, M.: Robot Control: The Task Function Approach. Oxford University Press, Oxford (1991)
44. Scholz, J.P., Schöner, G.: The uncontrolled manifold concept: identifying control variables for a functional task. Exp. Brain Res. **126**(3), 289–306 (1999)
45. Seth, A., Sherman, M., Eastman, P., Delp, S.: Minimal formulation of joint motion for biomechanisms. Nonlinear Dyn. **62**(1), 291–303 (2010)
46. Siciliano, B., Slotine, J.J.E.: A general framework for managing multiple tasks in highly redundant robotic systems. In: Fifth International Conference on Advanced Robotics, 1991. 'Robots in Unstructured Environments', 91 ICAR, pp. 1211–1216 vol. 2 (June 1991)
47. Standing, R.J., Maulder, P.S.: A comparison of the habitual landing strategies from differing drop heights of parkour practitioners (traceurs) and recreationally trained individuals. J. Sports Sci. Med. **14**(4), 723–731 (2015)
48. Stasse, O., Flayols, T., Budhiraja, R., Giraud-Esclasse, K., Carpentier, J., Del Prete, A., Souères, P., Mansard, N., Lamiraux, F., Laumond, J.-P., Marchionni, L., Tome, H., Ferro, F.: TALOS: a new humanoid research platform targeted for industrial applications. working paper or preprint (March 2017)
49. Venture, G., Ayusawa, K., Nakamura, Y.: Motion capture based identification of the human body inertial parameters. In: 2008 30th Annual International Conference of the IEEE Engineering in Medicine and Biology Society, pp. 4575–4578 (2008)
50. Wakai, M., Linthorne, N.P.: Optimum take-off angle in the standing long jump. Hum. Mov. Sci. **24**(1), 81–96 (2005)
51. Winter, D.A.: Biomechanics and Motor Control of Human Movement, 4th edn. Wiley, New York (2009)
52. Wu, G., Siegler, S., Allard, P., Kirtley, C., Leardini, A., Rosenbaum, D., Whittle, M., D'Lima, D.D., Cristofolini, L., Witte, H., Schmid, O., Stokes, I.: ISB recommendation on definitions of joint coordinate system of various joints for the reporting of human joint motionPart I: ankle, hip, and spine. J. Biomech. **35**(4), 543–548 (2002)
53. Wu, G., van der Helm, F.C.T., Veeger, H.E.J., Makhsous, M., Van Roy, P., Anglin, C., Nagels, J., Karduna, A.R., McQuade, K., Wang, X., Werner, F.W., Buchholz, B.: ISB recommendation on definitions of joint coordinate systems of various joints for the reporting of human joint motion. Part II: shoulder, elbow, wrist and hand. J. Biomech. **38**(5), 981–992 (2005)
54. Zatsiorsky, V.M., Seluyanov, V.N.: The mass and inertia characteristics of the main segment of human body. Biomechanics, pp. 1152–1159 (1983)

Multibody Optimisations: From Kinematic Constraints to Knee Contact Forces and Ligament Forces

Raphael Dumas, Laurence Cheze and Florent Moissenet

Abstract Musculoskeletal models are widely used in biomechanics today to better understand muscle and joint function. Musculo-tendon forces as well as joint contact forces and ligament forces can be estimated within an inverse dynamics computational framework. Using a musculoskeletal model of the lower limb, this chapter presents the different optimisations required to drive the model with experimental data and to compute these forces and their interactions. In these optimisations, the development of anatomical constraints representing, for example, the medial and lateral tibiofemoral contacts or the cruciate ligaments is crucial both to inverse kinematics and to inverse dynamics. Some emblematic results are presented for knee contact forces and ligament forces during gait, illustrating the couplings between joint degrees of freedom and the interactions between forces acting both in muscles and in joints.

List of symbols

T	matrix transpose
$\dagger$	matrix pseudo-inverse
$\mathbf{E}_{3\times3}$	identity matrix
i	index for segment
j	index for skin or virtual marker (in inverse kinematics) or muscle (in inverse dynamics)
$\mathbf{u}_i, \mathbf{v}_i, \mathbf{w}_i$	anterior, superior and lateral axes of segment
$\mathbf{r}_{P_i}, \mathbf{r}_{D_i}$	positions of the proximal (P_i) and distal (D_i) endpoints

R. Dumas (✉) · L. Cheze
LBMC UMR_T9406, Univ Lyon, Université Claude Bernard Lyon 1, IFSTTAR,
F69622 Lyon, France
e-mail: raphael.dumas@ifsttar.fr

F. Moissenet
Centre National de Rééducation Fonctionnelle et de Réadaptation—Rehazenter, Laboratoire
d'Analyse du Mouvement et de la Posture (LAMP), Luxembourg, Luxembourg

© Springer International Publishing AG, part of Springer Nature 2019
G. Venture et al. (eds.), *Biomechanics of Anthropomorphic Systems*, Springer Tracts
in Advanced Robotics 124, https://doi.org/10.1007/978-3-319-93870-7_4

$$\left(P_i, \mathbf{u}_i, \underbrace{\left(\mathbf{r}_{P_i} - \mathbf{r}_{D_i} \right)}_{\mathbf{v}_i}, \mathbf{w}_i \right) \quad \text{non-orthonormal segment coordinate system}$$

$(P_i, \mathbf{X}_i, \mathbf{Y}_i, \mathbf{Z}_i)$ orthonormal segment coordinate system

$\mathbf{B}_i$ transformation matrix from the non-orthonormal to the orthonormal segment coordinate system

$\alpha_i, \beta_i, \gamma_i$ constant angles between the axes of the non-orthonormal segment coordinate system

L_i segment length (between proximal and distal endpoints)

$\mathbf{Q}_i$ natural coordinates (2 position and 2 direction vectors)

$\mathbf{\Phi}_i^r$ rigid body constraints

$\mathbf{r}_{M_i^j}, \mathbf{r}_{V_i^j}$ position of skin or virtual marker (M_i^j or V_i^j)

$(n_i)_u, (n_i)_v, (n_i)_w$ coordinates in the non-orthonormal segment coordinate system

$\mathbf{N}_i$ interpolation matrix

$\mathbf{\Phi}^k$ kinematic constraints

d, θ model parameter (i.e. ligament length, angle between hinge axes)

$\mathbf{\Phi}^m$ driving constraints

$\mathbf{G}$ matrix of generalised masses

$\mathbf{Q}, \dot{\mathbf{Q}}, \ddot{\mathbf{Q}}:$ vectors of generalised positions, velocities and accelerations for all segments

$\mathbf{K}$ Jacobian matrix of the constraints

λ vector of Lagrange multipliers

$\mathbf{R}$ vector of generalised ground reaction

$\mathbf{Z}_{\mathbf{K}_2^T}$ orthogonal basis of the null space of $\mathbf{K}_2^T$ (corresponding to a subset of the constraints)

$\mathbf{P}$ vector of generalised weights

$\mathbf{L}$ matrix of generalised muscular lever arms

$\mathbf{f}$ vector of musculo-tendon forces

f, J objective functions

$\mathbf{W}$ optimisation weights

$\mathbf{F}_0^R, \mathbf{M}_0^R$ ground reaction force and moment vectors at the centre of pressure (P_0)

$f_u^{\mathbf{M}_0^R}, f_v^{\mathbf{M}_0^R}, f_w^{\mathbf{M}_0^R}$ forces applied about the axes of the non-orthonormal foot coordinate system representing the ground reaction moment.

1 Introduction

Musculoskeletal models are widely used in biomechanics to better understand muscle and joint function [1–5]. A powerful tool for exploring motion analysis data in clinical, sports or ergonomic contexts as well as for simulating synthetic data, musculoskeletal models rely on a representation of the muscle and joint anthropomorphic geometries and on the equations of motion of the multibody system under analysis. The computational framework, whether based on forward or inverse dynamics, involves several optimisations enabling the musculo-tendon forces for a given motor task to be estimated.

This chapter presents an overview of the theoretical and numerical aspects of the development and performance of a musculoskeletal model of the lower limb based on an inverse dynamics computational framework. Two constrained optimisations, designed for inverse kinematics and inverse dynamics, are performed to compute segment parameters, musculo-tendon forces, joint contact forces and ligament forces based on experimental data. A third computation, using a pseudo-inverse method instead of optimisation, is performed to compute the individual muscle contributions to joint contact forces and ligament forces.

Within this computational framework, two specific bottlenecks arise: the so-called soft tissue artefact (i.e. the relative movement between the skin markers and the underlying bones), which jeopardises the kinematics estimation, and the muscular redundancy, which makes the dynamics an indeterminate problem. This chapter therefore focuses on defining anatomical constraints, the Jacobian of which plays a strategic role in the couplings between joint degrees of freedom and in the interactions between the forces acting both in muscles and in joints.

2 Computational Framework

Figure 1 summarises the computational framework. The musculoskeletal model (see Sect. 3) is driven by experimental data: skin marker trajectories and ground reaction forces and moments. The inverse kinematics step (see Sect. 4) computes segment positions and accelerations according to the model constraints through a first optimisation (i.e. minimisation) process. The inverse dynamics step (see Sect. 5) computes musculo-tendon forces, joint contact forces and ligament forces according to the Jacobian matrix of the constraints and other model quantities through a second optimisation (i.e. minimisation) process. The contribution of each muscle to joint contact forces and ligament forces is further computed using a pseudo-inverse method (see Sect. 6) also based on the Jacobian matrix of the constraints.

3 Musculoskeletal Model of the Lower Limb

A 3D lower limb musculoskeletal model consisting of five segments (i.e. pelvis, thigh, patella, shank and foot) and 5 joint degrees of freedom is driven by the trajectories of 15 skin markers fixed on anatomical landmarks of the pelvis (i.e. the right and left anterior and posterior superior iliac spines) and the right lower limb (i.e. the great

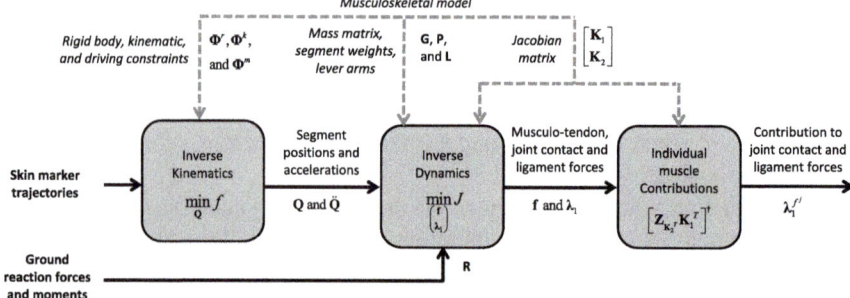

Fig. 1 Computational framework—musculoskeletal model driven by experimental data to compute musculo-tendon forces, joint contact forces, ligament forces and their interactions (the different vectors, matrices and functions are defined in the next sections)

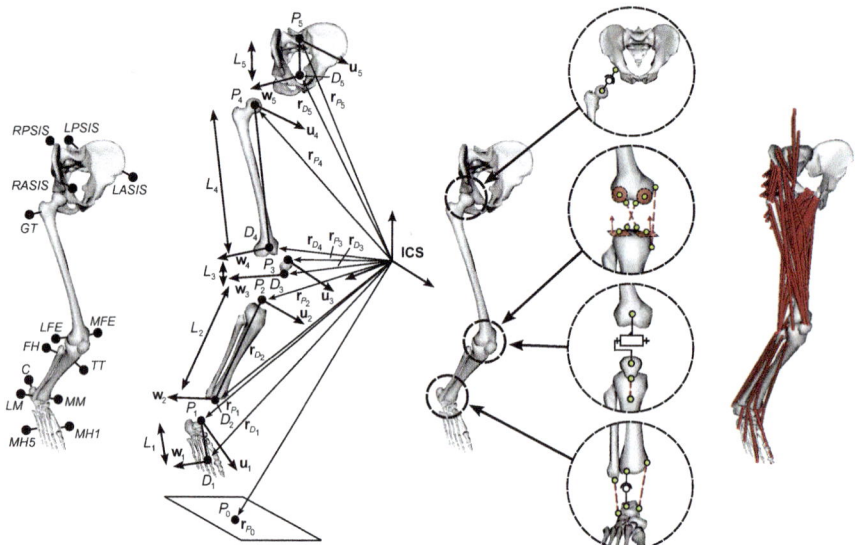

Fig. 2 Lower limb musculoskeletal model—skin markers, segment parameters, kinematic constraints and muscle lines of action

trochanter, medial and lateral epicondyles, peroneal head, anterior tibial tuberosity, medial and lateral malleoli, calcaneus, first and fifth metatarsal heads) and by the ground reaction forces and moments measured at the centre of pressure (Fig. 2). The centre of pressure, measured under one foot, is the point of the force plate surface where the ground reaction moment is perpendicular to the surface. It coincides with the zero moment point classically defined in robotics during the single support phase of gait.

Body segment inertial parameters are estimated by regression equations [6]. Muscle lever arms are computed using a muscular geometry derived from that proposed by Klein Horsman et al. [7] and adjusted by van Arkel et al. [8]. This geometry consists of 129 muscular lines of action representing 38 muscles. No activation dynamics (i.e. Hill-type muscle model) is introduced. The musculoskeletal model is scaled by homothety to fit the segment lengths of the subject under analysis. In other words, to all the model parameters corresponding to a segment, and typically muscle origins and insertions, is applied a scaling factor defined as the ratio between the model and the subject segment length. The subject pelvis, thigh, shank and foot length are estimated from skin markers.

3.1 Parameter Set

The position and orientation of the segments is parameterised using natural coordinates [9, 10] yielding, for each segment i, 2 position vectors (proximal P_i and distal D_i endpoints) and 2 unitary direction vectors (anterior $\mathbf{u}_i$ and lateral $\mathbf{w}_i$ axes):

$$\mathbf{Q}_i = \begin{pmatrix} \mathbf{u}_i \\ \mathbf{r}_{P_i} \\ \mathbf{r}_{D_i} \\ \mathbf{w}_i \end{pmatrix},$$

with $i = 1, 2, 3, 4$, and 5 for the foot, shank, patella, thigh, and pelvis, respectively. All position and direction vectors are expressed in an inertial coordinate system (ICS). The construction of the position vectors $\mathbf{r}_{P_i}$ and $\mathbf{r}_{D_i}$ and direction vectors $\mathbf{u}_i$ and $\mathbf{w}_i$ from skin markers is detailed in Duprey et al. [11].

In a forward kinematics scheme, the position of any point embedded in the segment (e.g. position $\mathbf{r}_{M_i^j}$ of the jth skin marker embedded in the ith segment, M_i^j) can be deduced linearly from $\mathbf{Q}_i$ using a constant interpolation matrix:

$$\mathbf{r}_{M_i^j} - \mathbf{r}_{P_i} = \left(n_i^{M_i^j} \right)_u \mathbf{u}_i + \left(n_i^{M_i^j} \right)_v \left(\mathbf{r}_{P_i} - \mathbf{r}_{D_i} \right) + \left(n_i^{M_i^j} \right)_w \mathbf{w}_i$$

$$\mathbf{r}_{M_i^j} = \underbrace{\left[\left(n_i^{M_i^j} \right)_u \mathbf{E}_{3\times3} \left(1 + \left(n_i^{M_i^j} \right)_v \right) \mathbf{E}_{3\times3} - \left(n_i^{M_i^j} \right)_v \mathbf{E}_{3\times3} \left(n_i^{M_i^j} \right)_w \mathbf{E}_{3\times3} \right]}_{\mathbf{N}_i^{M_i^j}} \mathbf{Q}_i ,$$

with $\left(\begin{array}{c}\left(n_i^{M_i^j}\right)_u \\ \left(n_i^{M_i^j}\right)_v \\ \left(n_i^{M_i^j}\right)_w\end{array}\right)$ the coordinates of M_i^j in the non-orthonormal segment coordinate

system $\left(P_i, \mathbf{u}_i, \underbrace{\left(\mathbf{r}_{P_i} - \mathbf{r}_{D_i}\right)}_{\mathbf{v}_i}, \mathbf{w}_i\right)$ and with $\mathbf{E}_{3\times3}$ the identity matrix. The same inter-

polation applies for the segment centre of mass and for virtual markers V_i^j standing, for example, for ligament origins and insertions (see Table 1). The orientation of any virtual direction $\mathbf{n}_i^j$ (see Table 1) embedded in the segment can be deduced in the same way. Virtual markers, as opposed to skin markers, refer to internal anatomical landmarks and, typically, ligament and muscle origins and insertions. Similarly, virtual directions refer to anatomical directions such as the normal to anatomical planes.

Note that using the non-orthonormal segment coordinate system $\left(P_i, \mathbf{u}_i, \underbrace{\left(\mathbf{r}_{P_i} - \mathbf{r}_{D_i}\right)}_{\mathbf{v}_i}, \mathbf{w}_i\right)$ can be convenient in biomechanics to include straight-forwardly segment lengths (i.e. $L_i = \left\|\mathbf{r}_{P_i} - \mathbf{r}_{D_i}\right\|$ which are used for model scaling) and functional axes of several joints (i.e. knee and ankle flexion-extension axis given by the vector $\mathbf{w}_i$). Moreover, this non-orthonormal coordinate system is directly related to the more classical orthonormal coordinate system $(P_i, \mathbf{X}_i, \mathbf{Y}_i, \mathbf{Z}_i)$, as standardised by the *International Society of Biomechanics* [12]:

$$\left[\mathbf{X}_i\ \mathbf{Y}_i\ \mathbf{Z}_i\right] = \left[\mathbf{u}_i\ \underbrace{\mathbf{r}_{P_i} - \mathbf{r}_{D_i}}_{\mathbf{v}_i}\ \mathbf{w}_i\right][\mathbf{B}_i]^{-1}$$

$$\mathbf{B}_i = \begin{bmatrix} 1 & L_i \cos \gamma_i & \cos \beta_i \\ 0 & L_i \sin \gamma_i & \frac{\cos \alpha_i - \cos \beta_i \cos \gamma_i}{\sin \gamma_i} \\ 0 & 0 & \sqrt{1 - (\cos \beta_i)^2 - \left(\frac{\cos \alpha_i - \cos \beta_i \cos \gamma_i}{\sin \gamma_i}\right)^2} \end{bmatrix},$$

with α_i, β_i, and γ_i the constant angles between the axes of the non-orthonormal segment coordinate system.

Since segment position and orientation are defined by 12 parameters, 6 rigid body constraints have to be considered. The rigid body constraints represent the fact that the norms of the axes of the non-orthonormal segment coordinate system equal 1

Table 1 Anatomical definition of the kinematic constraints Φ^k

Constraint	Segment	Virtual marker or direction	Anatomical definition	Parameter	Associated force (see Sect. 5)	References
Tibiotalar spherical joint condition	Shank	V_2^1	Position of joint centre in shank		Contact force applied from the tibia to the talus about the three axes of the orthonormal foot coordinate system	[13]
	Foot	V_1^1	Position of joint centre in foot			
		$\mathbf{X}_1, \mathbf{Y}_1, \mathbf{Z}_1$	Anterior, superior and lateral orthonormal axes of foot			
Tibiocalcaneal ligament length constancy	Shank	V_2^2	Origin of ligament in shank	Ligament length d_A^1	Ligament force applied along the ligament line of action	
	Foot	V_1^2	Insertion of ligament in foot			
Calcaneofibular ligament length constancy	Shank	V_2^3	Origin of ligament in shank	Ligament length d_A^2	Ligament force applied along the ligament line of action	
	Foot	V_1^3	Insertion of ligament in foot			
Sphere-on-plane medial contact condition	Thigh	V_4^1	Position of femur condyle centre in thigh	Condyle radius d_K^1	Contact force applied from the medial femur condyle to the medial tibia plateau about the plateau normal	[14]

(continued)

Table 1 (continued)

Constraint	Segment	Virtual marker or direction	Anatomical definition	Parameter	Associated force (see Sect. 5)	References
	Shank	V_2^1	Position of tibia plateau in shank			
		$\mathbf{n}_2^1$	Orientation of tibia plateau in shank			
Sphere-on-plane lateral contact condition	Thigh	V_4^2	Position of femur condyle centre in thigh	Condyle radius d_K^2	Contact force applied from the lateral femur condyle to the lateral tibia plateau about the plateau normal	
	Shank	V_2^5	Position of tibia plateau in shank			
		$\mathbf{n}_2^2$	Orientation of tibia plateau in shank			
Anterior cruciate ligament length constancy	Thigh	V_4^3	ACL origin in thigh	Ligament length d_K^3	Ligament force applied along the ligament line of action	
	Shank	V_2^6	ACL insertion in shank			
Posterior cruciate ligament length constancy	Thigh	V_3^4	PCL origin in thigh	Ligament length d_K^4	Ligament force applied along the ligament line of action	
	Shank	V_2^7	PCL insertion in shank			
Medial collateral ligament length constancy	Thigh	V_4^5	MCL origin in thigh	Ligament length d_K^5	Ligament force applied along the ligament line of action	
	Shank	V_2^8	MCL insertion in shank			

(continued)

Table 1 (continued)

Constraint	Segment	Virtual marker or direction	Anatomical definition	Parameter	Associated force (see Sect. 5)	References
Patellofemoral hinge joint condition	Thigh	V_4^6	Position of joint axis in thigh		Contact force applied from the femur to the patella about the three axes of the orthonormal patella coordinate system	[15]
	Patella	V_3^1	Position of joint axis in patella			
		$\mathbf{X}_3, \mathbf{Y}_3, \mathbf{Z}_3$	Anterior, superior and lateral orthonormal axes of patella			
		$\mathbf{n}_3^1$	Orientation of joint axis in patella	Hinge angles θ_K^1, θ_K^2		
Patellar ligament length constancy	Shank	V_2^9	PL insertion in shank (origin is point D_3 in patella)	Ligament length d_K^6	Ligament force applied along the ligament line of action	
Iliofemoral spherical joint condition	Pelvis	V_5^1	Position of joint centre in pelvis		Contact force applied from the ilium to the femur about the three axes of the orthonormal thigh coordinate system	
	Thigh	$\mathbf{X}_4, \mathbf{Y}_4, \mathbf{Z}_4$	Anterior, superior and lateral orthonormal axes of thigh			

and L_i, respectively, and that the angles between these axes are constant. All rigid body constraints are quadratic in $\mathbf{Q}_i$:

$$
\boldsymbol{\Phi}_i^r =
\begin{pmatrix}
\mathbf{u}_i^2 - 1 \\
\mathbf{u}_i \bullet \left(\mathbf{r}_{P_i} - \mathbf{r}_{D_i} \right) - L_i \cos \gamma_i \\
\mathbf{u}_i \bullet \mathbf{w}_i - \cos \beta_i \\
\left(\mathbf{r}_{P_i} - \mathbf{r}_{D_i} \right)^2 - L_i^2 \\
\left(\mathbf{r}_{P_i} - \mathbf{r}_{D_i} \right) \bullet \mathbf{w}_i - L_i \cos \alpha_i \\
\mathbf{w}_i^2 - 1
\end{pmatrix}.
$$

3.2   Anatomical Constraints

In addition to the aforementioned rigid body constraints, kinematic constraints $\boldsymbol{\Phi}^k$ are introduced to specify the kinematic chain of the lower limb. These constraints represent an anatomical description of the joints. At the ankle, with 1 degree of freedom (DoF), the anatomical constraints represent the tibiotalar spherical joint condition, and the tibiocalcaneal and calcaneofibular ligament length constancies. At the knee, with 1 DoF, the anatomical constraints represent sphere-on-plane medial and lateral contact conditions, anterior and posterior cruciate ligament length constancies, medial collateral ligament length constancy, patellofemoral hinge joint condition, and patellar ligament length constancy. At the hip, with 3 DoFs, the anatomical constraints represent iliofemoral spherical joint condition. All kinematic constraints are built quadratic in $\mathbf{Q}_i$:

$$\Phi^k = \begin{pmatrix} \left(N_2^{V_1^1}Q_2 - N_1^{V_1^1}Q_1\right) \bullet N_1^{X_1}Q_1 \\ \left(N_2^{V_1^1}Q_2 - N_1^{V_1^1}Q_1\right) \bullet N_1^{Y_1}Q_1 \\ \left(N_2^{V_1^1}Q_2 - N_1^{V_1^1}Q_1\right) \bullet N_1^{Z_1}Q_1 \\ \left(N_2^{V_1^2}Q_2 - N_1^{V_1^2}Q_1\right)^2 - \left(d_A^1\right)^2 \\ \left(N_2^{V_1^3}Q_2 - N_1^{V_1^3}Q_1\right)^2 - \left(d_A^2\right)^2 \\ \left(N_4^{V_4^1}Q_4 - N_2^{V_2^2}Q_2\right) \bullet N_2^{n_2^1}Q_2 - d_K^1 \\ \left(N_4^{V_4^2}Q_4 - N_2^{V_2^5}Q_2\right) \bullet N_2^{n_2^2}Q_2 - d_K^2 \\ \left(N_4^{V_4^3}Q_4 - N_2^{V_2^6}Q_2\right)^2 - \left(d_K^3\right)^2 \\ \left(N_4^{V_4^4}Q_4 - N_2^{V_2^7}Q_2\right)^2 - \left(d_K^4\right)^2 \\ \left(N_4^{V_4^5}Q_4 - N_2^{V_2^8}Q_2\right)^2 - \left(d_K^5\right)^2 \\ \left(N_4^{V_4^6}Q_4 - N_3^{V_3^1}Q_3\right) \bullet N_3^{X_3}Q_3 \\ \left(N_4^{V_4^6}Q_4 - N_3^{V_3^1}Q_3\right) \bullet N_3^{Y_3}Q_3 \\ \left(N_4^{V_4^6}Q_4 - N_3^{V_3^1}Q_3\right) \bullet N_3^{Z_3}Q_3 \\ u_4 \bullet N_3^{n_3^1}Q_3 - \cos\theta_K^1 \\ \left(r_{P_4} - r_{D_4}\right) \bullet N_3^{n_3^1}Q_3 - L_4\cos\theta_K^2 \\ \left(r_{D_3} - N_2^{V_2^9}Q_2\right)^2 - \left(d_K^6\right)^2 \\ \left(N_5^{V_5^1}Q_5 - r_{P_4}\right) \bullet N_4^{X_4}Q_4 \\ \left(N_5^{V_5^1}Q_5 - r_{P_4}\right) \bullet N_4^{Y_4}Q_4 \\ \left(N_5^{V_5^1}Q_5 - r_{P_4}\right) \bullet N_4^{Z_4}Q_4 \end{pmatrix}.$$

The virtual markers V_i^j, virtual directions n_i^j, and model parameters involved in these kinematic constraints are detailed in Table 1. The corresponding coordinates, components and values can be found in the listed references.

As detailed in Table 1, anatomical constraints refer to the kinematic constraints Φ^k and can be associated with the relevant joint contact forces and ligament forces.

However, within rigid body constraints, the segment length constancy also encompasses an anatomical significance. This constraint is also associated with segment compression force (for the thigh and shank or, more precisely, for the femur and tibia) or segment traction force (for the patella).

3.3 Equations of Motion

The equations of motion of the lower limb can be written to introduce the musculo-tendon forces and the Lagrange multipliers associated with the above rigid body and kinematic constraints:

$$\mathbf{G}\ddot{\mathbf{Q}} + \mathbf{K}^T\boldsymbol{\lambda} = \mathbf{R} + \mathbf{P} + \mathbf{Lf}$$

where $\mathbf{G}$ is the matrix of generalised masses, $\ddot{\mathbf{Q}} = \begin{pmatrix} \vdots \\ \ddot{\mathbf{Q}}_i \\ \vdots \end{pmatrix}$ is the vector of generalised

accelerations for all segments, $\mathbf{K}$ is the Jacobian matrix of the constraints, $\boldsymbol{\lambda}$ is the vector of Lagrange multipliers, $\mathbf{R}$ is the vector of generalised ground reaction (i.e. including the 3D force $\mathbf{F}_0^{\mathbf{R}}$ and moment $\mathbf{M}_0^{\mathbf{R}}$ at the centre of pressure P_0), $\mathbf{P}$ is the vector of generalised weights, $\mathbf{L}$ is the matrix of generalised muscular lever arms, and $\mathbf{f}$ is the vector of musculo-tendon forces. Details can be found in Dumas et al. and Moissenet et al. [16, 17].

Interestingly, the use of the natural coordinates $\mathbf{Q}_i$ leads to equations including a constant mass matrix and no centrifugal nor Coriolis terms [10]. Moreover, the lower limb musculoskeletal model consists of 5 segments but the equations of motion can be written excluding the pelvis (which is only required for the definition of the hip joint constraints and some muscle origins).

4 Inverse Kinematics

The musculoskeletal model is driven by skin markers placed on the subject, i.e., the

segment parameters $\mathbf{Q} = \begin{pmatrix} \vdots \\ \mathbf{Q}_i \\ \vdots \end{pmatrix}$ have to be computed from the trajectories $\mathbf{r}_{M_i^j}$

measured by a motion capture system. This step is generally described as inverse kinematics. The terminology of multibody kinematics optimisation or global optimisation is also used in the field of biomechanics [18]. The alternative of extended

Kalman filters has also been proposed. This inverse kinematics step has two specific aims: to compensate for the soft-tissue artefacts and to compute consistent accelerations for the further dynamics steps (i.e. inconsistent accelerations possibly result in spurious forces).

It is important to note that, when segment kinematics is computed directly from skin marker trajectories, as commonly done in gait analysis, apparent dislocations may take place in the joints [19]. This indicates that the soft tissue artefact has a deleterious effect on the estimation of bone motion. Errors are, in some cases, of the same order of magnitude as the motions at the joints being investigated. Nevertheless, the model-derived kinematics obtained by inverse kinematics, although physiologically consistent, remains different from the actual kinematics, especially for the tibiofemoral joint [20–23]. Indeed, the DoFs modelled in the inverse kinematics only capture gross joint motion.

4.1 Constrained Optimisation

Inverse kinematics minimises the sum of the squared differences between measured and model-derived skin marker trajectories. This minimisation of the Euclidean norm is standard in the field of biomechanics [18]. Based on the marker set, i.e. with several skin markers per segment, the optimisation is an over-determined problem. Moreover, because anatomical constraints are included in the musculoskeletal model leading to 5 joint DoFs, the optimisation is a constrained problem. Thus, the optimisation problem is:

$$\min_{\mathbf{Q}} f = \tfrac{1}{2} (\boldsymbol{\Phi}^m)^T \boldsymbol{\Phi}^m$$

$$\text{subject to} \begin{cases} \boldsymbol{\Phi}^k = 0 \\ \boldsymbol{\Phi}^r = 0 \end{cases},$$

with $\boldsymbol{\Phi}^r = \begin{pmatrix} \vdots \\ \boldsymbol{\Phi}^r_i \\ \vdots \end{pmatrix}$ the rigid body constraints for all segments. The driving constraints

$\boldsymbol{\Phi}^m = \begin{pmatrix} \vdots \\ \mathbf{r}_{M^j_i} - \mathbf{N}^{M^j_i}_i \mathbf{Q}_i \\ \vdots \end{pmatrix}$ represent the differences between measured and model-derived skin marker trajectories.

This optimisation problem is finally equivalent to a zero-search problem when using a Lagrange formulation:

$$
\mathbf{F}\begin{pmatrix} \mathbf{Q} \\ \lambda^k \\ \lambda^r \end{pmatrix} = \begin{pmatrix} [\mathbf{K}^m]^T\,(\mathbf{\Phi}^m) + \begin{bmatrix} \mathbf{K}^k & \mathbf{0} \\ \mathbf{0} & \mathbf{K}^r \end{bmatrix}\begin{pmatrix} \lambda^k \\ \lambda^r \end{pmatrix} \\ \mathbf{\Phi}^k \\ \mathbf{\Phi}^r \end{pmatrix} = \mathbf{0}
$$

with $\mathbf{K} = \frac{\partial \mathbf{\Phi}}{\partial \mathbf{Q}}$ the Jacobian of the constraints and λ the corresponding Lagrange multipliers. Recall here that all constraints are linear or quadratic in $\mathbf{Q}$. This zero-search can be solved, at each sampled instant of time, with the function *fsolve* in Matlab.

To ensure the consistency of accelerations, one possibility is to add $\dot{\mathbf{Q}}$ and $\ddot{\mathbf{Q}}$ as well as the corresponding constraints to the optimisation problem [24]. Another possibility, chosen for its expediency, is to compute the time derivatives (i.e. by central differences) and to project them on the null spaces of $\mathbf{K}$ and $\dot{\mathbf{K}}$ [17].

4.2   Initial Guess and Model Parameters

Crucial to the inverse kinematics are the initial guess for $\mathbf{Q}$ and the definition of the model parameters involved in the different constraints $\mathbf{\Phi}^m$, $\mathbf{\Phi}^k$, $\mathbf{\Phi}^r$. Nevertheless, the optimisation problem becomes the minimisation of a quadratic function f subject to quadratic constraints that will not be sensitive to the initial guess. Moreover, no upper and lower boundaries for $\mathbf{Q}$ are considered. The number of model parameters in this optimisation is quite high, 147, and the kinematic results are very sensitive to some of these parameters, typically ankle and knee ligament geometries [25]. Actually, the fact that the musculoskeletal model has 5 joint DoFs means that there are several DoF couplings. These couplings, such as the relation between tibiofemoral flexion and internal rotation angle, or tibiofemoral and patellofemoral flexion angle, depend directly on the anatomical constraints. As explained in Table 1, the model parameters involved in $\mathbf{\Phi}^k$ are obtained from the literature [13–15]. As for the parameters involved in $\mathbf{\Phi}^m$ and $\mathbf{\Phi}^r$, they can be deduced from the trajectories of the skin markers and averaged for all sampled instants of time as detailed in Sancisi et al. [26].

5   Inverse Dynamics

In addition to skin marker trajectories, the musculoskeletal model relies on the ground reaction forces and moments to estimate musculo-tendon forces, joint contact forces and ligament forces. This estimation is performed in an inverse dynamics procedure with the model-derived segment parameters $\mathbf{Q}$, $\dot{\mathbf{Q}}$ and $\ddot{\mathbf{Q}}$. This inverse dynamics step has two specific aims: to solve the force-sharing problem and to investigate the interactions between forces both in muscles and in joints. It is important to note that such interactions can be investigated only if musculo-tendon forces, joint contact forces and ligament forces can be optimised simultaneously. This becomes straightforward with the above equations of motion introducing the musculo-tendon forces and the Lagrange multipliers associated with kinematic constraints. In line with common biomechanics practice, the inverse dynamics computation assumes that all segments are rigid.

5.1 Constrained Optimisation

The principle of the force-sharing solution is to minimise the sum of the squared forces and, in the present case, the sum of musculo-tendon forces, joint contact forces and ligament forces. Musculo-tendon forces $\mathbf{f}$ contribute to the dynamics of the multibody system through lever arms $\mathbf{L}$. Similarly, joint contact forces and ligament forces, represented by a subset of Lagrange multipliers $\boldsymbol{\lambda}_1$, contribute to the dynamics of the multibody system through the Jacobian matrix $\mathbf{K} = \begin{bmatrix} \mathbf{K}_1 \\ \mathbf{K}_2 \end{bmatrix}$. Thus, the optimisation problem is:

$$\min_{\begin{pmatrix} \mathbf{f} \\ \boldsymbol{\lambda}_1 \end{pmatrix}} J = \tfrac{1}{2} \begin{pmatrix} \mathbf{f} \\ \boldsymbol{\lambda}_1 \end{pmatrix}^T \mathbf{W} \begin{pmatrix} \mathbf{f} \\ \boldsymbol{\lambda}_1 \end{pmatrix}$$

$$\text{subject to} \begin{cases} \mathbf{Z}_{\mathbf{K}_2^T} \begin{bmatrix} \mathbf{L} & -\mathbf{K}_1^T \end{bmatrix} \begin{pmatrix} \mathbf{f} \\ \boldsymbol{\lambda}_1 \end{pmatrix} = \mathbf{Z}_{\mathbf{K}_2^T} \left(\mathbf{G}\ddot{\mathbf{Q}} - \mathbf{P} - \mathbf{R} \right) \\ \begin{pmatrix} \mathbf{f}_{\min} \\ \boldsymbol{\lambda}_{1\min} \end{pmatrix} \leq \begin{pmatrix} \mathbf{f} \\ \boldsymbol{\lambda}_1 \end{pmatrix} \leq \begin{pmatrix} \mathbf{f}_{\max} \\ \boldsymbol{\lambda}_{1\max} \end{pmatrix} \end{cases}$$

with J the objective function, $\mathbf{W}$ a diagonal matrix composed of the optimisation weights associated with the unknowns $\begin{pmatrix} \mathbf{f} \\ \boldsymbol{\lambda}_1 \end{pmatrix}$, and $\mathbf{Z}_{\mathbf{K}_2^T}$ the orthogonal basis of the null space of $\mathbf{K}_2^T$ (i.e. $\mathbf{Z}_{\mathbf{K}_2^T} \mathbf{K}_2^T = \mathbf{0}$). The constraints of the optimisation are the dynamic equilibrium of the model and the upper and lower bounds of the musculo-tendon, joint contact forces and ligament forces. This constrained optimisation can be solved, at each sampled instant of time, with the function *fmincon* in Matlab.

5.2   Optimisation Weights and Boundaries

For this second constrained optimisation solving the force-sharing problem, the initial guess is simply a null vector. The optimisation weights in **W** are set at 1 for musculo-tendon forces, $1e^{-6}$ for ligament forces, 1 for ankle, patellofemoral and hip joint forces, 2 and 4 for medial and lateral tibiofemoral joint forces, respectively [27, 28]. Regarding upper and lower bounds, musculo-tendon forces and ligament forces can only be positive but some joint contact forces applied in the ankle, patella, femoral and ankle joints about the anterior and lateral axes of the relevant segments can change sign. Maximal forces are identified for muscles (e.g. based on the physiological cross-sectional areas reported in Klein Horsman et al. [7]) but remain unknown for joint contact forces and ligaments and are set at infinite.

The choice of objective function J (i.e. quadratic, cubic, etc.) and optimisation weights in **W** are typical of alterations used in the literature to obtain a more accurate estimation of joint contact forces [29]. For instance, these weights can be tuned to track measured joint contact forces (see Sect. 7), although this was not done here.

5.3   Multi-objective Optimisation

Instead of using arbitrary weights to define objective function J, a multi-objective optimisation can be performed. For this purpose, different concurrent objective functions are defined. The constraints and initial guess of the optimisation remain the same. The same principle is applied to solve the force-sharing problem, namely the minimum of the sum of the squared forces. However, in this case, the optimisation no longer introduces what is called an a priori articulation of preferences, i.e. the relative importance of each objective specifically defined through weights:

$$\min_{\begin{pmatrix} \mathbf{f} \\ \boldsymbol{\lambda}_1 \end{pmatrix}} \max \begin{pmatrix} J_1 = \frac{1}{n_1} \mathbf{f}^T \mathbf{f} \\ J_2 = \frac{1}{n_2} \left(\boldsymbol{\lambda}_1^c \right)^T \boldsymbol{\lambda}_1^c \\ J_3 = \frac{1}{n_3} \left(\boldsymbol{\lambda}_1^l \right)^T \boldsymbol{\lambda}_1^l \end{pmatrix}$$

with $\boldsymbol{\lambda}_1^c$ and $\boldsymbol{\lambda}_1^l$ the joint contact forces and ligament forces and n_1, n_2 and n_3 the number of forces involved in J_1, J_2 and J_3. This multi-objective optimisation can be solved, at each sampled instant of time, with the function *fminmax* in Matlab.

The rationale for considering these three concurrent objectives is to investigate muscle and joint functions, and their interactions [4, 27, 30, 31]. Naturally, the amplitude of compression forces (i.e. dealing with articular contacts) and of shear and torsion forces (i.e. dealing with ligaments) will depend on the distribution of musculo-tendon forces, and usually on the amount of co-contraction. Similarly, the distribution of musculo-tendon forces between mono- and bi-articular muscles will have an impact on the compression force in the bones.

6   Individual Muscle Contributions

A complementary approach to elucidating muscle and joint function is to compute the contributions of musculo-tendon forces to joint contact forces and ligament forces. This computation is another indeterminate problem but it can be solved using a pseudo-inverse method [32–34].

For this purpose, the generalised ground reaction $\mathbf{R}$ is conveniently re-written to replace the ground reaction moment $\mathbf{M}_0^{\mathbf{R}}$ by three equivalent forces:

$$
\mathbf{R} = \mathbf{L}^{\mathbf{R}}
\begin{pmatrix}
\mathbf{F}_0^{\mathbf{R}} \\
f_u^{\mathbf{M}_0^{\mathbf{R}}} \\
f_v^{\mathbf{M}_0^{\mathbf{R}}} \\
f_w^{\mathbf{M}_0^{\mathbf{R}}}
\end{pmatrix}.
$$

with the forces $\begin{pmatrix} f_u^{\mathbf{M}_0^{\mathbf{R}}} \\ f_v^{\mathbf{M}_0^{\mathbf{R}}} \\ f_w^{\mathbf{M}_0^{\mathbf{R}}} \end{pmatrix}$ applied about the axes of the non-orthonormal foot coordinate system representing the ground reaction moment. Then the contributions of musculo-tendon forces to ground reaction are computed by applying each individual musculo-tendon force in isolation (i.e. with $\mathbf{P} = \mathbf{0}$ and all $f^j = 0$ except one). The computation is based on a Moore-Penrose pseudo-inverse (denoted †):

$$
\begin{pmatrix}
\ddot{\mathbf{Q}}^{f^j} \\
\mathbf{F}_0^{\mathbf{R}, f^j} \\
f_u^{\mathbf{M}_0^{\mathbf{R}}, f^j} \\
f_v^{\mathbf{M}_0^{\mathbf{R}}, f^j} \\
f_w^{\mathbf{M}_0^{\mathbf{R}}, f^j}
\end{pmatrix}
= \left[\mathbf{Z}_{\mathbf{K}^T}\mathbf{G} \ -\mathbf{Z}_{\mathbf{K}^T}\mathbf{L}^{\mathbf{R}} \right]^\dagger \mathbf{Z}_{\mathbf{K}^T}\mathbf{L}
\begin{pmatrix}
\vdots \\
0 \\
\vdots \\
f^j \\
\vdots \\
0 \\
\vdots
\end{pmatrix}
$$

with $\ddot{\mathbf{Q}}^{f^j}$ the contribution of the jth muscle to generalised accelerations, $\mathbf{F}_0^{\mathbf{R},f^j}$ and

$\begin{pmatrix} f_u^{\mathbf{M}_0^{\mathbf{R}},f^j} \\ f_v^{\mathbf{M}_0^{\mathbf{R}},f^j} \\ f_w^{\mathbf{M}_0^{\mathbf{R}},f^j} \end{pmatrix}$ the contributions to ground reaction forces and moments and $\mathbf{Z}_{\mathbf{K}^T}$ the

orthogonal basis of the null space of $\mathbf{K}$ (i.e. $\mathbf{Z}_{\mathbf{K}^T}\mathbf{K}^T = \mathbf{0}$). The null space of the
Jacobian $\mathbf{K}$ is used to more straightforwardly link musculo-tendon forces and ground
reaction forces. However, in the second computation, it is the null space of $\mathbf{K}_2$ that
is used to link musculo-tendon forces, joint contact forces and ligament forces:

$$\boldsymbol{\lambda}_1^{f^j} = \left[\mathbf{Z}_{\mathbf{K}_2^T}\mathbf{K}_1^T\right]^{\dagger}\mathbf{Z}_{\mathbf{K}_2^T}\left(-\mathbf{G}\ddot{\mathbf{Q}}^{f^j} + \mathbf{L}^{\mathbf{R}}\begin{pmatrix} \mathbf{F}_0^{\mathbf{R},f^j} \\ f_u^{\mathbf{M}_0^{\mathbf{R}},f^j} \\ f_v^{\mathbf{M}_0^{\mathbf{R}},f^j} \\ f_w^{\mathbf{M}_0^{\mathbf{R}},f^j} \end{pmatrix} + \mathbf{L}\begin{pmatrix} \vdots \\ 0 \\ \vdots \\ f^j \\ \vdots \\ 0 \\ \vdots \end{pmatrix}\right).$$

with $\boldsymbol{\lambda}_1^{f^j}$ the contribution of the jth muscle to joint contact forces and ligament forces.

7 Knee Joint Kinematics, Contact Forces and Ligament Forces

This section presents results from retrospective experimental measurements (i.e.
skin marker trajectories, ground reaction forces and moments) on an asymptomatic
subject (male, 30 years old, 65 kg, 165 cm) [34] and one subject implanted with an
instrumented knee prosthesis (male, 86 years old, 75 kg, 180 cm) [35] walking at
comfortable speed. In both cases, five gait cycles are analysed. Inverse kinematics
and inverse dynamics are processed and the results time-normalised and averaged.

7.1 Joint Angles and Displacements

Model-derived segment parameters $\mathbf{Q}$ computed from the skin marker trajectories
measured during the asymptomatic subject's five gait cycles are used to obtain the

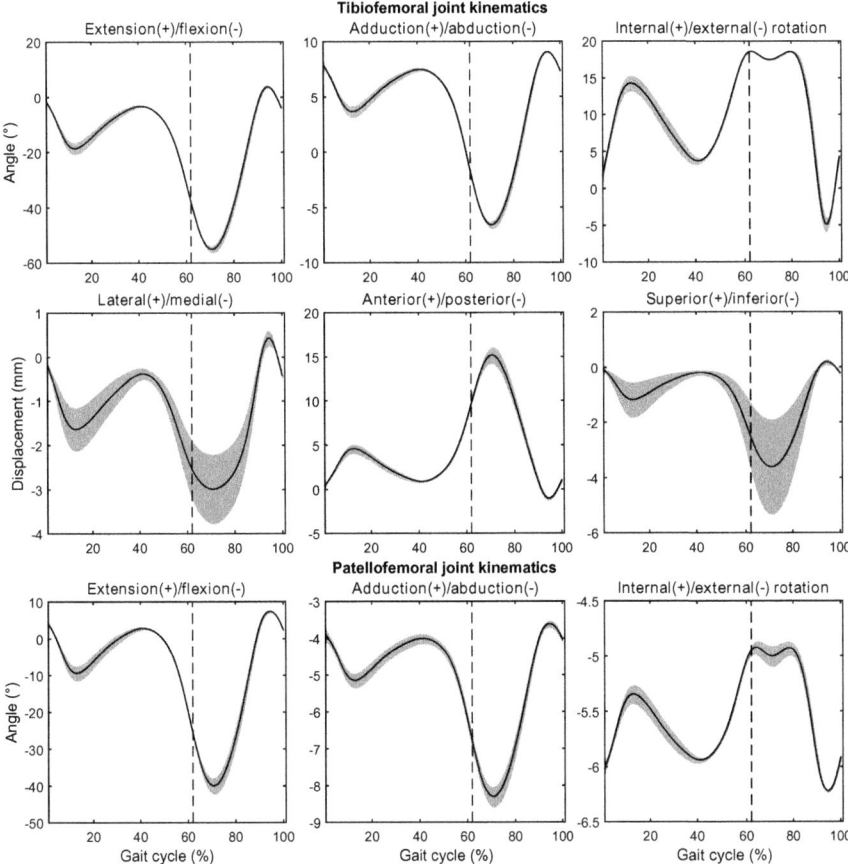

Fig. 3 Tibiofemoral joint angles and displacements, patellofemoral joint angles—mean and standard deviation on 5 gait cycles of the asymptomatic subject (toe-off is indicated by a vertical dotted line)

joint kinematics [36]. Figure 3 presents the tibiofemoral joint angles and displacements and the patellofemoral joint angles for the asymptomatic subject.

The curve patterns in Fig. 3 indicate clearly that the DoFs are not independent. Actually, the musculoskeletal model includes 11 anatomical constraints and, emblematically, 4 ligament constraints, so that shank and patella movement with respect to the thigh has only one independent DoF. Notably, the knee kinematics presents significant tibiofemoral internal rotation and anterior displacement depending on tibiofemoral flexion, which appear consistent with normal knee kinematics measured in vivo during dynamic activities [37]. The knee kinematics also presents coupled patellofemoral flexion.

7.2   *Joint Contact Forces and Ligament Forces*

Segment accelerations are then combined with ground reaction forces and moments
to compute musculo-tendon forces, joint contact forces and ligament forces. Figure 4
presents the tibiofemoral and patellofemoral joint contact forces, cruciate and patellar
ligament forces, and the influence of the main contributing muscles. As presented
in Table 1, according to the kinematic constraints definition, joint contact forces
are straightforwardly expressed in the coordinate system of the distal segment and
ligament forces along the ligament line of action.

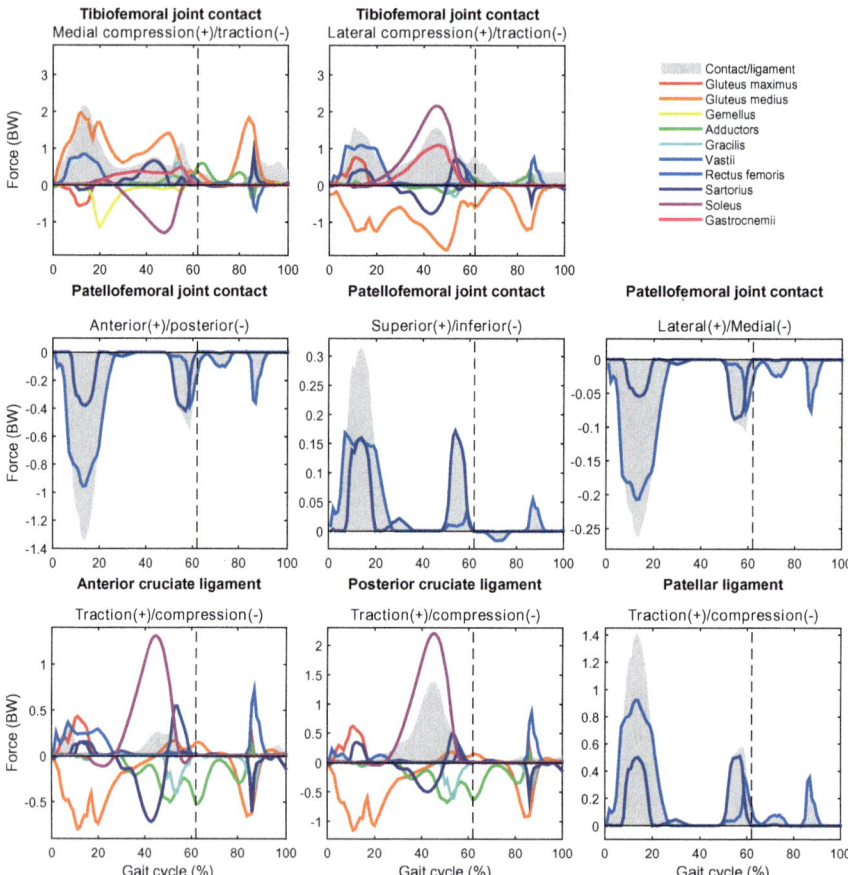

Fig. 4 Tibiofemoral and patellofemoral joint contact forces, cruciate and patellar ligament forces,
and individual muscle contributions—mean over 5 gait cycles of asymptomatic subject (all forces
are expressed in body weight (BW), toe-off is indicated by a vertical dotted line)

Contact forces reach 1.5–2 BW for the tibiofemoral and patellofemoral joints. Ligament forces also reach 1.5 BW. All forces show two main peaks during stance, at 15 and 45% of the gait cycle.

For both tibiofemoral medial and lateral contact forces, the vastii and gastrocnemii contribute to the first and the second peak, respectively. This is consistent with the contributions reported in the literature [38–40]. Moreover, the gluteus medius increases tibiofemoral medial contact force and decreases lateral contact force, while the soleus makes the opposite contribution. Sartorius, gracilis, and adductors, having an insertion on the medial side of the tibia, also increase medial contact force and decrease lateral contact force. For the patellofemoral joint, only the vastii and rectus femoris contribute to contact force.

For the anterior cruciate ligament, vastii and gluteus maximus contribute to the first force peak and soleus to the second force peak. Soleus also contributes principally to the second peak of posterior cruciate ligament force. Gluteus medius, sartorius, gracilis, and adductors tend to decrease cruciate ligament traction. For the patellofemoral ligament, only vastii and rectus femoris contribute to its traction.

7.3 Validation of Tibiofemoral Contact Forces

Similarly to the asymptomatic subject, tibiofemoral medial and lateral knee contact forces (as well as total contact force) are computed for the subject implanted with an instrumented knee prosthesis. Thus, these contact forces are validated against in vivo measurements. Figure 5 presents the results of this validation.

The amplitude of the estimated total force is in agreement with the in vivo measurements. The weighted optimisation generally results in slightly higher forces. The root mean square errors for total contact force are 0.59 BW with weighted optimisation and 0.57 BW with multi-objective optimisation.

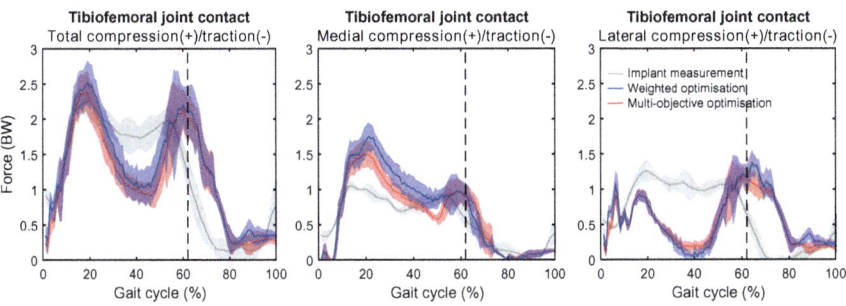

Fig. 5 Estimated total, medial, and lateral tibiofemoral joint contact forces for both weighted and multi-objective optimisations, and implant measurements—mean and standard deviation on 5 gait cycles of subject implanted with an instrumented knee prosthesis (all forces are expressed in body weight (BW), toe-off is indicated by a vertical dotted line)

However, an unbalanced ratio between medial and lateral contact forces is observed. The first peak is overestimated for medial contact force and underestimated for lateral force. This result may be due to insufficient similarity between subject and model joint geometries, which could be improved by customisation or personalisation of the model [41–43].

Furthermore, the pattern of the estimated contact forces reveals a late second peak and virtual unloading of lateral contact during stance (as already observed with other models [38, 41, 42, 44, 45]).

8 Conclusion

This chapter presents the different optimisations required to drive a musculoskeletal model of the lower limb with experimental data (i.e. skin marker trajectories and ground reaction forces and moments) and to compute musculo-tendon forces, joint contact forces and ligament forces as well as their interactions.

It is worthy of note that musculo-tendon forces, joint contact forces and ligament forces can be estimated simultaneously. Because anatomical constraints (e.g. medial and lateral contact conditions, cruciate and patellar ligament length constancies, segment length constancies) are introduced into the inverse kinematics and inverse dynamics steps, the biomechanical significance of the Lagrange multipliers associated with these constraints can be exploited. Depending on the context, the null space of the Jacobian matrix or Jacobian sub-matrix is also exploitable. This is useful, enabling consistent segment acceleration to be computed. Moreover, it allows multi-objective optimisations (i.e. with and without an a priori articulation of preferences) and the computation of individual muscle contributions. In the context of this book about the biomechanics of anthropomorphic systems, these anatomical constraints can be considered to be more representative of joint mechanics than simple spherical or hinge constraints.

Results for the proposed musculoskeletal model of the lower limb and optimisation methods are in agreement with results for other models, typically concerning joint contact forces, individual muscle contributions to them and to ground reaction forces (not presented here) [32, 38–40, 45, 46]. More original results presented here concern the forces in the cruciate ligaments, patellofemoral contact and patellofemoral ligaments, and individual muscle contributions to them [34]. Furthermore, the proposed musculoskeletal model is validated against in vivo measurements of knee contact forces with an instrumented prosthesis [28, 47] and compares favourably with other non-personalised models [29].

This musculoskeletal model of the lower limb offers promise for clinical applications such as osteoarthritis or ligament tears. Knee anatomical constraints can easily be personalised using bi-plane radiography or MRI [48–50]. Obviously, analysing interactions between forces acting both in muscles and in joints would have advantages for patients with articular or ligament pathologies.

References

1. Buchanan, T.S., Lloyd, D.G., Manal, K., Besier, T.F.: Neuromusculoskeletal modeling: estimation of muscle forces and joint moments and movements from measurements of neural command. J. Appl. Biomech. **20**(4), 367–395 (2004). https://doi.org/10.1123/jab.20.4.367
2. Chèze, L., Moissenet, F., Dumas, R.: State of the art and current limits of musculo-skeletal models for clinical applications. Mov. Sport Sci. **90**, 7–17 (2015). https://doi.org/10.1051/sm/2012026
3. Erdemir, A., McLean, S., Herzog, W., van den Bogert, A.J.: Model-based estimation of muscle forces exerted during movements. Clin. Biomech. **22**(2), 131–154 (2007). https://doi.org/10.1016/j.clinbiomech.2006.09.005
4. Pandy, M.G., Andriacchi, T.P.: Muscle and joint function in human locomotion. Annu. Rev. Biomed. Eng. **12**(1), 401–433 (2010). https://doi.org/10.1146/annurev-bioeng-070909-105259
5. Zajac, F.E.: Understanding muscle coordination of the human leg with dynamical simulations. J. Biomech. **35**(8), 1011–1018 (2002). https://doi.org/10.1016/S0021-9290(02)00046-5
6. Dumas, R., Chèze, L, Verriest, J.P.: Adjustments to McConville et al. and Young et al. body segment inertial parameters. J. Biomech. **40**(3), 543–553 (2007). https://doi.org/10.1016/j.jbiomech.2006.02.013
7. Klein Horsman, M.D., Koopman, H.F.J.M., Helm, F.C.T., Prosé, L.P., Veeger, H.E.J.: Morphological muscle and joint parameters for musculoskeletal modelling of the lower extremity. Clin. Biomech. **22**(2), 239–247 (2007). https://doi.org/10.1016/j.clinbiomech.2006.10.003
8. van Arkel, R.J., Modenese, L., Phillips, A.T.M., Jeffers, J.R.T.: Hip abduction can prevent posterior edge loading of hip replacements. J. Orthop. Res. **31**(8), 1172–1179 (2013). https://doi.org/10.1002/jor.22364
9. Dumas, R., Chèze, L.: 3D inverse dynamics in non-orthonormal segment coordinate system. Med. Biol. Eng. Compu. **45**(3), 315–322 (2007). https://doi.org/10.1007/s11517-006-0156-8
10. Garcia de Jalon, J., Unda, J., Avello, A.: Natural coordinates for the computer analysis of multibody systems. Comput. Methods Appl. Mech. Eng. **56**(3), 309–327 (1986). https://doi.org/10.1016/0045-7825(86)90044-7
11. Duprey, S., Cheze, L., Dumas, R.: Influence of joint constraints on lower limb kinematics estimation from skin markers using global optimization. J. Biomech. **43**(14), 2858–2862 (2010). https://doi.org/10.1016/j.jbiomech.2010.06.010
12. Wu, G., Siegler, S., Allard, P., Kirtley, C., Leardini, A., Rosenbaum, D., Whittle, M., D'Lima, D.D., Cristofolini, L., Witte, H., Schmid, O., Stokes, I.: ISB recommendation on definitions of joint coordinate system of various joints for the reporting of human joint motion–part I: ankle, hip, and spine. J. Biomech. **35**(4), 543–548 (2002). https://doi.org/10.1016/S0021-9290(01)00222-6. International Society of Biomechanics
13. Di Gregorio, R., Parenti-Castelli, V., O'Connor, J.J., Leardini, A.: Mathematical models of passive motion at the human ankle joint by equivalent spatial parallel mechanisms. Med. Biol. Eng. Compu. **45**(3), 305–313 (2007). https://doi.org/10.1007/s11517-007-0160-7
14. Feikes, J.D., O'Connor, J.J., Zavatsky, A.B.: A constraint-based approach to modelling the mobility of the human knee joint. J. Biomech. **36**(1), 125–129 (2003). https://doi.org/10.1016/S0021-9290(02)00276-2
15. Sancisi, N., Parenti-Castelli, V.: A new kinematic model of the passive motion of the knee inclusive of the patella. J. Mech. Rob. **3**(4), 041003–041007 (2011). https://doi.org/10.1115/1.4004890
16. Dumas, R., Moissenet, F., Gasparutto, X., Chèze, L.: Influence of joint models on lower-limb musculo-tendon forces and three-dimensional joint reaction forces during gait. Proc. Inst. Mech. Eng. [H] **226**(2), 146–160 (2012). https://doi.org/10.1177/0954411911431396
17. Moissenet, F., Chèze, L., Dumas, R.: Anatomical kinematic constraints: consequences on musculo-tendon forces and joint reactions. Multibody Sys. Dyn. **28**(1), 125–141 (2012). https://doi.org/10.1007/s11044-011-9286-3

18. Begon, M., Andersen, M.S., Dumas, R.: Multibody kinematics optimization for the estimation of upper and lower limb human joint kinematics: a systematized methodological review. J. Biomech. Eng. (2017) (Accepted)

19. Ojeda, J., Martínez-Reina, J., Mayo, J.: A method to evaluate human skeletal models using marker residuals and global optimization. Mech. Mach. Theory **73**, 259–272 (2014). https://doi.org/10.1016/j.mechmachtheory.2013.11.003

20. Andersen, M.S., Benoit, D.L., Damsgaard, M., Ramsey, D.K., Rasmussen, J.: Do kinematic models reduce the effects of soft tissue artefacts in skin marker-based motion analysis? An in vivo study of knee kinematics. J. Biomech. **43**(2), 268–273 (2010). https://doi.org/10.1016/j.jbiomech.2009.08.034

21. Clément, J., Dumas, R., Hagemeister, N., de Guise, J.A.: Can generic knee joint models improve the measurement of osteoarthritic knee kinematics during squatting activity? Comput. Methods Biomech. Biomedical Eng. **20**(1), 94–103 (2017). https://doi.org/10.1080/10255842.2016.1202935

22. Gasparutto, X., Sancisi, N., Jacquelin, E., Parenti-Castelli, V., Dumas, R.: Validation of a multi-body optimization with knee kinematic models including ligament constraints. J. Biomech. **48**(6), 1141–1146 (2015). https://doi.org/10.1016/j.jbiomech.2015.01.010

23. Richard, V., Cappozzo, A., Dumas, R.: Comparative assessment of knee joint models used in multi-body kinematics optimisation for soft tissue artefact compensation. J. Biomech. **62**, 95–101 (2017). https://doi.org/10.1016/j.jbiomech.2017.01.030

24. Andersen, M.S., Damsgaard, M., Rasmussen, J.: Kinematic analysis of over-determinate biomechanical systems. Comput. Methods Biomech. Biomedical Eng. **12**(4), 371–384 (2009). https://doi.org/10.1080/10255840802459412

25. El Habachi, A., Moissenet, F., Duprey, S., Cheze, L., Dumas, R.: Global sensitivity analysis of the joint kinematics during gait to the parameters of a lower limb multi-body model. Med. Biol. Eng. Comput. **53**(7), 655–667 (2015). https://doi.org/10.1007/s11517-015-1269-8

26. Sancisi, N., Gasparutto, X., Parenti-Castelli, V., Dumas, R.: A multi-body optimization framework with a knee kinematic model including articular contacts and ligaments. Meccanica **52**(3), 695–711 (2017). https://doi.org/10.1007/s11012-016-0532-x

27. Moissenet, F., Chèze, L., Dumas, R.: A 3D lower limb musculoskeletal model for simultaneous estimation of musculo-tendon, joint contact, ligament and bone forces during gait. J. Biomech. **47**(1), 50–58 (2014). https://doi.org/10.1016/j.jbiomech.2013.10.015

28. Moissenet, F., Chèze, L., Dumas, R.: Influence of the level of muscular redundancy on the validity of a musculoskeletal model. J. Biomech. Eng. **138**(2), 021019–021016 (2016). https://doi.org/10.1115/1.4032127

29. Moissenet, F., Modenese, L., Dumas, R.: Alterations of musculoskeletal models for a more accurate estimation of lower limb joint contact forces during normal gait: a systematic review. J. Biomech. **63**, 8–20 (2017). https://doi.org/10.1016/j.jbiomech.2017.08.025

30. Cleather, D.J., Bull, A.M.J.: An optimization-based simultaneous approach to the determination of muscular, ligamentous, and joint contact forces provides insight into musculoligamentous interaction. Ann. Biomed. Eng. **39**(7), 1925–1934 (2011). https://doi.org/10.1007/s10439-011-0303-8

31. Collins, J.J.: The redundant nature of locomotor optimization laws. J. Biomech. **28**(3), 251–267 (1995). https://doi.org/10.1016/0021-9290(94)00072-C

32. Lin, Y.C., Kim, H.J., Pandy, M.G.: A computationally efficient method for assessing muscle function during human locomotion. Int. J. Numer. Methods Biomed. Eng. **27**(3), 436–449 (2011). https://doi.org/10.1002/cnm.1396

33. Moissenet, F., Chèze, L., Dumas, R.: Contribution of individual musculo-tendon forces to the axial compression force of the femur during normal gait. Mov. Sport Sci. **93**, 63–69 (2016). https://doi.org/10.1051/sm/2015041

34. Moissenet, F., Chèze, L., Dumas, R.: Individual muscle contributions to ground reaction and to joint contact, ligament and bone forces during normal gait. Multibody Sys. Dyn. **40**(2), 193–211 (2017). https://doi.org/10.1007/s11044-017-9564-9

35. Fregly, B.J., Besier, T.F., Lloyd, D.G., Delp, S.L., Banks, S.A., Pandy, M.G., D'Lima, D.D.: Grand challenge competition to predict in vivo knee loads. J. Orthop. Res. **30**(4), 503–513 (2012). https://doi.org/10.1002/jor.22023
36. Dumas, R., Robert, T., Pomero, V., Chèze, L.: Joint and segment coordinate systems revisited. Comput. Methods Biomech. Biomed. Eng. **15**(sup1), 183–185 (2012). https://doi.org/10.108 0/10255842.2012.713646
37. Gasparutto, X., Moissenet, F., Lafon, Y., Chèze, L., Dumas, R.: Kinematics of the normal knee during dynamic activities: a synthesis of data from intracortical pins and biplane imaging. Appl. Bion. Biomech. **2017**, 9 (2017). https://doi.org/10.1155/2017/1908618
38. Ogaya, S., Naito, H., Okita, Y., Iwata, A., Higuchi, Y., Fuchioka, S., Tanaka, M.: Contribution of muscle tension force to medial knee contact force at fast walking speed. J Mech. Med. Biol. **15**(01), 1550002 (2015). https://doi.org/10.1142/S0219519415500025
39. Sritharan, P., Lin, Y.C., Pandy, M.G.: Muscles that do not cross the knee contribute to the knee adduction moment and tibiofemoral compartment loading during gait. J. Orthop. Res. **30**(10), 1586–1595 (2012). https://doi.org/10.1002/jor.22082
40. Winby, C.R., Lloyd, D.G., Besier, T.F., Kirk, T.B.: Muscle and external load contribution to knee joint contact loads during normal gait. J. Biomech. **42**(14), 2294–2300 (2009). https://do i.org/10.1016/j.jbiomech.2009.06.019
41. Lerner, Z.F., DeMers, M.S., Delp, S.L., Browning, R.C.: How tibiofemoral alignment and contact locations affect predictions of medial and lateral tibiofemoral contact forces. J. Biomech. **48**(4), 644–650 (2015). https://doi.org/10.1016/j.jbiomech.2014.12.049
42. Saliba, C.M., Brandon, S.C.E., Deluzio, K.J.: Sensitivity of medial and lateral knee contact force predictions to frontal plane alignment and contact locations. J. Biomech. **57**, 125–130 (2017). https://doi.org/10.1016/j.jbiomech.2017.03.005
43. Zeighami, A., Aissaoui, R., Dumas, R.: Knee medial and lateral contact forces in a musculoskeletal model with subject-specific contact point trajectories. J. Biomech. **69**, 138–145 (2018). https://doi.org/10.1016/j.jbiomech.2018.01.021
44. Kumar, D., Rudolph, K.S., Manal, K.T.: EMG-driven modeling approach to muscle force and joint load estimations: Case study in knee osteoarthritis. J. Orthop. Res. **30**(3), 377–383 (2012). https://doi.org/10.1002/jor.21544
45. Shelburne, K.B., Torry, M.R., Pandy, M.G.: Contributions of muscles, ligaments, and the ground-reaction force to tibiofemoral joint loading during normal gait. J. Orthop. Res. **24**(10), 1983–1990 (2006). https://doi.org/10.1002/jor.20255
46. Correa, T.A., Crossley, K.M., Kim, H.J., Pandy, M.G.: Contributions of individual muscles to hip joint contact force in normal walking. J. Biomech. **43**(8), 1618–1622 (2010). https://doi.o rg/10.1016/j.jbiomech.2010.02.008
47. Moissenet, F., Giroux, M., Chèze, L., Dumas, R.: Validity of a musculoskeletal model using two different geometries for estimating hip contact forces during normal walking. Comput. Methods Biomech. Biomed. Eng. **18**(sup1), 2000–2001 (2015). https://doi.org/10.1080/1025 5842.2015.1069596
48. Brito da Luz, S., Modenese, L., Sancisi, N., Mills, P.M., Kennedy, B., Beck, B.R., Lloyd, D.G.: Feasibility of using MRIs to create subject-specific parallel-mechanism joint models. J. Biomech. **53**, 45–55 (2017). https://doi.org/10.1016/j.jbiomech.2016.12.018
49. Clément, J., Dumas, R., Hagemeister, N., de Guise, J.A.: Soft tissue artifact compensation in knee kinematics by multi-body optimization: performance of subject-specific knee joint models. J. Biomech. **48**(14), 3796–3802 (2015). https://doi.org/10.1016/j.jbiomech.2015.09.0 40
50. Valente, G., Pitto, L., Stagni, R., Taddei, F.: Effect of lower-limb joint models on subject-specific musculoskeletal models and simulations of daily motor activities. J. Biomech. **48**(16), 4198–4205 (2015). https://doi.org/10.1016/j.jbiomech.2015.09.042

Creating Personalized Dynamic Models

G. Venture, V. Bonnet and D. Kulic

Abstract In human motion science, the dynamics plays an important role. It relates the movement of the human to the forces necessary to achieve this movement. It also relates the human and its environment through interaction forces. Estimating subject-specific dynamic models is a challenging problem, due to the need for both accurate measurement and modeling formalisms. In the past decade, we have developed solutions for the computation of the dynamic quantities of humans, based on individual (subject specific) models, inspired largely by Robotics geometric and dynamic calibration. In this chapter, we will present the state of the art and our latest advances in this area and show examples of applications to both humans and humanoid robots. With these research results we hope to contribute beyond the field of robotics to the fields of biomechanics and ergonomics, by providing accurate dynamic models of beings.

1 Introduction

Dynamic analysis is based on the equation of motion:

$$\mathbf{\Gamma} = \mathbf{M}_c(\boldsymbol{\theta})\ddot{\boldsymbol{\theta}} + \mathbf{C}_c\big(\boldsymbol{\theta}, \dot{\boldsymbol{\theta}}\big) + \mathbf{G}_c(\boldsymbol{\theta}) \tag{1}$$

G. Venture (✉)
Tokyo University of Agriculture and Technology, 2-24-16 Nakacho, Koganei, Tokyo, Japan
e-mail: venture@cc.tuat.ac.jp

V. Bonnet
University of Paris-Est, Créteil, France

D. Kulic
University of Waterloo, 200 University Avenue West, Waterloo, ON, Canada

© Springer International Publishing AG, part of Springer Nature 2019
G. Venture et al. (eds.), *Biomechanics of Anthropomorphic Systems*, Springer Tracts
in Advanced Robotics 124, https://doi.org/10.1007/978-3-319-93870-7_5

This equation describes the relationship between the internal force and moment of forces in the system joints $\boldsymbol{\Gamma}$ and the force and moment of force due to the motion $\boldsymbol{M}_c(\boldsymbol{\theta})\ddot{\boldsymbol{\theta}} + \boldsymbol{C}_c\left(\boldsymbol{\theta}, \dot{\boldsymbol{\theta}}\right)$ (acceleration, velocities) and due to the gravity $\boldsymbol{G}_c$. This relationship has been extensively used in robotics for more than 3 decades to model the kinematics of the robots, obtain the precise dynamics of the robots and control these robots [1]. It was first applied to manipulator robots, and then more recently to humanoid robots [2]. When applied to humanoid robots, Eq. (1) transforms into the well-known floating-base equation, where the floating-base is an arbitrary chosen body which motion is described specifically together with the kinematics chain as follows:

$$\begin{bmatrix} 0 \\ \boldsymbol{\Gamma} \end{bmatrix} + \begin{bmatrix} \boldsymbol{J}_b^T \\ \boldsymbol{J}^T \end{bmatrix} \boldsymbol{F} = \begin{bmatrix} \boldsymbol{M}_b & \boldsymbol{M}_{bc} \\ \boldsymbol{M}_{cb} & \boldsymbol{M}_c \end{bmatrix} \begin{bmatrix} \ddot{\boldsymbol{q}}_b \\ \ddot{\boldsymbol{\theta}} \end{bmatrix} + \begin{bmatrix} \boldsymbol{C}_b & \boldsymbol{C}_{bc} \\ 0 & \boldsymbol{C}_c \end{bmatrix} \begin{bmatrix} \dot{\boldsymbol{q}}_b \\ \dot{\boldsymbol{\theta}} \end{bmatrix} + \begin{bmatrix} \boldsymbol{G}_b \\ \boldsymbol{G}_c \end{bmatrix} \quad (2)$$

where $\boldsymbol{M}_c$, $\boldsymbol{C}_c$, $\boldsymbol{G}_c$, $\boldsymbol{M}_b$, $\boldsymbol{C}_b$, and $\boldsymbol{G}_b$, are the inertia, Coriolis and gravity matrices calculated at the articulated chain (subscript $_c$) and at the floating base level (subscript $_b$), respectively and function of $\boldsymbol{\theta}, \boldsymbol{q}_b, \dot{\boldsymbol{\theta}}, \dot{\boldsymbol{q}}_b, \ddot{\boldsymbol{\theta}}, \dot{\boldsymbol{\theta}}, \ddot{\boldsymbol{\theta}}$, and $\boldsymbol{\Gamma}$ are the joint angle, velocity, acceleration and torque vectors, respectively. $\boldsymbol{q}_b, \dot{\boldsymbol{q}}_b, \ddot{\boldsymbol{q}}_b$ are the 6 coordinates of the floating base and their first and second derivatives respectively. $\boldsymbol{J}^T$ is the Jacobian transpose matrix mapping the external forces to the joint space and $\boldsymbol{J}_b^T$ is the Jacobian transpose of the matrix mapping the external forces to the floating base frame. Finally, $\dot{\boldsymbol{q}}_b$, and $\ddot{\boldsymbol{q}}_b$ are the base Cartesian velocity and acceleration vectors, respectively. The first line of this equation describes the dynamics of the floating base, which is free to move in the world but is un-actuated, while the second line describes the dynamics of the articulated body in terms of the joint variables. Measuring movement and conducting dynamics analysis on robots is facilitated due to the availability of accurate measurements of the variables of interest: joint angles, joint velocities, joint accelerations and joint torques. On the other hand, to perform the equivalent analysis for humans is more complex, since there is no direct access to the data at the joint level. Indeed, to conduct similar computations on the human it is necessary to estimate these data. Leveraging the developments of motion capture technology together with kinematics and dynamics computations it is possible to estimate the joint variables and to obtain personalized models with or without contact [3, 4]. Until recently, even individualized model estimation relied on anthropomorphic tables interpolation to estimate the body segment inertial parameters (BSIP) [5–7]. Since these anthropomorphic tables only account for specific populations and assume a symmetry between the right side and the left side of the body, they are not adequate in many cases. Our recent results allow personalized estimation of BSIP without reliance on generic anthropometric tables. This computational personalization has been applied for a wide range of populations and in particular for populations that do not fit with the populations captured in anthropomorphic tables: children, elderly, paraplegic, obese populations (Fig. 1). However, inverse kinematics computations from marker data of photogrammetric systems are subject to error due mainly to the assumption of rigid body modelling, and solving for this error is an active field

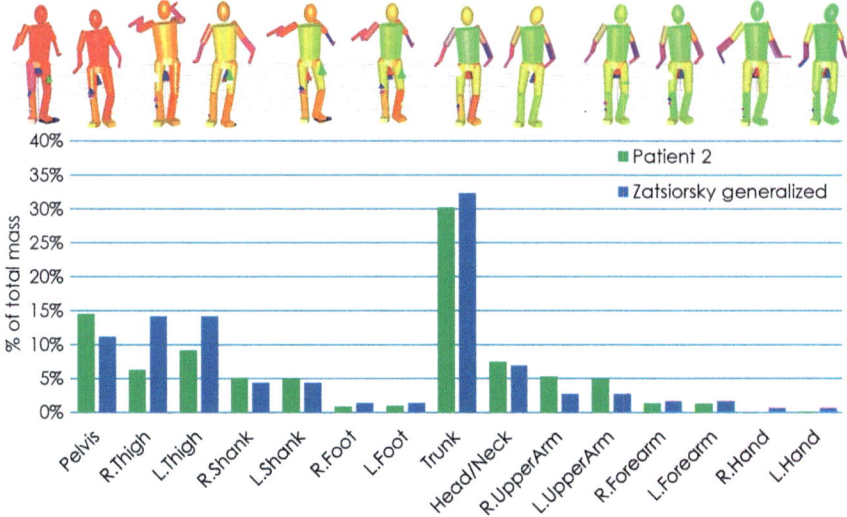

Fig. 1 Personalized dynamic model of a paraplegic patient, revealing the difference between the right side (paretic side) and the left side (healthy side) with identification (green), and not revealed with usual BSIP interpolation, here Zatsiorsky generalized [7]

of research in biomechanics [8]. Soft tissue artifacts are highly non-linear and are caused by motion of the skin and wobbling masses impacting bone pose estimates during the experimental measurement process, by as much as 15 degrees [9–11]. As the subsequent dynamic parameter estimates rely on the accuracy of the kinematics, accurate estimates of the kinematics are paramount to the estimation process. Finding a solution that can generalize the modelling of soft tissue artifacts remains an unsolved problem [12] and this will not be addressed in the present chapter.

2 Dynamically Consistent Inverse Kinematics [13]

The estimation of accurate kinematics, inter-segmental moments and external ground wrench (EGW) during the motion of a human is of crucial importance in countless applications [11, 14], for example when designing or adapting subject specific prosthetic or exoskeleton motion [15, 16]. The estimation of inter-segmental moments from inverse dynamics requires the knowledge of joint angles, velocities, and accelerations; geometric parameters and the body segment inertial parameters (BSIP). In human motion analysis, the classical method to estimate joint angles is to create a frame at each segment, from at least three markers, and then to calculate the relative linear and angular displacements between adjacent frames. Subsequently, from the resultant kinematics, recursive (bottom-up) methods can be used to propagate the measured EGW in order to estimate internal joint loads. Despite its ease of use, this method does not consider any kinematic constraints, i.e. the model does not have fixed segment lengths. Moreover, this method is sensitive to marker mis-placements and/or soft tissue artifacts [17]. To cope with this issue, in the literature, inverse kinematics is increasingly performed by minimizing the least-square differ-ence between the whole set of measured marker positions and their estimates from a forward kinematics model [18, 19, 20]. This approach is referred to as multi-body optimization in the biomechanics literature as opposed to single-body optimization [20, 21] that was widely used in the past. Multi-body optimization methods [18] can also estimate the segment lengths and the joint center positions using a multi-level optimization process. Because these multi-body optimization methods [18, 19] are solely based on kinematic data, the resultant joint trajectories might not be consistent with dynamic laws governing the system's whole body motion. To improve tracking of dynamic information, previous studies [19, 20] have proposed to use the available inconsistent positions and EGW measurements to adjust acceleration estimates. To do so, a weighted least-squares optimization approach was used to provide the most consistent optimal acceleration according to force data. However, the obtained joint accelerations were not necessarily consistent with the derivation rules between joint positions and velocities. In order to limit the errors due to the inverse kinematics when the final goal is to perform inverse dynamics, we have proposed to solve the inverse kinematics while taking into account the inverse dynamics with a two-layer or bi-level optimization process [5, 22]. Instead of using solely the motion data obtained from the markers, we add the force information collected with the force plates to drive the solution of the inverse kinematics to be dynamically consistent. Figure 2 presents a flowchart of the proposed method. In the first optimization step, indicated in blue, both the measured kinematic **Mk**$_{meas}$ and the measured dynamic **EGW**$_{meas}$ data are tracked at the same time in the cost function. This step allows finding the joint angles θ, velocities $\dot{\theta}$, and accelerations $\ddot{\theta}$ that track the marker trajectories and that are dynamically consistent with the external ground wrench measured by the force plates. In the second optimization step, indicated in green, the geometric parameters **L**, **Px** and **Py** and the BSIP #$_{dyn}$ are identified sequentially using the output of the first optimization step. The geometric parameters refer to the segment lengths **L** and local coordinates of the markers **Px** and **Py** in each segment frame. The BSIP refer to the mass, the coordinates of the center of mass, and the inertias of each segment.

We validated the method with 8 subjects performing a squatting task. From a standing position with the arms along the body, the subjects performed a squat and returned to their initial posture. Kinematic data were recorded using a motion capture system (MX VICON, 8 cameras, 100 Hz). The markers were located on the subjects' body surface according to the Plug-In-Gait marker template. Dynamometric data were collected with a force-plate (BERTEC, 100 Hz).

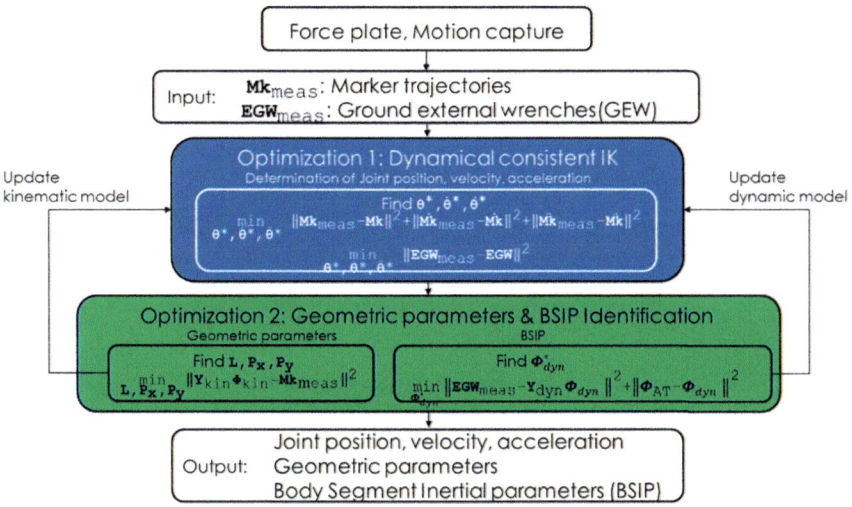

Fig. 2 Outline of the proposed dynamically consistent inverse kinematics algorithm

The new dynamically consistent inverse kinematics method, which minimizes the differences between both kinematic and dynamic estimates and measurements, is used to estimate joint trajectories, geometric parameters and BSIP. As shown in the figures the estimated joint angles are consistent and produce a good fitting of the kinematic data and an excellent tracking of the EGW measured by a force plate, for a detailed analysis please refer to [13]. Regarding the RMS of marker tracking in Fig. 3, the RMS values are consistent with the literature on soft-tissue-artefacts especially at the knee, thigh, and pelvis markers with RMSE of 15–20 mm [8–10]. The classical inverse kinematics method provides a better tracking of the markers, but embeds de facto in the resultant kinematics the error due to soft tissue artifacts, which subsequently influence the accuracy of the inverse dynamics when using these polluted joint trajectories. The RMSE of external forces of the classical method is two to five times larger than with the proposed bi-level optimization method as shown in Fig. 4. Solely minimizing the residue of the markers is not a sufficient criterion when performing inverse kinematics for dynamics computations. Finally, the error of the estimate of the EGW is also reduced thanks to the subject specific identification of BSIP. The knowledge of the individual segment mass, CoM and inertias pave the way to future monitoring tools that could account for segment changes during a rehabilitation process or along life.

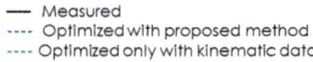

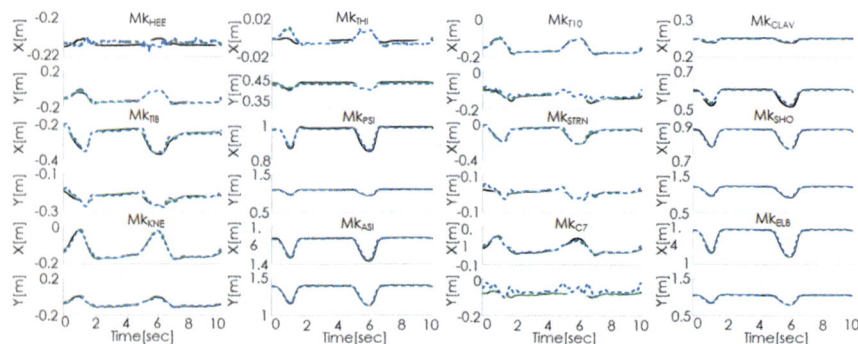

Fig. 3 Marker trajectories: measured data with photogrammetric system (black); estimated with the proposed dynamically consistent method (blue dotted line); estimated with conventional inverse kinematics (green dotted line), with x the sagittal direction, y the vertical direction and z the medio-lateral direction

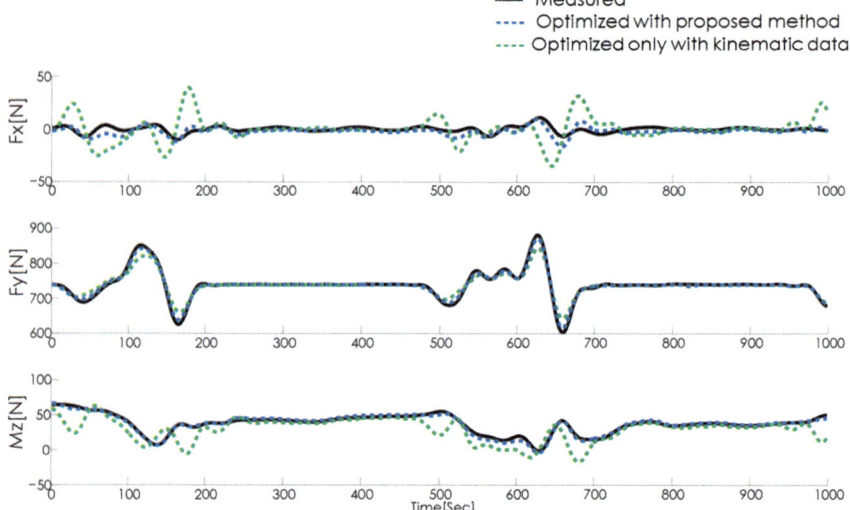

Fig. 4 External ground wrench: measured data with force plate (black); estimated with the proposed dynamically consistent method (blue dotted line); estimated with conventional inverse kinematics (green dotted line), with x the sagittal direction, y the vertical direction and z the medio-lateral direction

3 Contact Wrench Distribution Computation [23]

Standing up from a chair is a basic but crucial task of daily living. A decrease in this ability, experienced by elderly or individuals with sensory-motor deficiencies (Parkinson Disease, medullar lesion, post-stroke hemiplegia), limits their independence and can lead to their institutionalization. The inability to realize correctly a sit-to-stand (STS) task has been correlated with an increased risk of falling [24, 25] and hip fracture. When rising up from a chair, the hip joint torques can be larger than during other activities of daily living, such as walking and stair-climbing [26]. Consequently, assessing lower body strength, and thus functional performance through the STS test has been considered a reliable way to differentiate between subjects with different functional levels [27]. While in clinical practice typically only the kinematics of the STS is analyzed [25–28], dynamics parameters such as the internal joint loads [29], muscle forces [30], external wrench (EW) distribution, or the trajectory of the center of mass or center of pressure [27] have been the target of recent research. The proposed method relies on two main steps to identify, without force plate measurements, the external wrench distribution as described in Fig. 5. Step 1 identifies the subject-specific BSIP during an initial calibration phase. Step 1 is based on the use of a stereophotogrammetric system, and a force sensor under the subject's contact with the environment. Thanks to this identification procedure a more reliable estimate of the EW and of the joint torques can be obtained in Step 2. Step 2 needs only the kinematics data (i.e. the stereo-photogrammetric system). This second step solves in real-time an optimization problem in order to determine the optimal distribution of the EW between all the contact points. As described in by Bonnet et al. [23]. The minimized cost function is a hybrid function that includes the squared sum of all the joint torques $\Gamma(i)$ and the squared sum of the EW at each point k $F_{X,Yk}$, C_{Zk}:

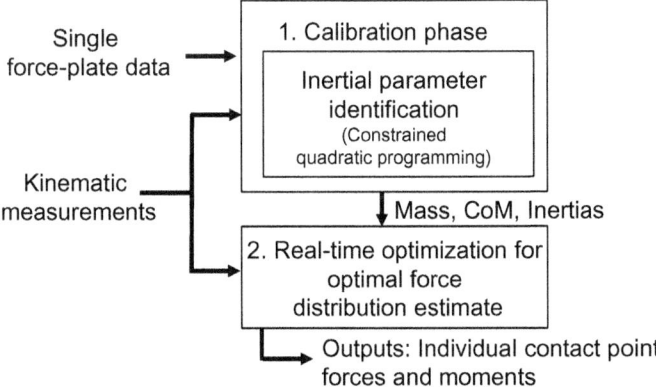

Fig. 5 Flowchart of the proposed method for contact wrench distribution computation

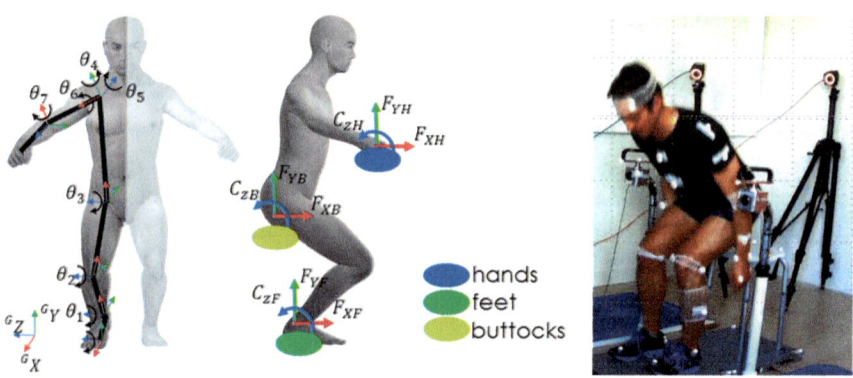

Fig. 6 External ground wrench: measured data with force plate (black); estimated with the proposed dynamically consistent method (blue dotted line); estimated with conventional inverse kinematics (green dotted line)

$$\text{min}C = \sum_{i=1}^{7} \left(\frac{\Gamma(i)}{\Gamma_{max}(i)} \right)^2 + \alpha \sum_{k=1}^{Nc} F_{X,Yk}^2 + \beta \sum_{k=1}^{Nc} C_{Zk}^2 \qquad (3)$$

where α and β are weights accounting for different quantities and to be tuned [23].

Besides their use for the calibration phase in Step 1, the force sensors used in this study are only for validation purposes.

In order to assess the accuracy of the proposed method in identifying the BSIP and in estimating the optimal EW distribution, an experiment with seven healthy volunteers (3 males and 4 females, age $= 31.6 \pm 13.4$ years, mass $= 65.5 \pm 5.9$ kg, stature $= 1.7 \pm 0.04$ m) was performed. Two sets of motions were collected, one for each part of the procedure (calibration / real-time observation). For the calibration part, volunteers were first asked to reproduce a screen-displayed set of exciting motions including oscillations of the arms and of the trunk at different frequencies, and ten STS tasks (Fig. 6). The motions were selected to be a good trade-off between excitability and reproducibility and non-physically challenging. This was done with the objective of using such motions with people who have reduced muscular strength and/or a deficit in attention. These motions lasted for 60 s and were used for the identification of the BSIP. In addition, three subjects performed the identification process with an artificially added mass located on the lower trunk, to assess the ability of the proposed method to detect and identify the individual segment mass.

The ability of the proposed method to estimate the EW distribution is depicted in Fig. 7, showing the horizontal (FX) and vertical (FY) ground reaction forces and the external couple (CZ) for each of the STS motions. For all the observed motions, the external forces and couples were estimated within an average RMS of 18 ± 10 N and 6 ± 4 nm

As expected, the identified subject-specific model allows a better prediction of the EW than a model based on averaged anthropometric tables, with an RMS difference factor reduction of 1.5 and 1.3 for the horizontal and vertical ground reaction forces

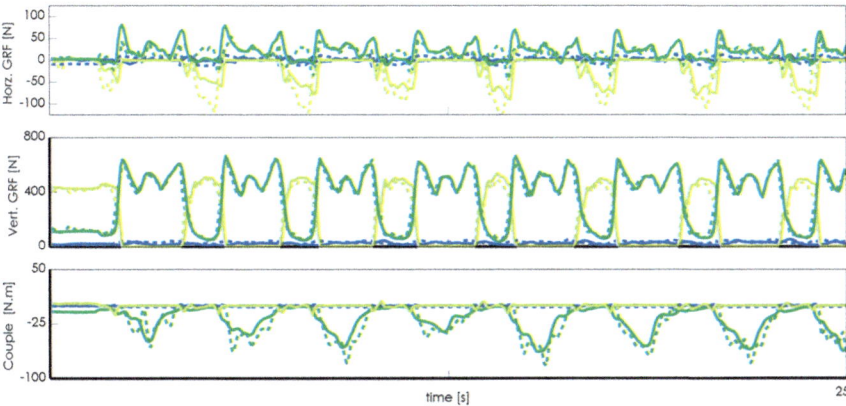

Fig. 7 External ground wrench: measured data with force plate (black); estimated with the proposed dynamically consistent method (blue dotted line); estimated with conventional inverse kinematics (green dotted line)

and 2.4 for the resultant couple. The identification process of the BSIP, using very simple and easily reproducible exciting motions, allows identifying the masses of each segment. This, as mentioned in our previous study [31], is extremely helpful to track the evolution of muscular mass during a long-term rehabilitation process. Since the investigated task is the sit-to-stand, solely the BSIP having a dynamical influence on this task need to be accurately identified.

4   Motion Segmentation with Inverse Optimal Control [32]

One other way to use dynamics calculation is in applying to the growing field of Inverse Optimal control (IOC). Indeed, the central nervous system, as the controller of the body, has the possibility to choose from an unlimited number of joint trajectories in order to carry out an action. Literature in human motor control over the last three decades has shown that the joint trajectory variance is limited to a much smaller subset, and could likely be the result of the minimization of a cost function [33]. Studies in biomechanics and human motion analysis have proposed many possible cost functions, such as minimizing time, joint velocity, acceleration or jerk [34]. Inverse optimal control [35] tries to estimate the cost function used to generate a motion given the data of this human motion trajectory and a set of cost function hypotheses. Previous studies [34, 36] have shown that different cost functions are used for different movements. Existing works typically segment continuous movement into discrete motion primitives and assume that the cost function does not change over the duration of a single primitive. However, our belief is that a continuous movement sequence may consist of multiple motion primitives, and each primitive may not necessarily share a common control strategy as seen in Fig. 8. Specifically,

Fig. 8 Illustration of optimal control in general, classic kinematics-based segmentation and the proposed segmentation approach based on inverse optimal control

we hypothesize that if the control strategy can be estimated as a function of the motion data, then a change in strategy may be used as an indication that the motion primitive being performed has changed, and this could be used to segment the motion in a new manner. To achieve this, we use a sliding window over the trajectory data rather than a window of the length of the motion primitive. This allows to determining the numerical weights of each of the cost function using IOC. These weights are averaged together to form a time varying feature of the motion trajectory. A threshold can then be applied to this feature to perform motion segmentation. The details of the method are given in [32]. Only the experimental validation is presented here.

The segmentation was performed on an experimental dataset of 8 healthy participants [37] with an average age of 30 ± 5 years old performing an average of 10 squats each. The data was collected using a VICON motion capture system. A 10-marker set was used, providing joint Cartesian position data. Joint angles were calculated from the cross products between markers, using a 3 DOF planar kinematic model as depicted in Fig. 9.

Figure 9 also shows the results of the weight recovery for a random subject. It shows clearly that the proposed method delineates between the squat motion with a high weight on the cartesian accelerations weight (greens), and the resting periods with high weights in power (blues), and can be used for segmentation. While the percentage allocated to minimizing power or cartesian accelerations may change, as indicated from the large differences in the shape of the bars, they predominate the other basis functions, indicating that they are more important in the squatting movement strategy from the basis functions used in the optimization. This finding is similar to previous findings, where acceleration [35] and power [38] have been found to be important basis functions.

The approach accepts arbitrary lengths of trajectories and estimates the underlying basis function weights for successive windows of that trajectory using inverse optimal control. A method to reject low-quality weight estimates by examining the residual

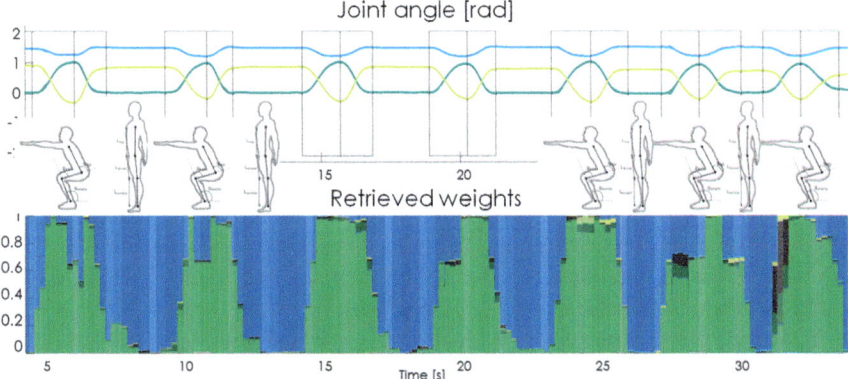

Fig. 9 Illustrated segmentation results: the top panel shows the hip, knee and ankle joint trajectories during the performance of the squat task, while the bottom shows the recovered IOC weights

norm is proposed, and the algorithm is demonstrated in both simulation and with real data. The basis weights of a set of squat tasks suggests that humans optimize for power and Cartesian acceleration during rest and movement, respectively, and that this choice of cost functions is consistent across the 8 healthy subjects in the dataset. It was also shown that a threshold-based segmentation method on the power basis weight achieved 84% in balanced accuracy.

5   Conclusion

In this chapter, we have presented an overview of personalization methods for biome-chanical models and some examples where this personalization is of importance and how these models can be used to improve the computation of inverse kinematics when dynamics computations are the final goal (Sect. 2); to utilize the results of the dynamics and geometric calibration for the estimation of the external ground wrench without force plates (Sect. 3); and finally for segmentation of motion using inverse optimal control (Sect. 4).

Many challenges are still ahead. The measurement technologies used for motion capture inherently introduce some error such as soft tissue artifacts, in addition the modelling techniques used introduce errors, for example assumptions about joint center positioning. The dynamically consistent inverse kinematics computation method presented in Sect. 2 improves performance because the inverse kinematics is not perfectly accurate. In the future if an accurate measure of the kinematics becomes possible such methods will become obsolete. However, the work presented in Sects. 3 and 4 are two important milestones that will be of increasing importance in the future. Indeed, being able to estimate forces without force sensors is of crucial importance in the development of in-house and low cost systems, as well as wide range systems.

Such computation allows for the development of personalized dynamic analysis. On the other hand, inverse optimal control is a field that is gaining more importance both in robotics and in biomechanics with the increasing ability to solve complex optimization problems. There is still much to investigate and understand, but using inverse optimal control provides a novel modality to segment human motion data, which is not based on kinematics information but on some more intricate data: the weights of the different cost functions of the controller. There is still no clear evidence that humans actually use optimality to control their motions and if they do how. This is just a small step in understanding human motion and one can expect that in the next few years many new advances will be made.

If it is now possible to build individually precise dynamics models that are simple to use and adapt, and require little computation for multiple applications, there are still a few next steps to investigate such as optimal and automatic human body model generation and to couple model based and learning approaches for better results.

References

1. Siciliano, B., Khatib, O. (eds.): Springer Handbook of Robotics. Springer, Berlin (2016)
2. Ayusawa, K., Venture, G., Nakamura, Y.: Identifiability and identification of inertial parameters using the underactuated base-link dynamics for legged multibody systems. Int. J. Rob. Res. **33**(3), 446–468 (2013). https://doi.org/10.1177/0278364913495932
3. Venture, G., Ayusawa, K., Nakamura, Y.: Real-time identification and visualization of human segment parameters. Conf. Proc. Int. Conf. IEEE Eng. Med. Biol. Soc. **2009**, 3983–3986 (2009). https://doi.org/10.1109/IEMBS.2009.5333620
4. Ayusawa, K.: *Scaling Kinematic Chains in the Air–Identification of Floating Systems Using Dynamics Constraint of the Baselink without Force Measurement*, pp. 19–25 (2011)
5. de Leva, P.: Adjustments to zatsiorsky-seluyanov's segment inertia parameters. J. Biomech. **29**(9), 1223–1230 (1996)
6. Dumas, R., Cheze, L., Verriest, J.-P.: Adjustments to McConville et al. and Young et al. body segment inertial parameters. J. Biomech. **40**(3), 543–553 (2007)
7. Zatsiorsky, V.: The mass and inertia characteristics of the main segments of the human body. Biomechanics, 1152–1159 (1983)
8. Andersen, M.S., Benoit, D.L., Damsgaard, M., Ramsey, D.K., Rasmussen, J.: Do kinematic models reduce the effects of soft tissue artefacts in skin marker-based motion analysis? An in vivo study of knee kinematics. J. Biomech. **43**, 268–273 (2010)
9. Peters, A., Galna, B., Sangeux, M., Morris, M., Baker, R.: Quantification of soft tissue artifact in lower limb human motion analysis: a systematic review. Gait Posture **31**, 1–8 (2010)
10. Zemp, R., List, R., Gülay, T., Elsig, J.P., Naxera, J., Taylor, W.R., Lorenzetti, S.: Soft tissue artefacts of the human back: Comparison of the sagittal curvature of the spine measured using skin markers and an open upright MRI. Plos One **9** (2014)
11. Hara, R., Sangeux, M., Baker, R., McGinley, J.: Quantification of pelvic soft tissue artifact in multiple static positions. Gait Posture **39**, 712–717 (2014)
12. Bonnet, V., Richard, V., Camomilla, V., Venture, G., Cappozzo, A., Dumas, R.: Multi-body optimisation and extended Kalman filter embedding a kinematic-driven STA model. J. Biomech. (2017) (in press)
13. Futamure, S., Bonnet, V., Dumas, R., Kulic, D., Venture, G.: Dynamically consistent inverse kinematics framework using optimizations for human motion analysis. In: IEEE-RAS International Conference on Humanoid Robot, pp. 436–441 (2016). https://doi.org/10.1109/humanoids.2016.7803312

14. Kulić, D., Venture, G., Yamane, K., Demircan, E., Mizuuchi, I., Mombaur, K.: Anthropomorphic movement analysis and synthesis: a survey of methods and applications. IEEE Trans. Robo **32** (2016)
15. Mandery, C., Borràs, J., Jöchner, M., Asfour, T.: Analyzing whole-body pose transitions in multi-contact motions. In: International Conference on Humanoid Robots, pp. 1020–1027 (2015)
16. Kolev, S., Todorov, E.: Physically consistent state estimation and system identification for contacts. In: International Conference on Humanoid Robots, pp. 1036–1043 (2015)
17. Leardini, A., Chiari, L., Croce, U.D., Cappozzo, A.: Human movement analysis using stereophotogrammetry Part 3. Soft tissue artifact assessment and compensation. Gait Posture **21**, 212–225 (2005)
18. Lee, J., Flashner, H., McNitt-Gray, J.L.: Estimation of multibody kinematics using position measurements. J. Comput. Nonlinear Dyn. **6**, 1–9 (2011)
19. Lu, T.-W, O'Connor, J.J.: Bone position estimation from skin marker co-ordinates using global optimisation with joint constraints. J. Biomech. Eng. **32**, 129–134 (1999)
20. Chèze, L., Fregly, B.J., Dimnet, J.: A solidification procedure to facilitate kinematic analyses based on video system data. J. Biomech. **28**(7), 879–884 (1995)
21. Söderkvist, I., Wedin, P.A.: Determining the movements of the skeleton using well-configured markers. J. Biomech. **26**(12), 1473–1477 (1993)
22. Bonnet, V., Daune, G., Joukov, V., Dumas, R., Fraisse, P., Kulić, D., Seilles, A., Andary, S., Venture, G.: A constrained extended Kalman Filter for dynamically consistent inverse kinematics and inertial parameters identification. In: International Conference on Biomedical Robotics and Biomechatronics (2016)
23. Bonnet, V., Azevedo Coste, C., Robert, T., Fraisse, P., Venture, G.: Optimal external wrench distribution during a multi-contact sit-to-stand task. IEEE Trans. Neural Syst. Rehabil. Eng. **25** (2017). https://doi.org/10.1109/tnsre.2017.2676465
24. Topper, A.K., Maki, B.E., Holliday, P.J.: Are activity-based assessments of balance and gait in the elderly predictive of risk of falling and/or type of fall ? J. Am. Geriatr. Soc. **41**, 479–487 (1993)
25. Papa, E., Cappozzo, A.: Sit-to-stand motor strategies investigated in able-bodied young and elderly subjects. J. Biomech. **33**, 1113–1122 (2000)
26. Rodosky, M.W., Andriacchi, T.P., Andersson, G.B.J.: The influence of chair height on lower-limb mechanics during rising. J. Orthop. Res. **7**, 266–271 (1989)
27. Cheng, P.T., Liaw, M.Y., Wong, M.K., Tang, F.T., Lee, M.Y., Lin, P.S.: The sit-to-fast deterstand movement in stroke patients and its correlation with falling. Arch. Phys. Med. Rehab. **79**, 1043–1046 (1998)
28. Millor, N., Lecumberri, P., Gomez, M., Martinez-Ramirez, A., Izquierdo, M.: Kinematic parameters to evaluate functional performance of sit-to-stand and stand-to-sit transitions using motion sensor devices: a systematic review. IEEE Trans. Neural. Syst. Rehab. Eng. **22**, 926–936 (2014)
29. Riley, P.O., Schenkman, M.L., Mann, R.W., Hodge, R.A.: Mechanics of a constrained chair rise. J. Biomech. **24**, 77–85 (1991)
30. Scarborough, M.D., Krebs, D.E., Harris, B.A.: Quadriceps muscle strength and dynamic stability in elderly persons. Gait Posture **10**, 10–20 (1999)
31. Bonnet, V., Venture, G.: Fast determination of the planar body segment inertial parameters using affordable sensors. IEEE Trans. Neural Syst. Rehabil. Eng. **23**, 628–635 (2015)
32. Lin, J.F.-S., Bonnet, V., Panchea, A.M., Ramdani, N., Venture, G., Kulic, D.: Human motion segmentation using cost weights recovered from inverse optimal control. In: IEEE-RAS International Conference on Humanoid Robot, pp. 1107–1113. Cancun, Mexico (2016)
33. Alexander, R.M.: The gaits of bipedal and quadrupedal animals. Int. J. Robot. Res. **3**, 49–59 (1984)
34. Flash, T., Hogan, N.: The coordination of arm movements: an experimentally confirmed mathematical model. J. Neurosci. **5**, 1688–1703 (1985)
35. Mombaur, K., et al.: From human to humanoid locomotion an inverse optimal control approach. Auton Robot **28**, 369–383 (2010)

36. Todorov, E.: Optimality principles in sensorimotor control. Nat. Neurosci. **7**, 907–915 (2004)
37. Bonnet, V., et al.: A least-squares identification algorithm for estimating squat exercise mechanics using a single inertial measurement unit. J. Biomech. **45**, 1472–1477 (2012)
38. Berret, B., et al.: Evidence for composite cost functions in arm movement planning: an inverse optimal control approach. PLoS Comput. Biol. **7**, 1–18 (2011)

Optimality and Modularity in Human Movement: From Optimal Control to Muscle Synergies

Bastien Berret, Ioannis Delis, Jérémie Gaveau and Frédéric Jean

Abstract In this chapter, we review recent work related to the optimal and modular control hypotheses for human movement. Optimal control theory is often thought to imply that the brain continuously computes global optima for each motor task it faces. Modular control theory typically assumes that the brain explicitly stores genuine synergies in specific neural circuits whose combined recruitment yields task-effective motor inputs to muscles. Put this way, these two influential motor control theories are pushed to extreme positions. A more nuanced view, framed within Marr's tri-level taxonomy of a computational theory of movement neuroscience, is discussed here. We argue that optimal control is best viewed as helping to understand "why" certain movements are preferred over others but does not say much about how the brain would practically trigger optimal strategies. We also argue that dimensionality reduction found in muscle activities may be a by-product of optimality and cannot be attributed to neurally hardwired synergies *stricto sensu*, in particular when the synergies are extracted from simple factorization algorithms

B. Berret
CIAMS, Université Paris-Sud, Université Paris-Saclay, 91405 Orsay, France

B. Berret
CIAMS, Université d'Orléans, 45067 Orl éans, France

B. Berret (✉)
Institut Universitaire de France (IUF), Paris, France
e-mail: bastien.berret@u-psud.fr

I. Delis
Department of Biomedical Engineering,
Columbia University, New York, NY 10027, USA
e-mail: bastien.berret@u-psud.fr

I. Delis
School of Biomedical Sciences, University of Leeds, Leeds LS2 9JT, UK

J. Gaveau
INSERM U1093-CAPS, Université Bourgogne Franche-Comté,
UFR des Sciences du Sport, 21000 Dijon, France

F. Jean
Unité de Mathématiques Appliquées, ENSTA ParisTech,
Université Paris-Saclay, 91120 Palaiseau, France

© Springer International Publishing AG, part of Springer Nature 2019
G. Venture et al. (eds.), *Biomechanics of Anthropomorphic Systems*, Springer Tracts in Advanced Robotics 124, https://doi.org/10.1007/978-3-319-93870-7_6

applied to electromyographic data; their putative nature is indeed strongly dictated by the methodology itself. Hence, more modeling work is required to critically test the modularity hypothesis and assess its potential neural origins. We propose that an adequate mathematical formulation of hierarchical motor control could help to bridge the gap between optimality and modularity, thereby accounting for the most appealing aspects of the human motor controller that robotic controllers would like to mimic: rapidity, efficiency, and robustness.

1 Introduction

The vision neuroscientist David Marr, in his posthumous book, distinguished three levels of analysis in the field of computational neuroscience [97]. In computational motor control, the higher level ("theory") addresses questions such as: why do humans displace their limbs the way they do? what hidden goal(s) do they try to achieve? [133]. Essentially, this level of analysis seeks to explain why human movement trajectories have certain characteristics and what makes the elected movement better than another. By intuition and also by analogy with the principle of least action in classical mechanics, some researchers (including Marr himself) presumed that human behavior may be optimal in a sense that remains to be specified. Formulated in this way, the question turns out to be an *inverse optimal control* problem, a class of problems that mathematicians started to tackle a long time ago [e.g. [83]] and has been recently applied to reaching [57], locomotion [27] or even flying [1]. Precisely, inverse optimal control is the process of recovering the optimality criterion (or cost function) according to which a bunch of *ex hypothesi* optimal trajectories are indeed optimal. It is an ill-posed problem and, in fact, harder than direct optimal control [8, 117]. Direct optimal control consists in a priori guessing the cost function and computing the corresponding optimal trajectories, which can be a tricky mathematical problem in itself in many occasions. Importantly, at this level of analysis, no assumption is made about how the brain could manage to generate these optimal trajectories or whether it constantly solves optimal control problems from scratch.[1] This type of question is left to the second level of analysis called "algorithm" which investigates how observably optimal trajectories could be generated by the sensorimotor system. To this aim, several motor control theories have been proposed such as the ones based on active inference which do not even require the specification

---

[1] A useful analogy from classical mechanics is the principle of least action. For instance, trajectories of conservative systems are *extrema* of the *Action*, i.e. the time integral of the Lagrangian (kinetic minus potential energies), while it is hardly arguable that objects explicitly "optimize" their trajectories on purpose. In fact, finding whether a Lagrangian exists for a given system of differential equations has been the topic of numerous investigations in physics which date back to the works of Maupertuis, Euler or Lagrange. This refers to the inverse problem of calculus of variations [47] and can be seen as the analog problem of inverse optimal control. Notably, inverse calculus of variations has been used in the context of motor control to investigate the origin of the two-thirds power law [93].

of a cost function to explain movement generation [54]. Active inference, however, crucially requires prior knowledge about limb trajectories. As inverse optimal control precisely seeks to provide the rationale about why certain trajectories may be more valuable than others a priori, it can complement active inference models by informing why certain priors are used. Conceivably, through evolution and life-span development, the central nervous system (CNS) may have learned statistical descriptions of movement distributions or may have found simplified ways to trigger optimal or, say, good enough movements. By storing synergies (also called building blocks, primitives or modules[2]), it has been proposed that the CNS might have found clever ways to group and coordinate different degrees of freedom (joints or muscles) so that only a few task-related variables would need to be tuned in order to produce adequate motor patterns and efficient movements.

Modularity can be assumed at different levels: kinematic, dynamic, muscular or neural. When focusing on the muscle activation level, as in the present chapter, we talk about the muscle synergy hypothesis whose main appeal is to simplify neural motor control through timely activation of precoded groups of muscles. However, what these "building blocks" are and according to what rules they are combined remains elusive. Very often these building blocks may take the form of muscle weightings or temporal patterns, and are identified via dimensionality reduction such as PCA (principal component analysis), NMF (non-negative matrix factorization) or ICA (independent component analysis), i.e. unsupervised machine learning techniques applied to approximate motor signals (electromyographic data, aka. EMG). Currently there is still a gap between theories investigating the structure of muscle activities and theories assuming that cost functions determine limb trajectories. Yet, the crucial pivot to both approaches is musculoskeletal modeling. In inverse and direct optimal control, a model of the musculoskeletal system is required because optimally driving a system requires some dynamical description of it (e.g. rigid body dynamics with more or less advanced models of musculotendon complexes). Musculoskeletal models are also required to test the modular control hypothesis in muscle space, especially for assessing the effectiveness of the extracted synergies (from EMG) in controlling the musculoskeletal system; i.e. feeding back the EMG-based synergies into the controlled system, which is rarely done in motor control studies.

A growing number of studies are interested in investigating the links between optimality - at the highest level of analysis - (according to Marr's taxonomy) and modularity - at the second level of analysis - in order to evaluate whether optimal (or at least good enough) trajectories can be obtained from the combination of a limited number of genuine motor modules. Whether or not the algorithm used by the CNS to trigger "good enough trajectories" truly relies on this kind of synergies is an open question but evidence is usually sought by tackling Marr's third level of analysis, that is, the "implementation" level. In particular, finding synergies of a neural origin

---

[2]In this chapter, the terms synergy, primitive, module or building block are loosely treated as synonyms and will be used interchangeably. In the literature, a precise mathematical definition specifying the exact nature of each term is generally lacking. Different authors may thus have their own conception regarding the meaning of each term.

would strongly support the theory. However, the nature and shape of the building blocks to be found within the CNS critically depends on the hypotheses made by the framework used to infer them. Therefore, research at the implementation level is linked to the algorithmic level; i.e. studies of the neurophysiological underpinnings of modular structures are conducted according to a predefined model of modularity.

In this chapter, we will review and discuss optimal and modular control theories and underline their strengths and limitations. It is worth noting that the three levels of analysis put forward by Marr are complementary but address conceptually different questions. Therefore, the interaction and links between the first two levels will be discussed. In Sect. 2, we will review the inverse optimal control approach which aims at deciphering the underlying high-level principles of motor control. In Sect. 3, we will review works related to the muscle synergy hypothesis that aims to assess the structure of muscle activity patterns from a dimensionality reduction perspective. Section 4 is dedicated to discussing the links between the two approaches and, finally, perspectives for future research are given in Sect. 5. A tutorial example illustrating the notions of modularity and optimality is provided in Appendix.

2 Optimal Control Hypothesis

2.1 Direct and Inverse Optimal Control Approaches to Motor Control

The inverse optimal control problem was first considered in a seminal paper by Kalman for linear-quadratic problems [83] and extended in [108]. In motor control, few studies employed such an inverse approach to tackle the motor planning problem until recently. This is not to say that optimal control has not been used; on the contrary it has been applied extensively mostly since the 80s and has emerged as a leading theory in the human movement control literature [49, 144]. The work of [106] is for instance characteristic of the classical (direct) approach that was initially employed: several costs were tested and compared to find the best descriptor/predictor of human motion data. Then, the researcher suggested that the elected cost constituted an explanatory principle of biological motion. Since that time the computational theory of motor control has developed and become very popular. An account of all the main ideas and concepts like "optimal feedback control" and "internal models" can be found in specialized reviews [45, 127, 129, 130, 144]. The success of this computational theory of motor control can be explained by two main reasons: first, once a cost function is defined, it captures everything about the possible trajectories of the system, thereby implementing an elegant dimensionality reduction from the infinite number of potential trajectories to cost functions [10]; second, it is readily implementable in artificial systems because the very same language is used by control theoretists and roboticists. The point we want to make here is that most studies tended to a priori choose a cost function and to test its predictions subsequently. Generally, they tried to match the standard LQR (linear-quadratic regulator) or LQG (linear-quadratic Gaussian) designs to make the problem easier to solve [84, 145]. Interestingly, these simple models were sufficient to explain several motor control phenomena (two-thirds power law, Fitts's law, [66]). However, the true nature of the question "why do we move like that?" better conforms to a reversed process, which brings *inverse* optimal control into play. An illustration of the difference between inverse and direct optimal control is given below (Eq. 1):

$$
\begin{aligned}
\text{observed trajectories} \; &\xrightarrow[\text{control approach}]{\text{inverse optimal}} \; && \text{cost function} \\
\text{cost function} \; &\xrightarrow[\text{control approach}]{\text{direct optimal}} \; && \text{compare observed vs.} \\
& && \text{predicted trajectories}
\end{aligned}
\tag{1}
$$

The need for inverse optimal control arises from the need to identify the most plausible (among all possible) cost functions. Mathematically, a cost function may take the form $J(u) = \int_0^T h(x, u, t)dt$ where x is the controlled variable (e.g. system state such as velocity, position etc.), u the control variable and t is time.[3] If a cost function h accounts for some data, nothing precludes another cost of a different nature to perform equally well or even better. This has been at the origin of some controversy because in many occasions divergent costs were found to account for planar point-to-point reaching movements. Innovative paradigms are thus required to disambiguate these candidate costs. So, the following question arises: which of these candidate costs is really relevant to human motor control? This situation is exemplified in Fig. 1.

Inverse optimal control is a difficult problem because it is primarily ill-posed unless very careful conditions are imposed. Therefore, to successfully study inverse optimal control problems, not only mathematical sophistication but also smart experimental designs are needed to distinguish putative cost functions.

2.2 Example of Inverse Optimal Control Results

In 2008, a necessary and sufficient condition of optimality for fast enough arm reaching movements was identified [12, 13, 57]. Namely, periods of simultaneous inactivation of opposing muscles - during the movement - were shown to be equivalent to the minimization of a cost including a term like the integral of the absolute power of muscle torques (termed "absolute work"). The strength of this result was the mathematical proof of an "equivalence" between a singular motion feature (muscular inactivation) and a cost feature (non-smoothness of the cost). Initially, the authors used a direct approach by guessing what could be a physically-relevant cost function [13]. Later, the mathematical analysis allowed to establish a powerful inverse optimal control result, whereby the authors could infer certain properties of the cost function

---

[3]In this chapter, we consider only integral costs for simplicity but we could easily add a terminal cost in all the optimal control problems.

Point-to-point reaching   Point-to-bar reaching

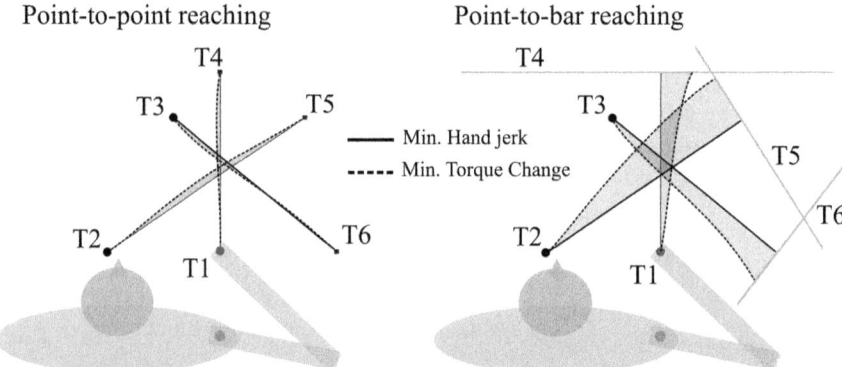

Fig. 1 Predicted hand trajectories for two influential motor costs during planar reaching movements. *Left panel* when reaching to a target point, the minimum jerk [53] and torque change [159] models predict highly similar and therefore equally plausible trajectories, whereas the two models are very different in nature (kinematic cost for the former and dynamic for the latter). *Right panel* when replacing the point with a line or a bar, the two costs lead to different trajectories, and may thus be distinguished empirically. Shaded areas indicate the difference between the trajectories predicted by the two models. Taken from [15]

right from experimental observations [57]. In practice, the work was guided by a different intriguing experimental result: the kinematics of vertical arm movements was shown to depend on its orientation relative to gravity vector, therefore suggesting a potential imprint of gravity on human movement [58, 63, 113]. Since then, a series of papers has shown that upward movements differed from downward movements of equal duration and amplitude (including one-degree-of-freedom motions during which only the sign of gravity effects changes with movement direction; i.e. assists/resists the acceleration of downward/upward movements respectively) in that the acceleration duration was shorter in the upward compared to the downward direction [32, 58, 60, 61, 63, 73, 92, 114, 128, 151, 165]. These directional asymmetries have been observed by independent groups of researchers so consistently, including in microgravity experiments where they were actually shown to progressively vanish, that we could now employ the term "law of asymmetries" to refer to this phenomenon. In particular, minimizing the absolute work of muscle torques has been shown to robustly reproduce these vertical asymmetries for fast enough movements, despite differences in initial postures [12], upper limb's segment [60] or gravito-inertial context [61]. This led the authors to suggest an optimal integration of gravity force during motion planning where gravity torque can be utilized to drive the limb during rapid movements. This interpretation is corroborated by EMG analyses showing that, during a rapid movement, the (phasic) activity of anti-gravity muscles is clearly lower than the (tonic) activity that would be needed to maintain a static arm posture in the same position, in certain movement phases[4] (e.g. [19, 38, 39, 52, 124]).

---

[4]It is generally assumed that the muscle torque τ acting at a given joint can be split into two terms such that $\tau = \tau_{stat} + \tau_{dyn}$, where τ_{stat} is a static term which only depends on the system position

Previously, the absolute work was shown to be one of the potential ingredients of a more general cost function underlying human movement [12]. Various researchers had already proposed other cost functions as well as the idea of composite costs which was explicitly tested in [15]. The idea of composite costs proposes that multiple complementary criteria shape human trajectories. For instance, motion smoothness and energy expenditure are complementary criteria as minimizing one may be detrimental to the other. This mixed nature of cost functions has been observed in various tasks such as reaching [15, 61, 162], landing after a jump [167] and walking [166]. At this point, it is important to classify cost functions in two categories: the first category, "subjective" costs, depends on the subject's choice; the second category, "objective" costs, depends on the task specification.[5] In general both objective and subjective costs are relevant to motor control. Forcing a subject to reach very fast to the target, tracking an imposed trajectory or freezing a joint are examples of objective costs. Yet, even with such constraints, most motor tasks are still redundant because there is an infinity of ways, i.e. motor control solutions, to perform them. Subjective costs come into play to resolve all residual redundancy and provide the rationale about why the task is eventually performed as it is. Here, we want to underline this last statement. Reproducing or trying to explain empirical data with an optimal control model does not compulsorily mean that the nervous system solves an optimal control problem, especially not constantly and from scratch. It only means that the motor solutions the nervous system has developed, whatever the time scale, are advantageous in a sense that has been delineated.

According to the objective/subjective dichotomy, one may suggest that, in order to eliminate the effects of objective costs in the identification process, experimenters should give maximal freedom to the participant and put as little task constraints as possible. In other words, considering highly redundant tasks with very few instructions given by the experimenter may help understanding the fundamental principles humans prefer to rely on. In goal-directed tasks, there are two ways to make a task more redundant: by adding intrinsic or extrinsic degrees of freedom. First, let us consider a pointing task where the only objective constraint is to point with the fingertip

---

(Footnote 4 continued)

and τ_{dyn} is a dynamic term which depends on its velocity and acceleration [7, 72]. Gravitational torque is part of the static term which may also include other terms like elastic forces. On this basis, researchers have proposed to split EMG activity into tonic and phasic components (e.g. [52]). To clarify our purpose, let us consider a single-joint upward movement here. If the static torques were to be compensated at all times, a phasic activity of the agonist muscle should come on top of its tonic activity during the entire motion duration. On the contrary, if the agonist EMG signal is found to be below its corresponding tonic level, it may suggest that gravity is not just counteracted but utilized as a driving force. This lack of tonic activity, already observed - but not fully considered - in several studies, actually echoes the inactivation principle mentioned in the main text. If observing proper inactivation may be tricky due to multiple factors such as the noisiness of EMGs, the predicted briefness of the phenomenon and the requirement of being under well-suited conditions of speed and amplitude, this lack of compensation of gravity torques, clearly apparent in EMG data, is additional evidence for an energy-related use of gravity in fast reaching movements.

[5]This terminology is borrowed from [88]. In [127], the terms internal and regularization are used for subjective costs while the term task-based is used for objective costs.

toward the target. The target location defines 3 constraints (in the 3-dimensional space) but to make the task very redundant we could consider adding other degrees of freedom. For instance, asking a participant to perform a whole-body reaching task would be advantageous to emphasize subjective motor decisions, however, it would also make the optimal control problem and musculoskeletal modeling harder to solve computationally. An other alternative is to reduce the constraints imposed by the target itself. Instead of a dot, asking a subject to point to a line, or a surface, makes the task redundant even with a simple two-link arm model (e.g. Fig. 1). The appeal of this second approach is to make explicit the *choice* problem faced by a subject when planning a movement[6] (in this type of task, the subject has to select an end point on his/her own). How a subject moves when asked to point toward a line (not a point) is instructive and has been studied in a couple of papers [16, 17, 105, 150]. This type of protocol has been used to investigate eye-hand coordination [17], posture-movement coordination [69] and the use of interaction torques and velocity dependence of cost functions [161, 162].

Hitherto, all the above-mentioned modeling studies have considered that movement time was known.[7] However, motor redundancy is not only spatial but also temporal. Why are some subjects faster than others or what determines someone's movement vigor[8] are questions that were addressed more recently [11, 76, 132, 134]. It turns out to be an inverse optimal control issue as well, if one considers the existence of a cost of time [11]. Assuming a linear separability of the temporal and spatial cost functions, i.e. $J(u) = \int_0^T h(x, u, t)dt = \int_0^T g(t) + l(x, u)dt$, it was shown that it is possible to accurately recover $g(t)$ for different times t in a free-time optimal control setting [11]. In particular, this approach requires having knowledge of both the subjective and objective terms of the trajectory cost $l(x, u)$. Given these assumptions and different simple models of the musculoskeletal plant, it was also found that the cost of time exhibited a sigmoidal growth. This, therefore, raised questions such as: how would the CNS proceed to depart from its spontaneous speed choice when speed instructions are given to a subject (such as "move fast")? A complex interplay between trajectory and time costs, but also between objective and subjective costs, may be hypothesized. Indeed, it was shown that speed instructions were most plausibly captured by adding objective trajectory criteria to $l(x, u)$ [82] rather than modifying $g(t)$, and that subjective trajectory costs appeared to be quite insensitive to speed instructions [162]. Overall, time, energy/effort, smoothness but also accuracy may be relevant to biological motion planning and control. The relative weights associated to each cost element may depend on the task characteristics. The multivariate aspect of the cost function and the addition of task-dependent costs,

---

[6]Remarkably, motor control has been conceived as a true (motoric) decision-making problem recently [164].

[7]When we say that movement time is known, modeling-wise, we mean that time is set by the user (often it is taken from experimental data). Therefore, time is an input to the model. Note, however, that time can also be a free variable that emerges from optimization just as the limb's trajectory does [138].

[8]Vigor loosely refers to the speed, extent or frequency of movement [48]. It is often characterized by relationships between amplitude and velocity or duration.

however, make the identification problem quite hard, even though these results may truly capture some fundamental high-level goals of the motor system.

2.3 Remarks Regarding Optimal Control Theory

In general, posing an inverse optimal control problem remains difficult as it requires several choices. Indeed, both devising the class of cost functions and the dynamical system to use is a modeling choice. First, the choice of which functional space of cost functions to consider is mostly left to the experimenter. Once such an infinite dimensional space is chosen, however, a sort of discretization is required to make the problem numerically tractable and to only have a finite number of parameters to be inferred in practice.[9] One may distinguish basis costs that are chosen for mathematical reasons (e.g. polynomials) from those that are chosen for their physical/biological meaning (e.g. energy, effort…). While the former costs can be good for fitting or reproducing anthropomorphic motion, they do not allow to explain it and, thus, do not answer Marr's top level question which constitutes the main appeal of the use of optimal control for motor neuroscience. Second, the choice of the (sensorimotor) level of investigation is a related concern. It specifies the dynamical system under consideration, thereby constraining the variables that can be included in the cost function. Overall, the CNS has a hierarchical organization from task space to muscle space. Thus, the CNS may first care about what happens in task space and work with a simplified model of the musculoskeletal dynamics. Accordingly, cost functions could be defined at a kinematic level, at a dynamic level or could even attempt to minimize the overall motoneuronal activity. What level of description/investigation is best suited remains uncertain. Nevertheless, it seems that, in many cases, we can get reasonably good predictions of human movement trajectories by simple models that capture the essence of the system being controlled. For example, rigid body dynamics with some very basic muscle dynamics implementation are sufficient to capture several important motor phenomena as discussed above. Yet more involved models of the musculoskeletal apparatus have been considered too [94]. While this is a valuable and complementary approach, this raised questions regarding the confidence one can put in the "optimality" of the solutions and in the dependence of the solutions on the relative uncertainty about the model parameters such as muscle time constants, pennation angles, length/velocity force dependencies, the physiological cross-sectional area and so on, and which are difficult to know precisely for a given individual. Solving an optimal control problem can actually be a tricky task (doing the optimal synthesis[10] even for simple problems may illustrate this [117]).

---

[9]For example, a researcher might decide to work in the space of cost functions that depend on position and speed variables, or might wish to include acceleration variables (e.g. [27]). Other assumptions could be made such as working with polynomials (e.g. [115, 139]). However, a numerical implementation would necessitate restricting to some degree n or working with a finite number of basis costs belonging to the function space under consideration.

[10]Formally, this is the set of all the optimal trajectories joining any initial state to any terminal one.

Quite often, we may only find sub-optimal solutions and get stuck in local minima when using numerical tools. Therefore, working with very high-dimensional and non-linear systems may be challenging also for this reason.

We must mention that, like with any theory, the question of falsifiability must be asked. Because inverse optimal control is a form of data fitting (at least its numerical implementation), we shall always find some functions that fit a given set of data. Thus, special care must be taken regarding overfitting: Occam's razor principle should be applied for numerical inverse optimal control especially when composite costs (i.e. combinations of multiple cost functions) are considered. If the CNS truly relies on certain composite costs and if one imposes the constraint that a limited number of costs should explain a variety of tasks, falsifiability could then be addressed. This would suppose that the same cost functions should be relevant to a variety of tasks in the sense of generalization and cross-validation. If, in contrast, a given mixture of costs allows to account for motor performance in a given task but not in other tasks, then it would mean that the model must be modified or even reconsidered more globally.

At last, some authors have argued that motor control is "good enough" instead of really optimal [96]. Actually, saying that a system is not behaving optimally can be considered a stronger claim than saying it is behaving optimally because it is always possible to find a cost replicating a given experimental trajectory (a trivial - admittedly meaningless - counterexample would be a cost tracking the specific trajectory to be reproduced). Yet, we agree that being good enough may be sufficient for the sensorimotor system, especially if one thinks of the existence of muscle synergies or primitives that restrain the repertoire of possible motor commands. In any case, saying that a behavior is good or favorable implies that it offers some advantages against others, which can be theoretically translated in terms of cost functions. Even though some behavioral strategy is a local, not global, optimum [55, 123], it is nevertheless an *extremum* of a certain cost function (e.g. satisfying the necessary conditions given by Pontryagin Maximum Principle, [117]), which would be useful to characterize. In the end, saying that something is good enough or even optimal (reciprocally not good enough or not optimal) makes no sense without adding "with respect to" some well-defined cost function.[11]

Other theories such as the passive motion paradigm [101] have also been opposed to optimal control. The core idea in this paradigm is to replace cost functions with endpoint force fields and assume that the mechanical system is moved by a virtual force acting at the level of the end-effector (e.g. like strings moving a puppet's limbs). This is seen as a means to maximally exploit the spontaneous dynamics of the musculoskeletal system. However the nature of the underlying force field has to be given (choosing some free parameters) and, once one is chosen, it could still be interpreted as arising from a certain optimality criterion (e.g. optimal feedback control arising from LQR/LQG settings typically leads to static or time-varying force fields, in Cartesian or joint space depending on the control system under investigation). The

---

[11] A useful biomechanical analogy would be to talk about the "moment of a force" without precising the fixed reference point with respect to which it is calculated.

same point could be made regarding dynamical system theory [87, 158] as, once a control law is defined, the musculoskeletal system may appear to be self-organized and governed by some ordinary differential equation. Inverse optimal control theory mainly seeks to justify why certain force fields or dynamical attractors would be utilized rather than others. Nonetheless, relying on learned dynamical patterns, or basis force fields, can be viewed as an efficient algorithmic way to solve the degrees of freedom problem and to generate effectively coordinated limb movements [103]. In our view, this level of investigation rather addresses Marr's second level of analysis (algorithm).

From a robotic perspective and towards an anthropomorphic motion factory, inverse optimal control is also appealing as it may allow roboticists to produce an infinity of human-like movements from a given cost function. Indeed, cost functions (together with a model of the plant) are able to plan movements that have not been tested or encountered previously. Yet, the difficulty to resolve quite involved optimal control problems in real time for robots with numerous degrees of freedom limits the appeal of its generalization power. Besides classical linear-quadratic formulations, we lack ways to quickly solve such problems. Model predictive control might partly resolve this issue. Although this is an increasingly popular approach, it remains however unknown whether its control architecture is biologically plausible [99]. Modularity may help efficiently solving optimal control problems in an ecological way, however further research efforts are needed to figure out how the CNS actually implements the control of limb movement.

3 Modular Control Hypothesis

3.1 Hierarchical Modular Control Approach to Motor Control

Another significant body of the motor control literature has focused on the idea of compositional or modular motor control, such as the muscle synergy hypothesis on which we focus in this chapter. This body of literature suggests that the CNS stores certain muscle synergies and is able to combine them adequately to generate a motor command that would allow accomplishing a given motor task. A useful metaphor to illustrate the concept of modularity for spatiotemporal motor signals is to depict movement generation as music playing. Music is created by combining "modules" such as a melody (notes) and a rhythm (tempo). Similarly, coordinated movement may be the outcome of the combination of such stereotyped modular structures. The idea of hierarchical modular control is illustrated in Fig. 2.

The computational appeal of this theory is that, if pre-coded invariant modules can be used as motor building blocks, the CNS would only have to coordinate them to execute consistent movements, thereby simplifying the control problem by, *de facto*, implementing a dimensionality reduction. Therefore, the aim is to understand

Hierarchical modular control

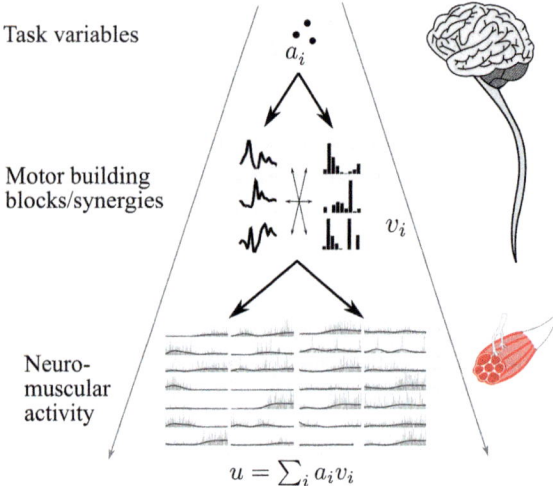

Task variables

Motor building
blocks/synergies

Neuro-
muscular
activity

$$u = \sum_i a_i v_i$$

Fig. 2 Illustration of the hierarchical modular control architecture at the muscle activation level. The scheme goes from task parameters (in a lower-dimensional space yielding some activations a_i), via combinations of precoded and stored motor modules (here loosely denoted by v_i), to neuromuscular inputs (in a higher-dimensional space, e.g. $u = \sum_i a_i v_i$ for a linear model). Dependencies on time/state and distinction between vector/scalar quantities are not specified on purpose. The nature of the modules and their combinations depend on the underlying model: they can be temporal waveforms, vectors of muscle activation ratios, spatiotemporal activity profiles, feedforward and/or feedback elementary control actions, and can be combined in a linear or nonlinear fashion etc. In all cases, their task-dependent modulation is assumed to account for the formation of genuine muscle patterns

both the representation of motor commands and according to which algorithm the CNS may set them up (the second Marr's level of analysis). More generally, at this level of analysis, one would want to elucidate how the CNS builds motor commands that generate good (enough) limb trajectories (which is not necessarily via muscle synergies but could relate to active inference or other means as mentioned above).

A first step in deciphering whether and how motor building blocks are stored within the CNS consists in defining a plausible modularity model. This is again a modeling choice. The model parameters can then be inferred from experimental data and correlated with their putative neural underpinnings. To implement this first step, several models of motor modularity have been proposed in the literature, mainly differing in their assumptions about (a) what quantities are stored by the motor system as invariant "modules" which can be reused in different movements and (b) what quantities are determined by the descending neural motor commands to recruit the modules in single trials [36, 79, 156]. Modules have been hypothesized to represent spatial, temporal or spatiotemporal invariant patterns in motor signals [5, 26, 43]. It is worth mentioning that modularity has been assumed to exist at different levels of the motor hierarchy (e.g. kinematic, [14, 126]; dynamic, [25, 124, 140];

neural, [28, 112]) with most of the studies placing modularity at the muscle activation level [34, 142, 143, 155]. Regarding the mathematical formulation, models of motor modularity are typically linear because it was found that the force fields resulting from the co-stimulation of two spinal loci were linear combinations of the individual force fields [20, 104]. Thus, the algorithms used to identify putative modules are commonly based on linear dimensionality reduction methods such as PCA or NMF and the extracted modules are assumed to be combined by feedforward motor mechanisms. Recently, techniques for the identification of modules of feedback nature have also been proposed for isometric tasks [118]. Hence, modeling choices include in general both the nature of the modules (spatial, temporal, feedforward, feedback etc.) and the associated combination rules (linear or not).

At the second step of this approach, motor signals recorded during a large number of motor tasks have been successfully fitted by modularity models. This was interpreted as evidence that performance of such motor behaviors relies on motor modules (e.g. reaching [26, 38, 43, 102, 157]; grasping [18, 110, 163]; walking [21, 46, 78–80, 91]; pedaling [74, 75]; reflex movements [22, 37, 116, 156]; postural tasks [29, 141, 152–154] etc.). Furthermore, modular structure in muscle activity has been shown to be preserved or adjusted after different types of brain injuries [23, 24, 30, 65, 120, 121].

At last, in order to probe the neural basis of modularity, the model parameters learned from the motor signals can be used as proxies of modular mechanisms, so their correlations with neural signals can be investigated. Several studies have identified potential neural bases of modules or their activations, both at the cortical level [71, 86, 111, 112] and in spinal structures [64, 67, 68, 125]. Crucially, this approach can serve to disentangle a) the nature and structure of modules, thereby informing the design of suitable modularity models and b) the level of the neural hierarchy where modules may be encoded, thereby addressing Marr's third level of analysis (the neural implementation). In this vein, Kargo and Giszter [85] showed that, at the spinal level of motor organization, premotor pulses (i.e. temporal modules) are more likely to be encoded than time-varying synergies (i.e. spatiotemporal modules). Also, Roh and collaborators [119] showed that medulla and spinal cord are sufficient for the expression of most (but not all) muscle synergies (i.e. spatial modules), which are likely activated by descending commands from supraspinal areas.

3.2 A Task-Space Perspective to Modularity

While many studies have tested modularity models based on whether they reconstruct the recorded muscle activation patterns for a number of task conditions using a limited number of invariant modules (input space assessment), recently a regain of interest to relate modularity to task space has been observed. A recent review of literature emphasizes this shift of paradigm [5]. As stated above, muscle synergies are typically extracted from recorded EMG data using unsupervised algorithms. The variance accounted for (VAF) or R^2 (R is the correlation coefficient) values are computed to evaluate the overall data approximation performed by the dimensionality reduction. However, this assessment has some limitations. First, VAF and R^2 values are somewhat arbitrary and defining an absolute threshold that indicates what is good fitting is a sensitive subject. Second, the musculoskeletal system being largely nonlinear, small errors in input space can lead to large errors in task space and undermine task achievement. Therefore, we should evaluate how putative motor synergies and their activations relate to task parameters. This idea has recently been put forward by many authors [6, 40–42, 135]. Mostly three approaches were taken: (1) using isometric tasks for which a virtual mapping from muscle space to task space can be defined by the experimenter, (2) quantifying the extent to which task identity is encoded in synergy-space so as to assess whether the way synergies are activated unequivocally determines the task at hand, as postulated by the theory, and (3) grounding on realistic musculoskeletal models to test whether experimentally-driven synergies can effectively be used for control. In a series of papers, de Rugy and collaborators evaluated "the usefulness of muscle synergies [...] in terms of errors produced not only in muscle space, but also in task space". They showed that even for what could appear as a relatively high VAF, control with muscle synergies could lead to unacceptable errors in task space. Delis and collaborators argued that single-trial task decoding/information techniques should be used to evaluate whether modularity can guarantee task performance in single trials [44]. The rationale was the following: if the performed movement cannot be discriminated in the reduced-dimension synergy space, this would cast serious doubts about the effectiveness of the proposed hierarchical modular control scheme (indeed, it is possible to get a large VAF with a low decoding score, which would invalidate a modular decomposition although it is doing a good job at reducing dimensionality). The authors concluded that complex and comprehensive data sets should be considered in general to conduct such analyses, in the spirit of large-scale neuroscience endeavors [56]. This is a relevant approach as it was shown that the number and efficiency of muscle synergies depends on the scope of the original database and on the complexity of the tasks under investigation [40, 136]. Another way to test the modular control hypothesis is to build an accurate musculoskeletal model, as done in [6, 107]. The authors of these studies applied this technique for walking and essentially showed that EMG-based synergies are rough starting point solutions that need to be fine-tuned to elicit adequate walking patterns.

Hence, the analysis of motor modularity should ideally consist of a closed loop between the recorded motor signals and their associated limb trajectories. This can be summarized as follows (Eq. 2):

$$
\begin{aligned}
&\text{real muscle patterns} \xrightarrow[\substack{\text{engineering}}]{\text{reverse}} \text{extracted muscle synergies} \\
&\text{extracted muscle synergies} \xrightarrow[\substack{\text{musculoskeletal dynamics}}]{\text{forward}} \text{reconstructed trajectories}
\end{aligned}
\tag{2}
$$

The first step allows to extract the potentially stored modules via some machine learning technique. To this aim, it is likely that switching from unsupervised to supervised learning algorithms taking into account the underlying biomechanics and trajectories in task space could lead to more advanced synergy models. However, the choice of unsupervised versus supervised learning algorithms is often neglected in practice because the simplicity of standard methods such as NMF is preferred. The second step consists in "playing back" the synergies into the musculoskeletal apparatus to test the produced motor behavior and evaluate the proposed control model in task space by analyzing the reconstructed limb trajectories.

3.3 Remarks Regarding Modular Control Theory

In reality, the questions of what the bricks of motor commands are, what model should be used to describe them and how motor commands are combined have not been resolved yet. It is likely that the models currently used are too simple to define a suitable framework from a control theory standpoint and to allow finding explicit neural correlates of the putative modules. In particular, most EMG-based identification models do not consider feedback processes whereas muscle synergies or coordination patterns may be posture-dependent as suggested in [118, 123] and, therefore, the state of the biomechanical system should possibly be taken into account during synergy identification. Another restriction is that current synergy extraction algorithms usually require equal movement durations across trials to simultaneously extract synergies from different movement conditions, which is not the case of real data. At last, except in some studies where multiple layers of modularity have been considered, thereby leading to nonlinear reductions of dimensionality [10], most existing approaches are linear, which might be a good first-order approximation but might be limiting in practice to control a nonlinear plant.

A major criticism of the modularity hypothesis is that, for a given task or set of tasks, it is always possible to reduce dimensionality to a certain extent or to account for muscle patterns if a sufficient set of modules is extracted. As a consequence, the extracted modules may be inherently task-dependent and may reflect the biomechanical constraints imposed by the human body (especially for neighboring muscles that control the same joint) [90, 160]. Moreover, if optimality drives human trajectory formation and/or optimization guides muscle pattern design, it is clear that, in general, small task variations (e.g. changing a target's position) will only produce small muscle activity variations. Overall, this would lead to certain commonalities in muscle patterns, which could always be isolated by machine learning or statistical techniques. Nevertheless, similarly to optimal control, generalization should be tested and special care should be taken with regard to the conclusions that can be reached via such EMG-driven analyses. In other words, while describing the organization of muscle patterns is valuable to provide synthetic views of how ensemble muscle patterns are structured, relating this structure to underlying active neural mechanisms is tricky. In practice, the existence of circumstantial evidence, as described above, combined with the possibility that any deviations from the theory can be attributed to either a coarse model of modularity or alternative neural pathways/mechanisms (e.g. allowing individual muscle control) make the modularity hypothesis hard to falsify [155].

Consequently, although the muscle synergy hypothesis is popular in human motor control, it remains rightly debated. At the core of the debate is whether EMG-based synergies are just a descriptive low-dimensional representation of expectedly well-structured motor outputs or whether they have a real neural basis. To better address this important point, we can hypothesize that such a representation in terms of synergies exists and examine what this would imply [35]. First, a control action achieving a given motor task is hypothetically built from the combined activation of a given set of synergies. Therefore, performing a new task would just require adjusting the way synergies are combined until a suitable one is found. If no combination of synergies allows to execute the task, it may be because it requires non-habitual muscle patterns that have not been experienced and stored before, thus new synergies may need to be learned.[12] This logical reasoning predicts that learning a task that is incompatible with a currently available set of synergies would be harder and take a longer time. d'Avella and collaborators nicely investigated the predictions of such a theory along these lines [9]. This approach of (a) formulating predictions of the modularity hypothesis and (b) designing experiments to critically (and, if possible, quantitatively) test them might provide more direct evidence either supporting or falsifying the theory even though the neural code is not directly accessed [2]. Studying behavior and its adaptation may actually represent a very valuable approach to motor neuroscience [89].

4 From Optimality to Modularity and Vice-Versa

4.1 Optimality with Modularity: Theoretical Ground

Optimality and modularity theories have often been treated separately yet some studies have attempted to combine the two concepts. Theoretical works are of particular interest among those studies. Interesting frameworks have been derived to understand if (and in what contexts) optimal controllers can be built from a limited set of elementary control actions (possibly optimal themselves). Most of the times, the mathematical analyses aiming at reconciling the two approaches were conducted from a control theory perspective. The link between optimality and modularity may be envisioned as follows (Eq. 3):

$$
\begin{aligned}
\text{Dynamics}: \quad \dot{x} &= f(x, u) \\
\text{Optimality}: \quad J(u) &= \int_0^T h(x, u, t)dt \\
\text{Modularity}: \quad u(x, t) &= \sum_{i \in \mathcal{I}} a_i v_i(x, t)
\end{aligned}
\tag{3}
$$

---

[12]In particular, this would be compatible with the claim that muscle patterns are habitual rather than optimal [123].

where u is the control action that can drive the system state x according to some dynamical constraints specified by $f(x, u)$ and which is built from certain motor primitives or basis modules $v_i(x, t)$,[13] h is the infinitesimal cost whose integral should be minimized. The main open question concerns the existence of motor building blocks allowing to effectively control the system for a given set of tasks. The basis modules may constitute a finite set of mixed feedforward and/or feedback control actions [3, 4, 109]. Linearity is often central to modularity studies because of experimental findings showing linear summation properties of spinal force fields (see Sect. 3). Interestingly, it can also be thought as a first-order approximation which simplifies the mathematical derivations. Notably, it allowed researchers to obtain elegant results for a restricted class of problems such as linear or feedback linearizable systems with quadratic costs or control-affine stochastic systems with control-quadratic costs that lead to a linear Bellman equation under suitable assumptions about noise [109, 146, 147]. These works showed that new optimal controls may be constructed from linear combinations of a finite number of elementary optimal control actions. However, the problem may be more complex when thinking of the human motor system as a whole because of the hierarchical nature of the neuromusculoskeletal system. Optimal control may indeed occur at different levels in the hierarchy (kinematic, dynamic, muscular or neural levels). It is, apropos, particularly remarkable how relatively simple optimal control models (e.g. minimum jerk model) capture the hand/joint kinematics although they neglect fine muscle properties (e.g. speed/length dependencies, concentric/eccentric contractions, slow/fast twitch muscle fibers etc.). Therefore, optimal control may conceivably apply in a low-dimensional space (task space or joint space) and lower level neuromuscular activity may subsequently conform to these higher level constraints already specifying the main spatiotemporal characteristics of the movement. In the spirit that approximate optimal motor commands are acceptable, a hierarchical control framework has been proposed [149], thereby providing a theoretical link between task parameters and motor synergies. Other approaches based on deep learning schemes have also been considered by acknowledging that an optimal control problem readily implements a (nonlinear) dimensionality reduction [10] which could lead to very effective movement representations in neural networks. Although defining motor building blocks in a compositional sense seems harder in those frameworks, they nevertheless nicely capture the idea of dimensionality reduction resulting from the concept of motor synergies and address important questions such as whether the monitoring of a restricted number of task variables can yield suitable coordination of the complex and nonlinear musculoskeletal plant.

---

[13]The basis modules $v_i(x, t)$ might be separated into spatial and temporal components $\sigma_i(t)w_i(x)$ such as in [103] or [95], and in a way which is reminiscent of the model proposed in [43]. In this case, spatial (state-dependent) modules, or muscle synergies, would be feedback-dependent as suggested in [118]. Analogously, this time-space separation is also apparent in the optimal control of finite-horizon LQR/LQG problems.

4.2 Optimality Versus Modularity: Paradox and Causality

At first sight, the coupling between optimality and modularity may seem paradoxical as the constraints imposed by modularity might severely compromise optimality. If one assumes existence of synergies, two seemingly competing questions arise: (1) whether modularity constrains and shapes the type or degree of optimality that can be attained in higher-level variables (i.e. end-effectors or joints) and (2) whether optimality naturally leads to an apparent modularity at the lower level (i.e. neuro-muscular). On one hand, a number of studies have analyzed the extent to which motor synergies may arise from optimal control. Since optimal control and/or opti-mization may give *de facto* a structure to the associated muscle patterns according to variations of the task demand, it can reasonably be expected that empirical motor synergies are just a byproduct of optimality conditions. Accordingly, motor synergies have been shown to emerge from optimal feedback control theory [148]. Further-more, numerical studies have provided interesting insights regarding what kind of time-varying synergies would result from optimal control policies applied to planar reaching movements [25]. On the other hand, the authors reversed the process to verify what kind of arm trajectories could be achieved via the extracted time-varying synergies [25]. Interestingly, these numerical experiments demonstrated that task constraints (reaching to targets in a plane) together with optimality objectives could be fulfilled with a small number of hypothetical motor synergies. In the same vein, other authors have investigated a similar issue for different tasks and subsystems. Computational studies based on musculoskeletal modeling showed that a simpli-fied construction of motor commands via modularity (either experimentally inferred using dimensionality reduction techniques or synthesized) could significantly affect optimality and lead to sub-optimal solutions in terms of effort costs during balance control [98] and to a limited ability to minimize energy but also to tune endpoint stiffness during an isometric upper-limb task [77]. It is worth mentioning that these studies relied on optimization rather than optimal control.[14] Other studies involving optimization techniques are also relevant here as they implemented the whole loop mentioned in Eq. 2 using EMG-driven virtual biomechanics [40, 123]. They could quantify how muscle synergies affected energy consumption and aiming errors in task-space. Numerical studies about modularity have also been conducted in robotics [3, 109, 137], in which errors in task-space were also evaluated. Overall, it may be concluded that the nature of the motor building blocks (feedback/feedforward) and the type of system and cost (linear-quadratic) is critical for effective and efficient motor control using modularity. As suggested by Neptune's studies using complex musculoskeletal models, muscle synergies (as extracted in EMG-based studies) may serve as a rough starting point for motor planning, which should then be refined through spinal and transcortical reflex pathways. As such, modularity may be favor-

---

[14]Optimization and optimal control should not be confused although they may be related when one comes to numerical resolution of optimal control problems. The former only deals with a standard function while the latter deals with a functional, i.e. a function of a function.

able to reduce the computational burden of behaving rigorously in an optimal way at the price of sacrificing some optimality.

Besides the question of the neural origin of muscle synergies, which is crucial to motor neuroscience (as already discussed) and proved difficult to answer [70], the chicken-and-egg situation of modularity versus optimality is of interest. Assuming that both optimality and modularity are present in the sensorimotor system, which one drives the other remains unclear. It is likely that developmental studies could help disentangling how optimality and modularity emerge during growth [46] and whether one constrained the other or vice-versa. It is possible that, if stored muscle synergies are hardwired in the CNS, they would limit the kind of optimality that can be attained (unless perhaps they have been precisely shaped to match the desired optimality conditions); this would suggest that motor development is the parallel process of finding both efficient limb trajectories and efficient representations of them, which would be partly dictated by evolutionary/homeostatic and neuroanatomical/biomechanical constraints. On a timescale of hours, it seems that motor learning is constrained by the existence of precoded/habitual solutions and that humans are not very good at finding global optima for new (never experienced) tasks [55, 100, 123]. It must however be noted here that recent studies actually undermine this temporal limitation [31, 50, 61, 131]. Yet, at least over long learning periods, a process of re-optimization may occur [81], possibly requiring the creation of new primitives, which would take a long time to acquire [9, 35].

5 Conclusion and Perspectives

Hierarchical motor control, in particular grounded on the concepts of optimality and modularity, is an appealing theory to explain the formation of muscle patterns from task parameters. Although it is unclear whether global optimality, in a strict sense is the immediate and primary goal of the human motor system when adapting to a new task on a short-time scale, daily life behaviors (such as reaching for a cup of coffee) undeniably display optimal-like signatures. Inverse optimal control offers a normative framework to formalize action selection and give the rationale for choosing one limb trajectory over another. The difficulties that the CNS faces in order to discover a global optimum when coping with a certain new motor task may be related to the way action planning is implemented; what is stored or learned versus what is specified on-the-fly. How the CNS seamlessly generates adequate motor commands in a fraction of a second might rely on modularity, i.e. the storage of muscle synergies that can be recalled and combined in a task-dependent manner to build genuine motor signals. If muscle activity conforms in some sense to higher-level optimality principles, it is not surprising that low-dimensional structures can be found in EMG-based studies. Classical empirical studies of muscle activity are undoubtedly useful for providing ensemble descriptions of muscle patterns, yet their implementation in neural networks (third Marr's level of analysis) remains putative despite remarkable efforts [68, 112]. At the same time we remain quite ignorant of what the (neural) nature of a motor building block (should it be called primitive, module or muscle synergy) is. The nature of the modules extracted from matrix factorization techniques applied to EMG data is often guided by the limitations of the method itself. For instance, NMF-based methods will only give rise to feedforward synergies. If synergies are posture-dependent as suggested in [118], this would mean that more advanced machine learning methods should be used. Towards an anthropomorphic motion factory and effective applications in robotics, we believe that there is still place for the proposal of new frameworks. The development of such a novel framework requires interdisciplinarity and will require advanced musculoskeletal models, neurophysiological data, psychophysical experiments and mathematics.

A generic framework capturing the essence of both optimality and modularity could be useful to advance our understanding of human movement control. Importantly, the mathematical formalism should be suited to port the main findings to robotics and engineering in order to improve the production of efficient movements in artificial systems in a rapid, robust and adaptive way. The idea of hierarchy being central and ubiquitous in motor control, it may be the cornerstone of such a framework. Based on previous experimental and computational results, a model relying on a cascade of optimal control problems could be envisioned. The problems could be solved recursively by the CNS: each problem could integrate a nominal trajectory coming from the previous problem and serving as a reference trajectory to the current level. An example of such a hierarchical framework connecting task-level to muscle level via skeleton level, could be as follows:

$$
\begin{aligned}
& x_{ref}(\cdot) \text{ built from task} \qquad\qquad \text{(task level)} \\
& \qquad\qquad \downarrow x_{ref} \\
& \min_{\tau(\cdot)} \int_0^T [h(q, \tau, t) + c(q, x_{ref})]dt \quad \text{(skeleton level)} \\
& \qquad\qquad \downarrow \tau_{ref} \\
& \min_{u(\cdot)} \int_0^T [H(\eta, u, t) + C(\eta, \tau_{ref})]dt \quad \text{(muscle level)} \\
& \qquad\qquad \downarrow u_{ref} \text{ (motoneurons)}
\end{aligned}
\tag{4}
$$

In Eq. 4, x would be the position/orientation of the end-effector in Cartesian space, x_{ref} would be a reference trajectory (either a fixed target location, a geometric path such as a geodesic or a full trajectory potentially coming from a least action principle or imposed by the task with a metronome). At the skeleton level, q would be the joint angles and τ the joint torques. A reference torque profile could be predicted by optimal control principles. At the muscle level, η would be the muscle forces and u the motoneuron inputs. An optimization problem could even be solved at this level if the spatiotemporal characteristics of the torque profiles are already set up. The core ingredient would be that each level would lead to a reference trajectory that could be tracked in the subsequent level. The tracking is implemented via the tracking costs $c(q, x_{ref})$ and $C(\eta, \tau_{ref})$, for instance by choosing at the skeleton level $c(q, x_{ref}) = c(\|\varphi(q) - x_{ref}\|)$ where $x = \varphi(q)$ is the forward kinematics (and something similar at the muscle level). Because of the existence of reference trajectories, the resolution of each optimal control problem may turn out to be faster and simplified (e.g. linearization or reduction of the search space). In this framework, we might insert other levels, introduce stochastic models and so on, but the structure is rich enough to incorporate a number of practical motor control problems.

Synergies may contribute to resolve each of these problems more efficiently, which could by the way explain why researchers have talked about modularity at various levels (kinematic, dynamic, muscular or even neural). Such a cascade of optimal control problems could be able to explain several experimental observations but this would remain to be investigated. This formulation may be reminiscent of an alternative formulation of hierarchical optimal control which has been proposed recently and in which a first optimal control problem giving rise to an infinity of solutions is solved before a subsequent optimal control problem is solved within the subspace of the optimal solutions of the previous level and so on [62, 122]. The present hierarchical framework may be flexible although it relies on the weighting of complementary objectives, thereby contrasting with the concept of a stack of tasks. Furthermore, this framework could account for the observation of kinematic persistence observed in some motor tasks, especially when controlling a visual cursor on a screen or adapting to microgravity, which may reflect the influence of task-space high-level goals in certain contexts [33, 59, 61, 100].

Acknowledgements This work is supported by a public grant overseen by the French National research Agency (ANR) as part of the Investissement d Avenir program, through the iCODE project funded by the IDEX Paris-Saclay, ANR-11-IDEX-0003-02.

Appendix

In this appendix, we provide a tutorial example illustrating the concepts of optimality and modularity that we discussed in the main text.

To this aim, we consider a simple controlled pendulum whose dynamic equation is

$$\ddot{\theta} = -\theta + u$$

where θ is the angular position and u the input torque. This may represent a simplified human arm model in the gravity field (with normalized anthropometry and small angle assumption).

We define an optimal control problem as follows: find the controller that drives the system from a given state $x_0 = (\theta_0, \dot{\theta}_0)$ to a final state $x_f = (\theta_f, \dot{\theta}_f)$ in time T and minimizes the cost function

$$C(u) = \int_0^T [u^2 + \theta^2 + \dot{\theta}^2] dt.$$

This is a linear-quadratic (LQ) problem of the form $\dot{x} = Ax + Bu$ (linear **dynamics**) and $C(u) = \int_0^T u^T Ru + x^T Qx dt$ (quadratic **optimality** criterion) where the matrices are identified as follows:

$$A = \begin{pmatrix} 0 & 1 \\ -1 & 0 \end{pmatrix}, \quad B = \begin{pmatrix} 0 \\ 1 \end{pmatrix}, \quad R = 1, \quad Q = I.$$

It can be shown that the optimal control can be written formally as (see [51] for mathematical proofs):

$$u(t) = -B^T P_+ e^{A_+ t} p - B^T P_- e^{A_-(t-T)} q \tag{5}$$

where P_+ and P_- are the maximal and minimal solutions of the associated Riccati equation $PA + A^T P - PBB^T P + Q = 0$. The matrices A_+ and A_- are defined as $A_+ = A - BB^T P_+$ and $A_- = A - BB^T P_-$, and p and q are some vectors depending on the initial/final states and on movement duration T. Importantly, the matrices $P_\pm$ and $A_\pm$ just depend on the optimal control problem specification, i.e. the matrices A, B, R, and Q.

The optimal control $u(t)$ can thus be written as a function of the eigenvalues of $A_\pm$. Simple computations show that the 4 eigenvalues are of the form $\pm\alpha \pm i\beta$ with $\alpha = \sqrt{(2\sqrt{2}-1)/2}$ and $\beta = \sqrt{(2\sqrt{2}+1)/2}$.

Therefore, the optimal control can be rewritten as the following linear combination:

$$u(t) = a_1 e^{-\alpha t}\cos(\beta t) + a_2 e^{\alpha t}\cos(\beta t) + a_3 e^{-\alpha t}\sin(\beta t) + a_4 e^{\alpha t}\sin(\beta t),$$

where the coefficients $(a_i)_{1 \le i \le 4}$ have to be adjusted depending on the constraints $x(0) = (\theta_0, \dot{\theta}_0)$ and $x(T) = (\theta_f, \dot{\theta}_f)$.

In summary, this simple example illustrates that for this system all optimal motor commands can be decomposed as follows:

$$u(t) = \sum_{i=1}^{4} a_i v_i(t)$$

where the functions $v_i(t)$ are time-varying primitives that are invariants of the problem and can thus be stored once for all (**modularity**). In contrast, the activation coefficients a_i must be set for each single movement in order to start/end in the adequate states and times. Note the similarity between the present equations and Eq. 3 in the main text. Storing invariant building blocks v_i and adjusting activation coefficients a_i in order to produce controllers that allow task achievement in an optimal fashion are the core concepts discussed in the present Chapter. It is worth noting that the same conclusion could actually be drawn for any well-defined LQ problem given the general form of Eq. 5.

References

1. Ajami, A., Gauthier, J.P., Maillot, T., Serres, U.: How humans fly. ESAIM: Control, Optimisation and Calculus of Variations **19**(4), 1030–1054 (2013)
2. Ajemian, R., Hogan, N.: Experimenting with theoretical motor neuroscience, (2010)
3. Alessandro, C., Nori, F.: Identification of synergies by optimization of trajectory tracking tasks. In: 2012 4th IEEE RAS & EMBS International Conference on Biomedical Robotics and Biomechatronics (BioRob), IEEE, pp. 924–930 (2012)
4. Alessandro, C, Carbajal, J.P., d'Avella, A.: A computational analysis of motor synergies by dynamic response decomposition. Front. Comput. Neurosci. **7** (2013a)
5. Alessandro, C., Delis, I., Nori, F., Panzeri, S., Berret, B.: Muscle synergies in neuroscience and robotics: from input-space to task-space perspectives. Front. Comput. Neurosci. **7**, 43 (2013b)
6. Allen, J.L., Neptune, R.R.: Three-dimensional modular control of human walking. J. Biomech. **45**(12), 2157–2163 (2012)
7. Atkeson, C.G., Hollerbach, J.M.: Kinematic features of unrestrained vertical arm movements. J. Neurosci. **5**(9), 2318–2330 (1985)
8. Bellman, R.E.: Dynamic Programming. Princeton, NJ (1957)
9. Berger, D.J., Gentner, R., Edmunds, T., Pai, D.K., d'Avella, A.: Differences in adaptation rates after virtual surgeries provide direct evidence for modularity. J. Neurosci. **33**(30), 12384–12394 (2013)
10. Berniker, M., Kording, K.P.: Deep networks for motor control functions. Front. Comput. Neurosci. **9** (2015)
11. Berret, B., Jean, F.: Why don't we move slower? the value of time in the neural control of action. J. Neurosci. **36**(4), 1056–1070 (2016)
12. Berret, B., Darlot, C., Jean, F., Pozzo, T., Papaxanthis, C., Gauthier, J.P.: The inactivation principle: mathematical solutions minimizing the absolute work and biological implications for the planning of arm movements. PLoS Comput. Biol. **4**(10), e100,0194 (2008a)
13. Berret, B., Gauthier, J.P., Papaxanthis, C.: How humans control arm movements. Proc. Steklov Inst. Mathematics **261**, 44–58 (2008b)
14. Berret, B., Bonnetblanc, F., Papaxanthis, C., Pozzo, T.: Modular control of pointing beyond arm's length. J. Neurosci. **29**(1), 191–205 (2009)
15. Berret, B., Chiovetto, E., Nori, F., Pozzo, T.: Evidence for composite cost functions in arm movement planning: an inverse optimal control approach. PLoS Comput. Biol. **7**(10), e1002,183 (2011a)
16. Berret, B., Chiovetto, E., Nori, F., Pozzo, T.: Manifold reaching paradigm: how do we handle target redundancy? J. Neurophysiol. **106**(4), 2086–2102 (2011b)
17. Berret, B., Bisio, A., Jacono, M., Pozzo, T.: Reach endpoint formation during the visuomotor planning of free arm pointing. Eur. J. Neurosci. **40**(10), 3491–3503 (2014)
18. Brochier, T., Spinks, R.L., Umilta, M.A., Lemon, R.N.: Patterns of muscle activity underlying object-specific grasp by the macaque monkey. J. Neurophysiol. **92**(3), 1770–1782 (2004)
19. Buneo, C.A., Soechting, J.F., Flanders, M.: Muscle activation patterns for reaching: the representation of distance and time. J. Neurophysiol. **71**(4), 1546–1558 (1994)
20. Caggiano, V., Cheung, V.C., Bizzi, E.: An optogenetic demonstration of motor modularity in the mammalian spinal cord. Sci. Rep. **6**(35), 185 (2016)
21. Cappellini, G., Ivanenko, Y.P., Poppele, R.E., Lacquaniti, F.: Motor patterns in human walking and running. J. Neurophysiol. **95**(6), 3426–37 (2006)
22. Cheung, V.C., d'Avella, A., Tresch, M.C., Bizzi, E.: Central and sensory contributions to the activation and organization of muscle synergies during natural motor behaviors. J. Neurosci. **25**(27), 6419–34 (2005)
23. Cheung, V.C., Piron, L., Agostini, M., Silvoni, S., Turolla, A., Bizzi, E.: Stability of muscle synergies for voluntary actions after cortical stroke in humans. Proc. Natl. Acad. Sci. U S A **106**(46), 19563–19568 (2009)

24. Cheung, V.C.K., Turolla, A., Agostini, M., Silvoni, S., Bennis, C., Kasi, P., Paganoni, S., Bonato, P., Bizzi, E.: Muscle synergy patterns as physiological markers of motor cortical damage. Proc. Natl. Acad. Sci. U S A **109**(36), 14652–14656 (2012)
25. Chhabra, M., Jacobs, R.A.: Properties of synergies arising from a theory of optimal motor behavior. Neural. Comput. **18**(10), 2320–2342 (2006)
26. Chiovetto, E., Berret, B., Delis, I., Panzeri, S., Pozzo, T.: Investigating reduction of dimensionality during single-joint elbow movements: a case study on muscle synergies. Front. Comput. Neurosci. **7**, 11 (2013)
27. Chittaro, F., Jean, F., Mason, P.: On the inverse optimal control problems of the human locomotion: stability and robustness of the minimizers. J. Math. Sci. **195**(3), 269–287 (2013)
28. Churchland, M.M., Cunningham, J.P., Kaufman, M.T., Foster, J.D., Nuyujukian, P., Ryu, S.I., Shenoy, K.V.: Neural population dynamics during reaching. Nature **487**(7405), 51–56 (2012)
29. Chvatal, S.A., Torres-Oviedo, G., Safavynia, S.A., Ting, L.H.: Common muscle synergies for control of center of mass and force in nonstepping and stepping postural behaviors. J. Neurophysiol **106**(2), 999–1015 (2011)
30. Clark, D.J., Ting, L.H., Zajac, F.E., Neptune, R.R., Kautz, S.A.: Merging of healthy motor modules predicts reduced locomotor performance and muscle coordination complexity post-stroke. J. Neurophysiol **103**(2), 844–857 (2010)
31. Cluff, T., Scott, S.H.: Apparent and actual trajectory control depend on the behavioral context in upper limb motor tasks. J. Neurosci. **35**(36), 12465–12476 (2015)
32. Crevecoeur, F., Thonnard, J.L., Lef èvre, P.: Optimal integration of gravity in trajectory planning of vertical pointing movements. J. Neurophysiol **102**(2), 786–796 (2009)
33. Danziger, Z., Mussa-Ivaldi, F.A.: The influence of visual motion on motor learning. J. Neurosci. **32**(29), 9859–9869 (2012)
34. d'Avella, A., Bizzi, E.: Shared and specific muscle synergies in natural motor behaviors. Proc. Natl. Acad. Sci. U S A **102**(8), 3076–3081 (2005)
35. d'Avella, A., Pai, D.K.: Modularity for sensorimotor control: evidence and a new prediction. J. Motor Behaviob **42**(6), 361–369 (2010)
36. d'Avella, A., Tresch, M.C.: Modularity in the motor system: decomposition of muscle patterns as combinations of time-varying synergies. In: Becker, S., Ghahramani, Z. (eds.) Dietterich TG, pp. 141–148. NIPS, MIT Press (2001)
37. d'Avella, A., Saltiel, P., Bizzi, E.: Combinations of muscle synergies in the construction of a natural motor behavior. Nat. Neurosci. **6**(3), 300–308 (2003)
38. d'Avella, A., Portone, A., Fernandez, L., Lacquaniti, F.: Control of fast-reaching movements by muscle synergy combinations. J. Neurosci. **26**(30), 7791–7810 (2006)
39. d'Avella, A., Fernandez, L., Portone, A., Lacquaniti, F.: Modulation of phasic and tonic muscle synergies with reaching direction and speed. J. Neurophysiol **100**(3), 1433–1454 (2008)
40. de Rugy, A., Loeb, G.E., Carroll, T.J.: Are muscle synergies useful for neural control? Front. Comput. Neurosci. **7**, 19 (2013)
41. Delis, I., Berret, B., Pozzo, T., Panzeri, S.: A methodology for assessing the effect of correlations among muscle synergy activations on task-discriminating information. Front. Comput. Neurosci. **7**, 54 (2013a)
42. Delis, I., Berret, B., Pozzo, T., Panzeri, S.: Quantitative evaluation of muscle synergy models: a single-trial task decoding approach. Front Comput. Neurosci. **7**, 8 (2013b)
43. Delis, I., Panzeri, S., Pozzo, T., Berret, B.: A unifying model of concurrent spatial and temporal modularity in muscle activity. J. Neurophysiol **111**(3), 675–693 (2014)
44. Delis, I., Panzeri, S., Pozzo, T., Berret, B.: Task-discriminative space-by-time factorization of muscle activity. Front Hum. Neurosci. **9**, 399 (2015)
45. Diedrichsen, J., Shadmehr, R., Ivry, R.B.: The coordination of movement: optimal feedback control and beyond. Trends Cogn. Sci. (2009)
46. Dominici, N., Ivanenko, Y.P., Cappellini, G., d'Avella, A., Mondí, V., Cicchese, M., Fabiano, A., Silei, T., Di Paolo, A., Giannini, C., Poppele, R.E., Lacquaniti, F.: Locomotor primitives in newborn babies and their development. Science **334**(6058), 997–999 (2011)

47. Douglas, J.: Solution of the inverse problem of the calculus of variations. Trans. Am. Mathe. Soc. **50**(1), 71–128 (1941)
48. Dudman, J.T., Krakauer, J.W.: The basal ganglia: from motor commands to the control of vigor. Curr. Opin. Neurobiol. **37**, 158–166 (2016)
49. Engelbrecht, S.: Minimum principles in motor control. J. Math. Psychol. **45**(3), 497–542 (2001)
50. Farshchiansadegh, A., Melendez-Calderon, A., Ranganathan, R., Murphey, T.D., Mussa-Ivaldi, F.A.: Sensory agreement guides kinetic energy optimization of arm movements during object manipulation. PLoS Comput. Biol. **12**(4), e1004,861 (2016)
51. Ferrante, A., Marro, G., Ntogramatzidis, L.: A parametrization of the solutions of the finite-horizon lq problem with general cost and boundary conditions. Automatica **41**, 1359–1366 (2005)
52. Flanders, M., Pellegrini, J.J., Geisler, S.D.: Basic features of phasic activation for reaching in vertical planes. Exp. Brain Res. **110**(1), 67–79 (1996)
53. Flash, T., Hogan, N.: The coordination of arm movements: an experimentally confirmed mathematical model. J. Neurosci. **5**(7), 1688–1703 (1985)
54. Friston, K.: What is optimal about motor control? Neuron **72**, 488–498 (2011)
55. Ganesh, G., Haruno, M., Kawato, M., Burdet, E.: Motor memory and local minimization of error and effort, not global optimization, determine motor behavior. J. Neurophysiol **104**(1), 382–390 (2010)
56. Gao, P., Ganguli, S.: On simplicity and complexity in the brave new world of large-scale neuroscience. Curr. Opin. Neurobiol **32**, 148–155 (2015)
57. Gauthier, J.P., Berret, B., Jean, F.: A biomechanical inactivation principle. Proc. Steklov Inst. Mathematics **268**, 93–116 (2010)
58. Gaveau, J., Papaxanthis, C.: The temporal structure of vertical arm movements. PLoS One **6**(7), e22,045 (2011)
59. Gaveau, J., Paizis, C, Berret, B., Pozzo, T., Papaxanthis, C.: Sensorimotor adaptation of point-to-point arm movements after space-flight: the role of the internal representation of gravity force in trajectory planning. J. Neurophysiol (2011)
60. Gaveau, J., Berret, B., Demougeot, L., Fadiga, L., Pozzo, T., Papaxanthis, C.: Energy-related optimal control accounts for gravitational load: comparing shoulder, elbow, and wrist rotations. J. Neurophysiol **111**(1), 4–16 (2014)
61. Gaveau, J., Berret, B., Angelaki, D.E., Papaxanthis, C.: Direction-dependent arm kinematics reveal optimal integration of gravity cues. eLife **5**, 16,394 (2016)
62. Geisert M, Del Prete A, Mansard N, Romano F, Nori F (2017) Regularized hierarchical differential dynamic programming. IEEE Trans Rob
63. Gentili, R., Cahouet, V., Papaxanthis, C.: Motor planning of arm movements is direction-dependent in the gravity field. Neuroscience **145**(1), 20–32 (2007)
64. Giszter, S.F., Hart, C.B.: Motor primitives and synergies in the spinal cord and after injury-the current state of play. Ann. N.Y. Acad. Sci. **1279**(1), 114–126 (2013)
65. Gizzi, L., Nielsen, J.F., Felici, F., Ivanenko, Y.P., Farina, D.: Impulses of activation but not motor modules are preserved in the locomotion of subacute stroke patients. J. Neurophysiol **106**(1), 202–10 (2011)
66. Harris, C.M., Wolpert, D.M.: Signal-dependent noise determines motor planning. Nature **394**(6695), 780–784 (1998)
67. Hart, C.B., Giszter, S.F.: Modular premotor drives and unit bursts as primitives for frog motor behaviors. J. Neurosci. **24**(22), 5269–5282 (2004)
68. Hart, C.B., Giszter, S.F.: A neural basis for motor primitives in the spinal cord. J. Neurosci. **30**(4), 1322–1336 (2010)
69. Hilt, P.M., Berret, B., Papaxanthis, C., Stapley, P.J., Pozzo, T.: Evidence for subjective values guiding posture and movement coordination in a free-endpoint whole-body reaching task. Sci. Rep. **6**(23), 868 (2016)
70. Hirashima, M., Oya, T.: How does the brain solve muscle redundancy? filling the gap between optimization and muscle synergy hypotheses. Neurosci. Res. **104**, 80–87 (2016)

71. Holdefer, R.N., Miller, L.E.: Primary motor cortical neurons encode functional muscle synergies. Exp. Brain Res. **146**(2), 233–43 (2002)
72. Hollerbach, J.M., Flash, T.: Dynamic interactions between limb segments during planar arm movement. Biol. Cybern. **44**(1), 67–77 (1982)
73. Hondzinski, J.M., Soebbing, C.M., French, A.E., Winges, S.A.: Different damping responses explain vertical endpoint error differences between visual conditions. Exp. Brain Res. **234**(6), 1575–1587 (2016)
74. Hug, F., Turpin, N.A., Guével, A., Dorel, S.: Is interindividual variability of emg patterns in trained cyclists related to different muscle synergies? J. Appl. Physiol. **108**(6), 1727–1736 (2010)
75. Hug, F., Turpin, N.A., Couturier, A., Dorel, S.: Consistency of muscle synergies during pedaling across different mechanical constraints. J. Neurophysiol **106**(1), 91–103 (2011)
76. Huh, D., Sejnowski, T.J.: Conservation law for self-paced movements. Proc. Natl. Acad. Sci. U S A **113**(31), 8831–8836 (2016)
77. Inouye, J.M., Valero-Cuevas, F.J.: Muscle synergies heavily influence the neural control of arm endpoint stiffness and energy consumption. PLoS Comput. Biol. **12**(2), e1004,737 (2016)
78. Ivanenko, Y.P., Grasso, R., Zago, M., Molinari, M., Scivoletto, G., Castellano, V., Macellari, V., Lacquaniti, F.: Temporal components of the motor patterns expressed by the human spinal cord reflect foot kinematics. J. Neurophysiol **90**(5), 3555–65 (2003)
79. Ivanenko, Y.P., Poppele, R.E., Lacquaniti, F.: Five basic muscle activation patterns account for muscle activity during human locomotion. J. Physiol. **556**(Pt 1), 267–282 (2004)
80. Ivanenko, Y.P., Cappellini, G., Dominici, N., Poppele, R.E., Lacquaniti, F.: Coordination of locomotion with voluntary movements in humans. J. Neurosci. **25**(31), 7238–7253 (2005)
81. Izawa, J., Rane, T., Donchin, O., Shadmehr, R.: Motor adaptation as a process of reoptimization. J. Neurosci. **28**(11), 2883–2891 (2008)
82. Jean, F., Berret, B.: On the Duration of Human Movement: From Self-paced to Slow/Fast Reaches up to Fitts's Law, pp. 43–65. Springer International Publishing, Cham (2017)
83. Kalman, R.: When is a linear control system optimal? ASME Transactions. J. Basic Eng. **86**, 51–60 (1964)
84. Kappen, H.J.: Optimal control theory and the linear bellman equation. In: Barber, D., Cemgil, A.T., Chiappa, S. (eds.), Bayesian Time Series Models, Cambridge University Press, pp. 363–387, Cambridge Books Online (2011)
85. Kargo, W.J., Giszter, S.F.: Individual premotor drive pulses, not time-varying synergies, are the units of adjustment for limb trajectories constructed in spinal cord. J. Neurosci. **28**(10), 2409–25 (2008)
86. Kargo, W.J., Nitz, D.A.: Early skill learning is expressed through selection and tuning of cortically represented muscle synergies. J. Neurosci. **23**(35), 11255–11269 (2003)
87. Kelso, J.S.: Dynamic patterns: the self-organization of brain and behavior. MIT press, Cambridge (1997)
88. Knill, D.C., Bondada, A., Chhabra, M.: Flexible, task-dependent use of sensory feedback to control hand movements. J. Neurosci. **31**(4), 1219–1237 (2011)
89. Krakauer, J.W., Ghazanfar, A.A., Gomez-Marin, A., MacIver, M.A., Poeppel, D.: Neuroscience needs behavior: correcting a reductionist bias. Neuron **93**(3), 480–490 (2017)
90. Kutch, J.J., Valero-Cuevas, F.J.: Challenges and new approaches to proving the existence of muscle synergies of neural origin. PLoS Comput. Biol. **8**(5), e1002,434 (2012)
91. Lacquaniti, F., Ivanenko, Y.P., Zago, M.: Patterned control of human locomotion. J. Physiol. **590**(Pt 10), 2189–2199 (2012)
92. Le Seac'h, A.B., McIntyre, J.: Multimodal reference frame for the planning of vertical arms movements. Neurosci. Lett. **423**(3), 211–215 (2007)
93. Lebedev, S., Tsui, W.H., Van Gelder, P.: Drawing movements as an outcome of the principle of least action. J. Math. Psychol. **45**(1), 43–52 (2001)
94. Li, W., Todorov, E.: Iterative linearization methods for approximately optimal control and estimation of non-linear stochastic system. Int. J. Control. **80**(9), 1439–1453 (2007)

95. Loeb, E., Giszter, S., Bizzi, P.S.E., Mussa-Ivaldi, F.: Output units of motor behavior: an experimental and modeling study. J. Cognit. Neurosci. **12**(1), 78–97 (2000)
96. Loeb, G.E.: Optimal isn't good enough. Biol. Cybern. **106**(11–12), 757–765 (2012)
97. Marr, D.: Vision: A Computational Investigation into the Human Representation and Processing of Visual Information. Henry Holt & Company, New York (1983)
98. McKay, J.L., Ting, L.H.: Optimization of muscle activity for task-level goals predicts complex changes in limb forces across biomechanical contexts. PLoS Comput Biol **8**(4), e1002,465 (2012)
99. Mehrabi, N., Razavian, R.S., Ghannadi, B., McPhee, J.: Predictive simulation of reaching moving targets using nonlinear model predictive control. Front. Comput. Neurosci. **10** (2016)
100. Mistry, M., Theodorou, E., Schaal, S., Kawato, M.: Optimal control of reaching includes kinematic constraints. J. Neurophysiol. **110**(1), 1–11 (2013)
101. Mohan, V., Morasso, P.: Passive motion paradigm: an alternative to optimal control. Front Neurorob. **5**, 4 (2011)
102. Muceli, S., Boye, A.T., d'Avella, A., Farina, D.: Identifying representative synergy matrices for describing muscular activation patterns during multidirectional reaching in the horizontal plane. J. Neurophysiol **103**(3), 1532–42 (2010)
103. Mussa-Ivaldi, F.A.: Nonlinear force fields: a distributed system of control primitives for representing and learning movements. In: Proceedings., 1997 IEEE International Symposium on Computational Intelligence in Robotics and Automation, 1997. CIRA'97., IEEE, pp. 84–90 (1997)
104. Mussa-Ivaldi, F.A., Giszter, S.F., Bizzi, E.: Linear combinations of primitives in vertebrate motor control. Proc. Natl. Acad. Sci. U S A **91**(16), 7534–7538 (1994)
105. Nashed, J.Y., Crevecoeur, F., Scott, S.H.: Influence of the behavioral goal and environmental obstacles on rapid feedback responses. J. Neurophysiol **108**(4), 999–1009 (2012)
106. Nelson, W.L.: Physical principles for economies of skilled movements. Biol. Cybern. **46**(2), 135–147 (1983)
107. Neptune, R.R., Clark, D.J., Kautz, S.A.: Modular control of human walking: a simulation study. J. Biomech. **42**(9), 1282–7 (2009)
108. Nori, F., Frezza, R.: Linear optimal control problems and quadratic cost functions estimation. In: 12th Mediterranean Conference on Control and Automation, MED'04. Kusadasi, Aydin, Turkey (2004)
109. Nori, F., Frezza, R.: A control theory approach to the analysis and synthesis of the experimentally observed motion primitives. Biol. Cybern. **93**(5), 323–342 (2005)
110. Overduin, S.A., d'Avella, A., Roh, J., Bizzi, E.: Modulation of muscle synergy recruitment in primate grasping. J. Neurosci. **28**(4), 880–92 (2008)
111. Overduin, S.A., d'Avella, A., Carmena, J.M., Bizzi, E.: Microstimulation activates a handful of muscle synergies. Neuron. **76**(6), 1071–1077 (2012)
112. Overduin, S.A., d'Avella, A., Roh, J., Carmena, J.M., Bizzi, E.: Representation of muscle synergies in the primate brain. J. Neurosci. **35**(37), 12615–12624 (2015)
113. Papaxanthis, C., Pozzo, T., Schieppati, M.: Trajectories of arm pointing movements on the sagittal plane vary with both direction and speed. Exp. Brain Res. **148**(4), 498–503 (2003)
114. Papaxanthis, C., Pozzo, T., McIntyre, J.: Kinematic and dynamic processes for the control of pointing movements in humans revealed by short-term exposure to microgravity. Neuroscience **135**(2), 371–383 (2005)
115. Pauwels, E., Henrion, D., Lasserre, J.B.: Inverse optimal control with polynomial optimization. In: 2014 IEEE 53rd Annual Conference on Decision and Control (CDC), IEEE, pp. 5581–5586 (2014)
116. Perreault, E.J., Chen, K., Trumbower, R.D., Lewis, G.: Interactions with compliant loads alter stretch reflex gains but not intermuscular coordination. J. Neurophysiol **99**(5), 2101–2113 (2008)
117. Pontryagin, L.S., Boltyanskii, V.G., Gamkrelidze, R.V., Mishchenko, E.F.: The Mathematical Theory of Optimal Processes. Pergamon Press, Oxford (1964)

118. Razavian, R.S., Mehrabi, N., McPhee, J.: A model-based approach to predict muscle synergies using optimization: application to feedback control. Front Comput. Neurosci. **9** (2015)
119. Roh, J., Cheung, V.C.K., Bizzi, E.: Modules in the brain stem and spinal cord underlying motor behaviors. J. Neurophysiol **106**(3), 1363–1378 (2011)
120. Roh, J., Rymer, W.Z., Perreault, E.J., Yoo, S.B., Beer, R.F.: Alterations in upper limb muscle synergy structure in chronic stroke survivors. J. Neurophysiol **109**(3), 768–781 (2013)
121. Roh, J., Rymer, W.Z., Beer, R.F.: Evidence for altered upper extremity muscle synergies in chronic stroke survivors with mild and moderate impairment. Front Hum. Neurosci. **9**, 6 (2015)
122. Romano, F., Del Prete, A., Mansard, N., Nori, F.: Prioritized optimal control: A hierarchical differential dynamic programming approach. In: 2015 IEEE International Conference on Robotics and Automation (ICRA), IEEE, pp. 3590–3595 (2015)
123. de Rugy, A., Loeb, G.E., Carroll, T.J.: Muscle coordination is habitual rather than optimal. J. Neurosci. **32**(21), 7384–7391 (2012)
124. Russo, M., D'Andola, M., Portone, A., Lacquaniti, F., d'Avella, A.: Dimensionality of joint torques and muscle patterns for reaching. Front Comput. Neurosci. **8**, 24 (2014)
125. Saltiel, P., Wyler-Duda, K., D'Avella, A., Tresch, M.C., Bizzi, E.: Muscle synergies encoded within the spinal cord: evidence from focal intraspinal nmda iontophoresis in the frog. J. Neurophysiol **85**(2), 605–19 (2001)
126. Santello, M., Flanders, M., Soechting, J.F.: Postural hand synergies for tool use. J. Neurosci. **18**(23), 10105–10115 (1998)
127. Schwartz, A.B.: Movement: how the brain communicates with the world. Cell **164**(6), 1122–1135 (2016)
128. Sciutti, A., Demougeot, L., Berret, B., Toma, S., Sandini, G., Papaxanthis, C., Pozzo, T.: Visual gravity influences arm movement planning. J. Neurophysiol **107**(12), 3433–3445 (2012)
129. Scott, S.H.: Optimal feedback control and the neural basis of volitional motor control. Nat. Rev. Neurosci. **5**(7), 532–546 (2004)
130. Scott, S.H.: The computational and neural basis of voluntary motor control and planning. Trends Cognitive Sci. **16**(11), 541–549 (2012)
131. Selinger, J.C., O'Connor, S.M., Wong, J.D., Donelan, J.M.: Humans can continuously optimize energetic cost during walking. Curr. Biol. **25**(18), 2452–2456 (2015)
132. Shadmehr, R.: Control of movements and temporal discounting of reward. Curr. Opin. Neurobiol. **20**(6), 726–730 (2010)
133. Shadmehr, R., Krakauer, J.W.: A computational neuroanatomy for motor control. Exp. Brain. Res. **185**(3), 359–381 (2008)
134. Shadmehr, R., Huang, H.J., Ahmed, A.A.: A representation of effort in decision-making and motor control. Curr. Biol. **26**(14), 1929–1934 (2016)
135. Sponberg, S., Daniel, T.L., Fairhall, A.L.: Dual dimensionality reduction reveals independent encoding of motor features in a muscle synergy for insect flight control. PLoS Comput. Biol. **11**(e1004), 168 (2015)
136. Steele, K.M., Tresch, M.C., Perreault, E.J.: The number and choice of muscles impact the results of muscle synergy analyses. Front Comput. Neurosci. **7** (2013)
137. Taïx, M., Tran, M.T., Souères, P., Guigon, E.: Generating human-like reaching movements with a humanoid robot: a computational approach. J. Comput. Sci. **4**(4), 269–284 (2013)
138. Tanaka, H., Krakauer, J.W., Qian, N.: An optimization principle for determining movement duration. J. Neurophysiol **95**(6), 3875–3886 (2006)
139. Terekhov, A.V., Pesin, Y.B., Niu, X., Latash, M.L., Zatsiorsky, V.M.: An analytical approach to the problem of inverse optimization with additive objective functions: an application to human prehension. J. Math. Biol. **61**(3), 423–453 (2010)
140. Thomas, J.S., Corcos, D.M., Hasan, Z.: Kinematic and kinetic constraints on arm, trunk, and leg segments in target-reaching movements. J. Neurophysiol **93**(1), 352–364 (2005)
141. Ting, L.H., Macpherson, J.M.: A limited set of muscle synergies for force control during a postural task. J. Neurophysiol **93**(1), 609–613 (2005)

142. Ting, L.H., McKay, J.L.: Neuromechanics of muscle synergies for posture and movement. Curr. Opin. Neurobiol **17**(6), 622–628 (2007)
143. Ting, L.H., Chiel, H.J., Trumbower, R.D., Allen, J.L., McKay, J.L., Hackney, M.E., Kesar, T.M.: Neuromechanical principles underlying movement modularity and their implications for rehabilitation. Neuron **86**(1), 38–54 (2015)
144. Todorov, E.: Optimality principles in sensorimotor control. Nat. Neurosci. **7**(9), 907–915 (2004)
145. Todorov, E.: Optimal control theory. In: Doya K (ed.) Bayesian Brain: Probabilistic Approaches to Neural Coding chap 12, pp. 269–298 (2006)
146. Todorov, E.: Compositionality of optimal control laws. Adv. Neural Inf. Process. Syst. **22**, 1856–1864 (2009a)
147. Todorov, E.: Efficient computation of optimal actions. Proc. Natl. Acad. Sci. U S A **106**(28), 11478–11483 (2009b)
148. Todorov, E., Jordan, M.I.: Optimal feedback control as a theory of motor coordination. Nat. Neurosci. **5**(11), 1226–1235 (2002)
149. Todorov, E., Li, W., Pan, X.: From task parameters to motor synergies: a hierarchical framework for approximately-optimal control of redundant manipulators. J. Robot Syst. **22**(11), 691–710 (2005)
150. Togo, S., Yoshioka, T., Imamizu, H.: Control strategy of hand movement depends on target redundancy. Sci. Rep. **7**(45), 722 (2017)
151. Toma, S., Sciutti, A., Papaxanthis, C., Pozzo, T.: Visuomotor adaptation to a visual rotation is gravity dependent. J. Neurophysiol **113**(6), 1885–1895 (2015)
152. Torres-Oviedo, G., Ting, L.H.: Muscle synergies characterizing human postural responses. J. Neurophysiol **98**(4), 2144–56 (2007)
153. Torres-Oviedo, G., Ting, L.H.: Subject-specific muscle synergies in human balance control are consistent across different biomechanical contexts. J. Neurophysiol **103**(6), 3084–98 (2010)
154. Torres-Oviedo, G., Macpherson, J.M., Ting, L.H.: Muscle synergy organization is robust across a variety of postural perturbations. J. Neurophysiol **96**(3), 1530–1546 (2006)
155. Tresch, M.C., Jarc, A.: The case for and against muscle synergies. Curr. Opin. Neurobiol **19**(6), 601–7 (2009)
156. Tresch, M.C., Saltiel, P., Bizzi, E.: The construction of movement by the spinal cord. Nat. Neurosci. **2**(2), 162–7 (1999)
157. Tresch, M.C., Cheung, V.C.K., d'Avella, A.: Matrix factorization algorithms for the identification of muscle synergies: evaluation on simulated and experimental data sets. J. Neurophysiol **95**(4), 2199–2212 (2006)
158. Turvey, M.T.: Coordination. Am. Psychol. **45**(8), 938 (1990)
159. Uno, Y., Kawato, M., Suzuki, R.: Formation and control of optimal trajectory in human multijoint arm movement. minimum torque-change model. Biol. Cybern. **61**(2), 89–101 (1989)
160. Valero-Cuevas, F.J., Venkadesan, M., Todorov, E.: Structured variability of muscle activations supports the minimal intervention principle of motor control. J. Neurophysiol **102**(1), 59–68 (2009)
161. Vu, V.H., Isableu. B., Berret, B.: Adaptive use of interaction torque during arm reaching movement from the optimal control viewpoint. Sci. Rep. **6** (2016a)
162. Vu, V.H., Isableu, B., Berret, B.: On the nature of motor planning variables during arm pointing movement: Compositeness and speed dependence. Neuroscience **328**, 127–146 (2016b)
163. Weiss, E.J., Flanders, M.: Muscular and postural synergies of the human hand. J. Neurophysiol **92**(1), 523–535 (2004)
164. Wolpert, D.M., Landy, M.S.: Motor control is decision-making. Curr. Opin. Neurobiol. **22**(6), 996–1003 (2012)
165. Yamamoto, S., Kushiro, K.: Direction-dependent differences in temporal kinematics for vertical prehension movements. Exp. Brain Res. **232**(2), 703–711 (2014)
166. Yandell, M.B., Zelik, K.E.: Preferred barefoot step frequency is influenced by factors beyond minimizing metabolic rate. Sci. Rep. **6** (2016)
167. Zelik, K.E., Kuo, A.D.: Mechanical work as an indirect measure of subjective costs influencing human movement. PLoS One **7**(2), e31,143 (2012)

Human Movements: Synergies, Stability, and Agility

Mark L. Latash

Abstract When people move, they organize large, abundant sets of elements (limbs, joints, digits, muscles, motor units, etc.) in a task–specific way by the central nervous system. Such organizations (synergies) ensure action stability, which is crucial given the varying internal body states and external forces. Action stability is controlled in a task-specific way. In particular, stability is reduced in a feed-forward manner (anticipatory synergy adjustment, ASA) if a person plans to perform a quick change of a salient performance variable. The importance of controlled stability for everyday movements is exemplified by studies of neurological patients who show deficits in both aspects of controlled stability: reduced stability during steady-state actions and small/delayed ASAs in preparation to a quick action. The physical approach to movement synergies has been developed using two theoretical frameworks. One of them is the idea of control with spatial referent coordinates (RCs) for salient variables. The other is the idea of intention-specific stability of redundant systems developed within the uncontrolled manifold (UCM) hypothesis. The RC and UCM hypotheses have been united into a single theory incorporating the idea of hierarchical control. This theory is able to account for results of a variety of studies that used perturbations of ongoing movements, analysis of variance across repetitive trials, and analysis of motor equivalence. Recent studies have provided links of this theory to neurophysiological structures and provide tools sensitive to impaired stability and agility of movements in patients with various neurological disorders.

1 Movements of Inanimate and Biological Systems

Movements of inanimate and living objects are strikingly different. If one knows the initial state of an inanimate object and all the forces acting on that object, its motion can be predicted with certainty based on the relevant laws of nature, such

M. L. Latash (✉)
Department of Kinesiology, The Pennsylvania State University, Rec.Hall-268N,
University Park, PA 16802, USA
e-mail: mll11@psu.edu

135
G. Venture et al. (eds.), *Biomechanics of Anthropomorphic Systems*, Springer Tracts in Advanced Robotics 124, https://doi.org/10.1007/978-3-319-93870-7_7

as the Newton Laws. This is not true for an inanimate object. Indeed, imagine an animal (for example, a mouse) and a toy, which is a perfect replica of that animal. If you place the toy on a table, walk out of the room, and then return, the toy will most likely be waiting for you exactly in the place where you left it (let us assume that no earthquake took place and no person walked into the room and took the toy). If you repeat this experiment with an animal, it will most likely be gone by the time you return to the room. Why do living objects, but not inanimate objects, frequently run uphill, swim against the current, and fly against the wind?

There are no reasons to suspect that living objects can violate basic laws of nature (physical laws). However, their behavior is not predictable based on those laws, at least at our current level of knowledge. If we do not invoke supernatural factors, this means that there are other, biologically specific, laws of nature that we may be unaware of at this time. Such biologically specific physical laws (BioPL) are likely based on chains of complex simpler interactions that ultimately can be traced down to basic physical laws common for all objects.

A physical law is typically expressed as an equation involving variables that are constrained by the law and parameters. For example, the Newton Second Law, $\mathbf{F} = m\mathbf{a}$, involves two variables that describe the object of interest, $\mathbf{F}$—force and $\mathbf{a}$—acceleration, and one parameter, m—mass (Fig. 1). A change in one of the two variables causes a change in the other variable. It is typically assumed that parameters in equations reflecting basic physical laws stay constant or change slowly as compared to changes in variables. To induce motion of an inanimate system from one state to another, one typically changes variables (such as force) thus causing changes in other variables constrained by physical laws, including the object's coordinates and their derivatives.

It is possible to induce changes in motion of an inanimate object by changing its parameters. Imagine a pendulum consisting of a weightless rope and a point mass at its end in the field of gravity. Motion of the pendulum will depend on several parameters such as length of the rod, L, and coordinates of its point of suspension, x_0 (Fig. 1). Imagine now that you can change these parameters, for example by changing the length of the rope or by grabbing the free end of the rope (suspension point) and moving it in space. Trajectory of the point mass will change with such manipulations. We will address this type of control as *parametric control*.

Fig. 1 Illustrations of variables and parameters of physical laws. Left: The Second Newton Law: $\mathbf{F} = m\mathbf{a}$. Right: Motion of a pendulum in the field of gravity (g). Parametric control: Changing parameters of the pendulum (L and x_0) leads to changes in its motion

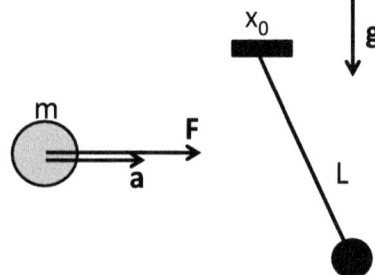

Now we are ready to introduce a definition of a living object (or of life) that could be useful to explore natural laws that define motion of such systems. Note that most definitions of life available in encyclopedias, Internet, and other resources enumerate distinguishing features of living systems such as homeostasis, metabolism, growth, reproduction, and a few others, instead of offering a concise definition. The definition accepted in this paper is based on the idea of parametric control:

A living system is a system able to: (1) unite basic physical laws into chains and clusters leading to new stable and pervasive relations among physical variables (addressed as BioPLs) and involving new parameters; and (2) modify these parameters to induce motion.

To illustrate this definition, consider the neural control of a single muscle (Fig. 2). Stretching a muscle from a short length leads to a small increase in its force (shown as a dashed line in Fig. 2) due to mechanical properties of the tissues. At some value of muscle length, the muscle shows first signs of electrical activation (λ in Fig. 2a); this length value is addressed as threshold of the stretch reflex [15, 17, 34]. Further stretch leads to a much steeper increase in muscle force (the solid curve to the right of λ); this is accompanied by an increase in the muscle activation level. Note that muscle behavior illustrated in Fig. 2 is seen in experiments when the animal cannot voluntarily change muscle activation (e.g., [18, 38]). Changes in descending signals from the brain lead to a nearly parallel translation of the F(L) curve along the length axis. Such changes can be accompanied by changes in muscle force, length, or both, depending on the external load. For example, the same change in λ, from λ_1 to λ_2 in Fig. 2b) would lead to a change in force in isometric conditions (length is fixed) and to a change in length in isotonic conditions (external load does not change).

Different functional forms can been used to describe the F(L) relation illustrated in Fig. 2. For example, $F = k_0 \exp(L - \lambda)$ when $L > \lambda$ and $F = 0$ for $L \leq \lambda$ [16]. This equation links two salient variables, F and L, with the help of two parameters, k_0 and λ. A number of experiments on both animals and humans have provided evidence that changes in descending signals can only change λ for a muscle but not k_0 [15, 18, 38]. So, the neural control of a single muscle can be described with changes in one parameter (λ) of a BioPL that links two salient state variables, force and length. Note that this BioPL is based on numerous processes that contribute to the shape of the F(L) dependence. These include processes of muscle deformation, reaction of spindle (and other) receptors to the deformation, generation of action potentials by the spindle sensory endings, transmission of those potentials along the afferent axons into the spinal cord, synaptic transmission, activation of alpha-motoneurons, conduction of action potentials along the axons to the muscle, neuromuscular transmission, excitation-contraction coupling, force generation, and muscle fiber contraction. In fact, each of the mentioned steps is, by itself, based on chains and clusters of more elemental processes, down to the basic physical laws.

The idea of parametric control can be generalized for whole-body actions. Imagine a hungry donkey (Fig. 3). If you place a carrot on a stick in front of the donkey's head, the carrot will define a target coordinate in the external space to which the donkey's head is attracted. It may be viewed as a referent coordinate (RC) for the head, and the donkey will try to move the head to that RC. When the donkey was

Fig. 2 **a** Stretching a muscle leads to a small increase in its force (the dashed line) until the threshold of the stretch reflex (λ). When active muscle force equilibrates the external load, the system achieves an equilibrium point (EP) characterized with a combination of force and length, $\{L_{EP}, F_{EP}\}$. **b** Changing λ (e.g., from λ_1 to λ_2) can lead to different mechanical consequences depending on the external load, e.g., a change in muscle force in isometric conditions and a change in muscle length in isotonic conditions

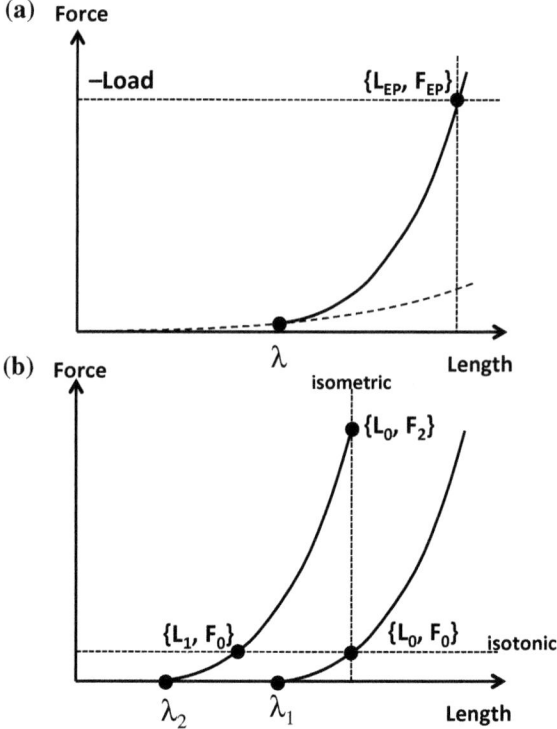

a little baby-donkey, it learned that to move the head to a target it had to move the legs. So, the carrot for the head is transformed to milli-carrots for the legs that have to follow trajectories compatible with the desired head motion. To move a leg, one has to move its joint. So, trajectories of each milli-carrot have to be transformed into trajectories of joint-specific micro-carrots. To move a joint, one has to activate muscles. So, motion of each micro-carrot produces trajectories of nano-carrots for individual muscles. As we already know from Fig. 2, nano-carrots are λs for the muscles.

The transformation from the task-specific carrot (RC) to RCs at lower hierarchical levels is illustrated in Fig. 3b. Note that, at each step of this transformation, a relatively low-dimensional set of RCs leads to a relatively higher-dimensional set. So, the involved transformations are few-to-many, which is commonly addressed as redundant. In the next part, I am going to suggest a better term for such transformations in living systems.

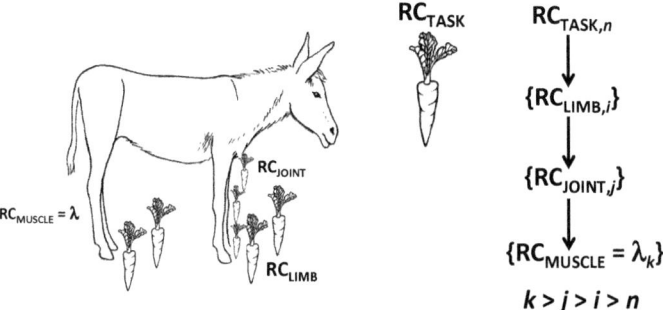

Fig. 3 Spatial location of a carrot defines a referent coordinate (RC_{TASK}) for the hungry donkey. To move to the carrot, the donkey's brain has to transform RC_{TASK} into time functions of referent coordinates at hierarchically lower levels, RC_{LIMBS}, RC_{JOINTS} and $RC_{MUSCLES}$ (which are identical to λs). These transformations are typically redundant

2   Redundancy and Abundance in Movements

State redundancy is a distinguishing feature of the design of the human body (reviewed in [32]). This term implies that any motor task can be accomplished with varying means because of the involvement of multiple elements (joints, digits, muscles, etc.). For example, there is an infinite number of joint configurations that can be used to place the fingertip to a point in space, there is an infinite number of individual finger forces that can be used to produce a desired net force on a hand-held object, and there is an infinite number of muscle force combinations to produce a desired torque in a joint. In other words, the central nervous system seems to be confronted with a computational problem: How to select a single, specific combination of elemental variables (those produced by elements) to match a desired common output specified by the task.

There is a related problem of trajectory redundancy. Indeed, even if one considers a single element that has to reach a desired value of its output over a certain time interval, there are an infinite number of trajectories leading to the desired state. Here we focus on the state redundancy only.

The notion of motor redundancy was central to the theory of the neural control of movement developed by Bernstein [7, 8]. In particular, he coined the phrase: *The main problem of motor control is the elimination of redundant degrees of freedom.* This formulation implies that, somehow, apparently redundant degrees-of-freedom are eliminated and a single solution is chosen and implemented. This view led to ideas of additional neural constraints and optimization approaches to problems of motor coordination (reviewed in Prilutsky and Zatsisorsky [48, 63].

An alternative view on the apparently redundant design of the human body has been developed as the principle of abundance. According to this principle, apparently redundant degrees of freedom are not eliminated but always used to ensure stable behavior with respect to salient, task-specific variables. This idea has been

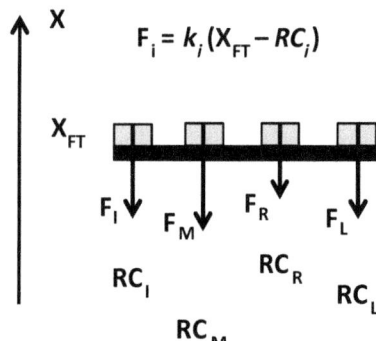

Fig. 4 To produce a desired level of total force while pressing with four fingers along a coordinate X one has to transform RC_{TASK} (equivalent to a combination of apparent stiffness k and referent coordinate RC_i) into a set of RC for the individual fingers. This is an abundant mapping

supported by many experiments showing relatively large inter-trial variability at the level of elements in combination with relatively low variability of salient performance variables (reviewed in [26]). Arguably, the first, most influential, experiment was run by Bernstein nearly 100 years ago on professional blacksmiths [5]. In that study, participants performed series of their professional labor movements—hitting the chisel with the hammer—and inter-trial variability was estimated at the level of joint rotations and at the level of hammer motion. The hammer trajectory was much less variable suggesting that joint rotations formed a synergy stabilizing the trajectory of the hammer.

The idea of abundance can also be illustrated using the notion of RCs introduced earlier. Imagine the task of pressing with the four fingers of a hand in isometric conditions along a coordinate X and reaching a certain total force level (Fig. 4). At the level of individual finger forces, high inter-trial variability is expected compared to the total force variability; this prediction has been confirmed in experiments (e.g., [45, 60]). In a linear approximation, force produced by a finger i (F_i) can be expressed as: $F_i = k_i \cdot (X_{FT} - RC_i)$, where X_{FT} is the fixed coordinate of the fingertip and k_i is a coefficient of apparent stiffness [31]. At this level of description, the production of even a single finger force is an abundant task because it involves two parameters, k_i and RC_i. Studies of single-finger tasks showed large inter-trial variability of both k_i and RC_i with strong co-variation leading to a low level of variability in F_i [1]. At the level of four fingers, the task involves eight parameters (four pairs of k_i and RC_i) that can co-vary to stabilize total force.

3 Stability of Biological Movement

Stability is a crucial feature of all functional biological movements. Indeed, as emphasized by Bernstein over 80 years ago [6], it is impossible to predict variations in external forces and intrinsic body states during natural behaviors. For example, stepping on a pebble or on an uneven surface during walking leads to an immediate effect on

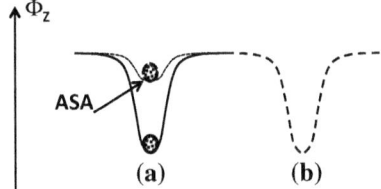

Fig. 5 Imagine an object in a potential field in a state of equilibrium (**a**). High stability corresponds to a deep potential well with steep slopes. To facilitate a quick action to a new location (**b**), it is advantageous to attenuate stability in preparation to the action: The object's location does not change but it is in a shallow potential well (dashed)

the movement kinematics resulting in immediate changes in muscle forces, which are length- and velocity-dependent. Changes in the state of spinal neurons come at a very short time delay (on the order of 15 ms) and affect muscle activations after about 30–40 ms. These time delays are much shorter than the shortest reaction time (≥ 100 ms). So, to avoid movement disruption at each and every step, the action has to be dynamically stable. This means that the system has a natural ability to return close to its original trajectory following a small, transient perturbation.

An important notion of *task-specific stability* of biological movements was introduced by Schöner [58]. It implies an ability of an abundant system to ensure dynamic stability with respect to different salient performance variables in a task-specific way. If one considers a multi-dimensional, abundant, space of elemental variables, different directions in that space are expected to show different stability and this structure of stability is expected to be specific for salient performance variables. For example, imagine walking down the hallway with a mug of coffee in the hand. Keeping the mug vertical is an important task-specific component, and joint rotations of the arm (and trunk) have to be coordinated to avoid spilling the coffee. If you step on an uneven surface or another person unexpectedly pushes your elbow, the joints are expected to yield in directions that keep the mug orientation relatively unaffected. Now imagine the same situation but with a hand-held object of about the same mass but no particular requirements to the object's orientation (e.g., a tennis ball). The same unexpected perturbation would likely lead to a different pattern of joint rotations not related to the orientation of the object.

Stability is important with respect to variables that have to be kept within a particular small range (such as the mug orientation in the former example). If one wants to change a variable quickly, high stability may become undesirable. Consider a system in a potential field organized along a task-specific coordinate Z (Fig. 5). If the system resides in a deep potential well with steep walls, it will be in a stable state (point A in Fig. 5). To move a system to another coordinate (point B), the potential field has to be changed with a new well at the new, desired location. Keeping in mind that changes in biological systems are never instantaneous, stability in point A may have to be attenuated in preparation to a quick action. This can be done by reducing the depth of the well and/or making its walls less steep (see the dashed well).

Consider the fact that all airplanes have a vertical tail fin (rudder), which helps the pilot to stabilize the plane, while no birds have a vertical tail fin. This may be seen as a reflection of the ambiguous role of stability in biological motion: Too much stability is detrimental for agility (maneuverability), which may have high evolutionary value. This issue will be discussed in a later section with an emphasis on the amazing ability of biological systems to adjust stability of specific performance variables depending on planned actions. In other words, we have retractable invisible rudders.

4 Synergies and the Uncontrolled Manifold Concept

Unfortunately, there is much jargon in the field of movement studies. A typical example is the term *synergy* that has been used in at least three different meanings. In clinical literature, this word has a strong negative connotation meaning uncontrolled stereotypical patterns of joint movement and/or muscle activation that interfere with purposeful movements [9, 12]. Such movements are commonly seen in stroke survivors. Another, broadly used, meaning of *synergy* is a group of variables showing parallel changes in time or with changes in task parameters. Such groups have been identified in spaces of kinematic, kinetic, and muscle activation variables [11, 20, 25, 53]. Organizing variables into groups has been seen as a step toward solving the problem of motor redundancy. We prefer to use the word *synergy* within the principle of abundance for neural mechanisms (mostly unknown) that ensure task-specific stability of salient performance variables. This definition explicitly identifies the purpose of synergies as ensuring stability of actions.

The last definition makes synergies directly related to the concept of task-specific stability [58]. Quantitative methods of estimating synergies stabilizing specific performance variables have been developed within the uncontrolled manifold (UCM) hypothesis [57]. UCM is defined as a sub-space within an abundant space of elemental variables where a specific performance variable does not change. Hence, if the only purpose of a hypothetical neural controller is to ensure a specific value of that variable, no control is needed as long as the abundant system resides somewhere within the corresponding UCM; this is the origin of the term *uncontrolled manifold*.

Figure 6 illustrates the UCM for the task of producing a desired magnitude of total force (F_{TOT}) in a two-finger pressing task. The solution space is linear, shown by the slanted solid line. F_{TOT} stays constant within the UCM and it changes along the orthogonal to the UCM sub-space (ORT, shown with the dashed slanted line). By definition, a F_{TOT}–stabilizing synergy should lead to high stability of the system in ORT and relatively low stability in UCM. This is illustrated in Fig. 6 with two potential fields, a shallow one along the UCM and a steep one along ORT.

Two methods have been developed to estimate relative stability along UCM and ORT. The first method is based on the following idea. Imagine that a person is performing the same task several times. Although the task is "the same", the initial conditions can never be perfectly matched. The difference in the initial conditions is expected to lead to diverging trajectories along unstable directions and converging

Fig. 6 An illustration of a task of pressing with two fingers and producing a desired magnitude of total force (F_{TOT}). The solution space (UCM) is shown by the slanted solid line. F_{TOT} changes along the orthogonal to the UCM sub-space (ORT, the dashed slanted line). A F_{TOT}–stabilizing synergy is characterized by a shallow potential field (Φ_{UCM}) along the UCM and a steep one along ORT (Φ_{ORT})

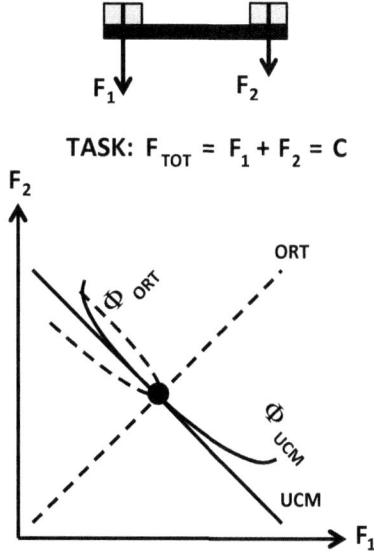

TASK: $F_{TOT} = F_1 + F_2 = C$

trajectories along stable directions. So, if one aligns the trials and then computes inter-trial variance at a selected phase of the action, higher variance is expected along less stable directions (along UCM) and lower variance is expected along more stable directions (along ORT). Note that this analysis is always performed with respect to a specific performance variable to which all elemental variables contribute, i.e. to a specific UCM. If the UCM is non-linear, for example during multi-joint movements, it can be locally linearized assuming relatively small inter-trial data scatter at a selected action phase. Then, analysis is performed not with respect to the UCM but with respect to its linear approximation, the null-space of the corresponding Jacobian matrix, **J**. The **J** matrix contains partial derivatives of the performance variable with respect to elemental variables. For the earlier example of the two-finger pressing task, $\mathbf{J} = [1\ 1]$ over the whole range of force values. For multi-joint movements, entries of **J** are trigonometric functions of joint angles and limb segment length values. As a result, **J** changes during movements with changes in the effector configuration.

The second method is based on an idea that a quick action always represents a perturbation for the pre-existent state of a system. As a result, a quick action (or correction of a perturbed action) is expected to lead to deviations along less stable directions, i.e. within the UCM for a potentially salient variable. Such deviations have been termed motor equivalent (ME); they have been reported in a series of studies of multi-finger, multi-joint, and multi-muscle action [39–41, 54]. Note that ME motion along the UCM is, by definition, wasteful if the purpose is to change the related performance variable. Hence, large magnitudes of motion along the UCM reported in the cited studies suggest that quick actions and reactions are not programmed in some optimal way but represent behaviors of a physical system defined by natural laws. The amount of ME motion can also be used as a proxy of stability. Note that non-

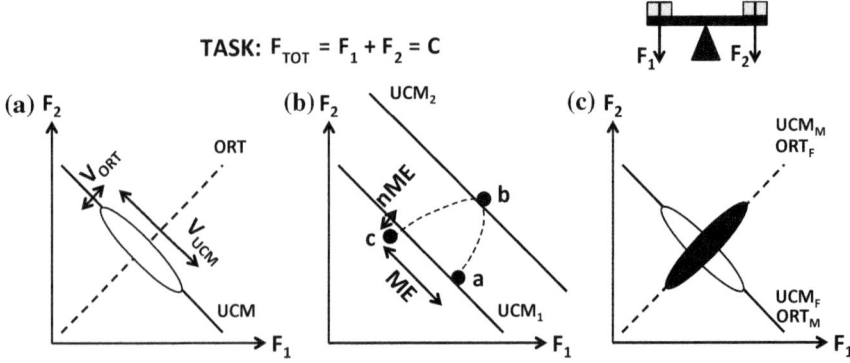

Fig. 7 The task of pressing with two fingers and producing a desired magnitude of total force (F_{TOT}). **a** Inter-trial cloud of data points shows large variance along the solution space (V_{UCM}) compared to variance in the orthogonal to the UCM space (V_{ORT}). **b** Motion to a higher force magnitude (point b) and back to the previous force magnitude (point c) is accompanied by a large motor equivalent motion along the UCM (ME) compared to the small motion along ORT (nME). **c** If the same task is performed in conditions of balancing with respect to a pivot, the cloud of data points rotates and becomes elongated along the UCM for the total moment of force (UCM_M), which is parallel to ORT for the total force (ORT_F)

motor equivalent, nME, motion (orthogonal to the UCM) is strongly task-dependent. For example, a movement from an initial state to a target and back is expected to lead to small nME motion, while a movement to a target without the return segment leads to large nME motion.

The two methods are illustrated in Fig. 7, which uses the example of two-finger accurate force production. If a person performs this task multiple times, variance along the UCM (V_{UCM}) is expected to be larger than variance along ORT (V_{ORT}). Note that the inequality $V_{UCM} > V_{ORT}$ has to be considered after proper normalization of the variance indices per dimension in both UCM and ORT. This is not a problem in Fig. 7a (both UCM and ORT are one-dimensional) but has to be done in more realistic tasks and systems. Note that that the inequality may or may not be true independently of the overall amount of inter-trial variance. In other words, a person may show highly accurate, stereotypical performance with $V_{UCM} = V_{ORT}$, which may be interpreted as no specific synergy stabilizing F_{TOT}, while another person may show a sloppier performance (higher V_{ORT}) but with a F_{TOT}-stabilizing synergy.

If the person is asked to increase F_{TOT} quickly to a new target level (point b in Fig. 7b) from an initial state (point a, in Fig. 7b), deviations both along UCM and along ORT are expected. The nME deviation along ORT is defined by the location of the target (its distance from the initial state), while the ME deviation along UCM is free to be of any magnitude. If the subject returns close to the initial F_{TOT} value (point c), the nME deviation is expected to be small, while the ME deviation can be large reflecting the lower stability along the UCM.

Since UCM is, by definition, specific to a particular performance variable, changing the task (or its interpretation by the actor) may lead to a different UCM. For example, if the F_{TOT}-production task is performed with an additional requirement to balance the forces about a pivot (as shown in the insert in Fig. 7c), stabilizing the total moment of force (M_{TOT}) may become more important than stabilizing F_{TOT}. Note that the UCM for M_{TOT} corresponds to positive finger force co-variation across trials and is orthogonal to the UCM for F_{TOT}, which requires negative finger force co-variation. As a result, this task has conflicting requirements. Indeed, most subjects in such tasks stabilize M_{TOT} at the expense of failing to stabilize F_{TOT} and show ellipses of data points elongated along the UCM for M_{TOT} [28]. This is an illustration of task-specific stability, which has also been shown for multi-joint reaching tasks and multi-muscle whole-body tasks [24, 56].

5 Feed-Forward Control of Stability

Animals frequently behave in a predictive fashion. For example, the fox does not run to the point where it sees a rabbit but to a point where it plans to intercept the rabbit. Predictive, feed-forward control is common during everyday human movements. Most frequently, it has been described as muscle activations with the purpose to accommodate anticipated future forces from the environment or from other parts of the body. For example, when a person prepares to lift a hand-held object, the gripping force shows an increase in anticipation of the motion to ensure adequate friction between the fingertips and the contact surface [23]. A quick motion of one of the joints of a limb is accompanied by activation of muscles acting at other joints with the purpose to counteract the expected change in expected, motion-dependent, interaction torques (e.g., [3, 52]). When a standing person makes a fast action (for example, extends the arms quickly), first changes in muscle activation levels are seen in the muscles of the leg and trunk [4], not in the muscles producing the intended action. These anticipatory postural adjustments (APAs, reviewed in [37]) generate forces that help to avoid postural perturbations expected from the intended action. In all these examples, feed-forward changes in muscle activations emerge during early development and show major disruptions in patients with various neurological disorders.

There are less specific examples of feed-forward control. For example, when a person performs an action during which an unexpected event, e.g., a force perturbation, may occur, frequently increased muscle co-contraction is seen leading to stronger peripheral muscle reactions to any perturbations. This is an example of the so-called *preflexes* [35], peripheral reactions of muscles to length changes that are not mediated by the central nervous system but can be modulated in advance by changing the background muscle activation level.

Another, subtle, example of feed-forward control involves changes in stability of an ongoing action. As illustrated earlier with the plane vs. bird example (also in Fig. 5), high stability may not always be desirable. In particular, it may be counterproductive in tasks that require a quick change in the salient performance variable. Humans possess an ability to adjust stability of salient variables in preparation to

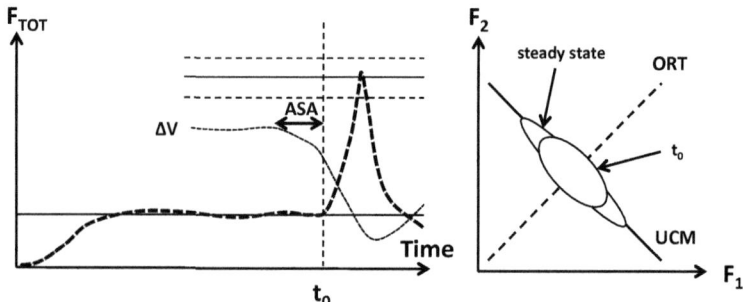

Fig. 8 An illustration of anticipatory synergy adjustment (ASA) in the task of producing a constant force level with two fingers followed by a quick force pulse. The index of synergy, ΔV (normalized difference between V_{UCM} and V_{ORT}) shows a drop about 300 ms prior to the force pulse initiation (t_0). In the force-force plane, this is reflected in a less elliptical data distribution (right panel)

quick actions, i.e., in a feed-forward way. Such anticipatory synergy adjustments (ASAs, [42]) are seen in young healthy persons as early as 300 ms prior to the action initiation. Using the previous example of two-finger force production, imagine that the subject in this experiment is asked to reach a level of F_{TOT} with visual feedback on F_{TOT} and then, in a self-paced manner, produce a very quick force pulse into a target (Fig. 8). This task is performed several times and all the trials are aligned by the force pulse initiation (t_0 in Fig. 8). During the steady-state phase of the task, strong synergies are expected stabilizing F_{TOT}. They are reflected in the structure of inter-trial variance: $V_{UCM} \gg V_{ORT}$. One can use the normalized difference between V_{UCM} and V_{ORT} as a metric of synergy, ΔV, which is illustrated in Fig. 8 with the thin dashed line. This index is expected to be positive during the steady-state phase of the task.

Starting about 300 ms prior to t_0, one can see a drop in ΔV, which may be associated with a drop in V_{UCM}, an increase in V_{ORT}, or both. This drop is the ASA. It is illustrated in Fig. 8 using ellipses of data distribution, which become less and less elongated along the UCM as the time of force pulse initiation approaches. The function of ASAs is to destabilize the variable that the person plans to change quickly and thus facilitate its change and avoid fighting one's own synergies stabilizing that variable.

ASAs and APAs share certain common features. In particular, they become much shorter or even disappear if an action is performed under the simple reaction time instruction, i.e., initiated as quickly as possible after an imperative signal. Both ASAs and APAs are reduced in older persons and in patients with various neurological disorders. Note that APAs involve changes in performance and elemental variables that can be seen in averaged across trials patterns. In contrast, ASAs are seen only in indices of across-trial co-variation reflecting changes only in action stability; as a result, no changes can be seen in averaged across trials patterns. Also, ASAs start much earlier, about 300 ms prior to action initiation, compared to APAs, which typically begin about 100 ms prior to action initiation. These differences suggest that neural mechanisms of APA and ASA are different. The similarities suggest, however, that the two types of adjustments may share neural mechanisms common across different feed-forward processes.

6   Models of Controlled Stability

So far, two approaches have been used to interpret the larger inter-trial variance along the UCM compared to ORT. The first is based on the idea of optimal feedback control [61–63]. It assumes that the action is organized to minimize a cost function, which involves two summands. The first reflects error in the salient performance variable, while the other reflects variance in effort, which is associated with variance in the space of elemental variables. Minimizing this cost function over action time leads to compensation of errors in the performance variable (i.e., deviations along ORT) without correcting deviations along UCM since such actions would not reduce the error-related component of the cost function while increasing the effort-related component. This model has been able to replicate some of the important features of actions involving abundant sets of elements including the structure of variance ($V_{UCM} > V_{ORT}$).

The model based on optimal feedback control has been highly influential within the computational approaches to the neural control of movements (reviewed in [13, 49]). This model, however, does not specify variables manipulated by the central nervous system and thus has unclear links to the neural processes associated with the control of multi-element movements. Also, assuming computations within the object of interest is not expected to help with understanding laws of nature that define behavior of this object. This is particularly questionable with respect to computation of integrals over movement time prior to movement initiation.

A model avoiding these problems has been based on the idea of short-latency back-coupling loops with modifiable gains illustrated in Fig. 9 [29]. This model was developed for a four-element system (such as four fingers pressing on individual sensors) with the task of producing a magnitude of their sum given Gaussian noise added at the level of sharing the desired output among the four elements. Figure 9 illustrates the RC at the task level shared among four neural structures ("neurons") with the outputs corresponding to RC at the level of elements. A change in RC for an element leads to a change in this element's contribution to the task. The feedback loops in the model are somewhat similar to the well-known system of Renshaw cells that are excited by branches of the axons of individual alpha-motoneurons and produce inhibitory feedback projections on all the alpha-motoneurons of the pool. Note that Fig. 9 shows back-coupling projections from one of the N1 neurons only, while such projections are assumed from all the neurons of the pool. If all the projections are negative, such a system produces inter-trials variance patterns similar to those observed in experiments with accurate four-finger total force production. Moreover, assuming realistic time delays in the system is able to replicate a non-trivial finding that when a person starts a four-finger force production task from a

Fig. 9 An illustration of the
idea of central
back-coupling. The referent
coordinate at the task level,
RC_{TASK}, is distributed
among RCs for the elements
(generated by hypothetical
neurons, N1). The RC of
each element contributes to
the task. Each N1 neuron
projects on a small neuron,
which projects back to all the
neurons of N1. Changing the
gain matrix (G) of those
projections controls stability
of the total output of the
system

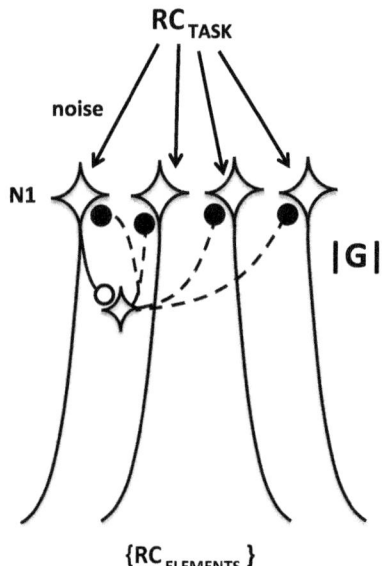

relaxed state, it takes a few tens of milliseconds for the structure of variance (V_{UCM} > V_{ORT}) to become observable [59].

Changing the gains of the back-coupling projections in the scheme in Fig. 9 can lead to stabilization of different variables or more than one variable at a time. For example, turning a few gains positive allows stabilizing both total force and total moment of force produced by the four fingers (as in [28, 55]). Hence, the scheme implies two groups of variables specified at a higher, task-related, level. One of them defines the desired level of performance (similar to the RC_{TASK} in Fig. 3), while the other defines the gain matrix, **G**, of the back-coupling loops. Changing RC_{TASK} without a change in **G** allows changing performance without a change in synergies stabilizing salient performance variables (e.g., [10]). Changing **G** without a change in RC_{TASK} is expected to lead to a change in the stability properties without a change in performance, as in the aforementioned phenomenon of ASA.

A similar back-coupling idea has been used in a scheme linking the ideas of performance-stabilizing synergies and the control with RCs [36]. Within that scheme, back-coupling loops from peripheral receptors have been added to allow linking inter-trial structure of variance to natural feedback from somatosensors and other receptors. For example, a few recent studies have shown the importance of visual feedback for multi-finger synergies, which disappear after turning the visual feedback off [46, 47].

7 Neurophysiological Mechanisms of Controlled Stability

Synergies can be organized at multiple levels of the central nervous system starting from the segmental levels in the spinal cord. For example, the aforementioned system of Renshaw cells can be seen as an example of a synergy stabilizing output of the alpha-motoneuronal pool. Indeed, if one of the active alpha-motoneurons unexpectedly stops generating action potentials, the corresponding Renshaw cell stop its inhibitory action of that alpha-motoneuron and other motoneurons of the pool. This contributes to the overall increase in the pool activation (increasing the frequency of firing of recruited neurons and/or recruiting new ones) thus compensating partly for the original error. Figure 10a illustrates schematically a pool of alpha-motoneurons projecting on a muscle with one Renshaw cell (RC) and its inhibitory projections onto each of the alpha-motoneurons. Note the similarity of this scheme and the one in Fig. 9.

Another low-level example of a synergy is the action of stretch reflex, which acts to stabilize the equilibrium state of the muscle acting against an external load (Fig. 10b). This scheme assumes that the input to a pool of motoneurons can be represented as a change in threshold of the stretch reflex (λ). Further, the output of each motoneuron leads to changes in activation of the corresponding motor unit and

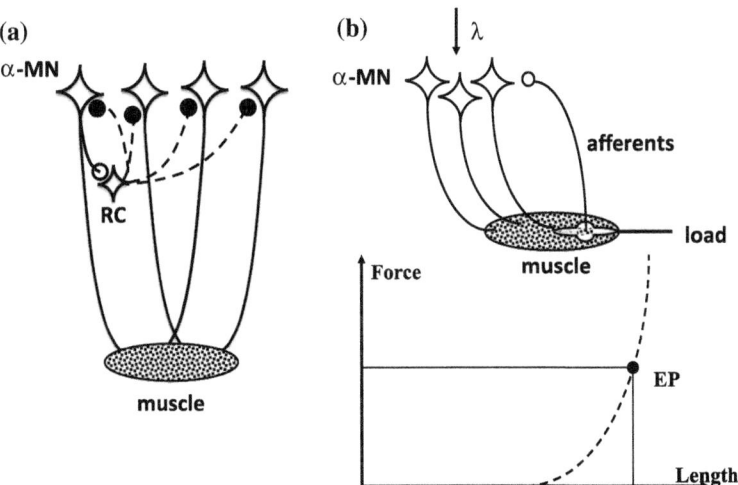

Fig. 10 Two illustrations of synergies at the segmental spinal level. **a** The system of Renshaw cells stabilizes the output of the alpha-motoneuronal pool. Each motoneuron of the pool contributes to muscle action and also excites its Renshaw cell, which has inhibitory projections to each of the motoneurons of the pool. **b** The action of the stretch reflex stabilizes equilibrium state (EP) of the system "muscle plus reflex loops plus external load". This scheme assumes that the input to a pool of motoneurons can be represented as a change in the threshold of the stretch reflex (λ). Further, the output of each motoneuron leads to changes in activation of the corresponding motor unit and its mechanical contribution to muscle action

Fig. 11 The time changes in the synergy index (ΔV_Z) in the task of producing a force pulse from a steady level of total force by four fingers. Note the lower steady-state ΔV in patients with Parkinson's disease (PD) and smaller anticipatory synergy adjustments (ASA). Stroke patients show unchanged steady-state ΔV but smaller ASA

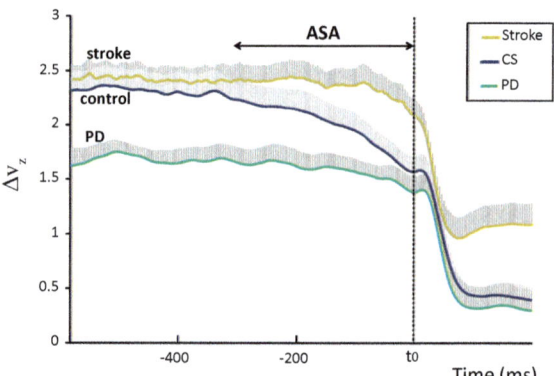

its mechanical contribution to muscle action. Indeed, if muscle force suddenly drops (e.g., a motor unit suddenly turns off), the external force becomes larger than the muscle force, the muscle is stretched, spindle sensory endings increase their level of activity and contribute to activation of the corresponding alpha-motoneuronal pool. This chain of events compensates partly for the original deviation from the equilibrium point. Overall, since most of the known reflex loops represent negative feedbacks, they contribute to stabilization of the equilibrium state of the motor system in its interaction with the environment.

Very little is known about supraspinal mechanisms of synergic control. So far, only one study [2] explored possible relations between the structure of motor variability and brain activation quantified using electroencephalography (EEG). The study documented correlations between EEG-based metrics and inter-trial variation of force sharing in a two-finger force production task.

Several important insights have come from studies of various synergies in patients with neurological disorders (reviewed in [30]). In particular, patients with Parkinson's disease show decreased indices of synergies (ΔV reflecting the normalized difference between V_{UCM} and V_{ORT}) and shorter/smaller ASAs. Figure 11 illustrates results of a study with the production of a quick force pulse starting from a steady-state force level. The figure shows clearly both main differences between the PD patients and age-matched control subjects: Note the lower ΔV in the PD group during steady-state force production and smaller/shorter APAs in preparation to the force pulse. The reduced ΔV index may be interpreted as reflecting reduced stability of actions, while smaller ASAs are signs of impaired agility. While these conclusions may seem contradictory, they match well the primary clinical signs of PD such as problems with postural stability and problems with action initiation, sometimes leading to episodes of freezing. Dopamine-replacement drugs, used commonly to treat PD, lead to an increase in both ΔV and ASA [44] indices suggesting that the original changes are indeed produced by the disease-related lack of dopamine production by the substantia nigra, one of the main nuclei of the basal ganglia.

The mentioned observations in PD patients were made using both multi-finger pressing and prehensile tasks and multi-muscle whole-body tasks. Changes in the ΔV and ASA indices are seen very early in PD. In fact, patients with clinical signs limited to only one side of the body (stage I according to [19]) show reduced ΔV and ASAs in both hands. Patients classified as having no problems with postural stability (stage II) show reduced ΔV and ASAs in postural tasks [14]. Moreover, a study of healthy professional welders, known for being at high risk for parkinsonism, has shown reduced ΔV in multi-finger tasks, and there was a significant correlation of ΔV with MRI signals in the globus pallidus, one of the main nuclei of the basal ganglia [33].

PD involves malfunctioning of one of the major circuits involving both subcortical (the basal ganglia and the thalamus) and cortical structures. So, the mentioned observations in PD patients and in persons at high risk for parkinsonism suggest crucial importance of subcortical loops for the organization of controlled stability. Other systemic disorders with involvement of subcortical structures, such as multi-system atrophy and multiple sclerosis, are also accompanied by reduced ΔV and ASA indices during multi-finger tasks [22, 43].

In contrast, a few studies of patients after a single cortical stroke resulted in conflicting findings. The first study reported no changes in the synergy index (similar to ΔV) during a multi-joint reaching task [50]. A follow-up [51] suggested that synergies after stroke may be workspace dependent and show an impairment for some tasks. A more recent study of multi-finger pressing tasks also failed to find significant differences between the stroke patients and healthy controls [21]. Its main results are illustrated in Fig. 11. Note the similar ΔV magnitudes in the stroke and control groups during the steady-state phase. There was, however, a significant drop in the ASA in the stroke participants, similar to the findings in PD. So, cortical stroke resulted in no significant impairment in action stability but a significant impairment in agility. Note that similar effects of stroke on ASAs were seen in the ipsilesional and contralesional hands; this is in contrast to typical clinical findings in stroke survivors who show much more pronounced impairment in the extremities contralateral to the injury site as compared to ipsilesional extremities that show only mild problems.

To summarize the results of clinical studies within the scheme shown in Fig. 9, cortical damage seems to have significant effects on processes involved in specification of the feedback matrix $\mathbf{G}$, but not the task-related input RC_{TASK}. In contrast, damage to loops involving subcortical structures affect the specification of both $\mathbf{G}$ and RC_{TASK}. Of course, these conclusions should be taken with a grain of salt given the sparse evidence and the involvement of both cortical and subcortical structures into the loops through the basal ganglia and the cerebellum.

8 Concluding Comments

Analysis of biological movements as consequences of laws of nature is the way to develop physics of living systems at the level of the sensory-motor function. This approach has incorporated naturally the ideas of hierarchical control, control with spatial RCs for salient variables, and task-specific stability (originally suggested within the framework of the UCM hypothesis). This theoretical approach allows accounting for results of a variety of studies with perturbations of ongoing movements, analysis of variance across repetitive trials, analysis of motor equivalence, and analysis of movements by patients with various neurological disorders. It has been applied to intentional voluntary movements and unintentional drifts in motor performance. Generalization of the main concepts has also been suggested for such functions as perception and language. While the latter applications have not been reviewed in this chapter, they have been covered in a recent review [27]. Synergies stabilizing performance can be organized at multiple levels of the central nervous system starting from the spinal segmental systems and up to the whole-brain mechanisms. Recent clinical studies have emphasized the importance of subcortical loops for motor synergies.

References

1. Ambike, S., Mattos, D., Zatsiorsky, V.M., Latash, M.L.: Unsteady steady-states: central causes of unintentional force drift. Exp. Brain Res. **234**, 3597–3611 (2016)
2. Babikian, S., Kanso, E., Kutch, J.J.: Cortical activity predicts good variation in human motor output. *Experimental Brain Research* (2017) (e-pub https://doi.org/10.1007/s00221-017-4876-9)
3. Bastian, A.J., Martin, T.A., Keating, J.G., Thach, W.T.: Cerebellar ataxia: abnormal control of interaction torques across multiple joints. J. Neurophysiol. **76**, 492–509 (1996)
4. Belenkiy, V.Y., Gurfinkel, V.S., Pal'tsev, Y.I.: Elements of control of voluntary movements. Biofizika **10**, 135–141 (1967)
5. Bernstein, N.A.: A new method of mirror cyclographie and its application towards the study of labor movements during work on a workbench. *Hygiene, Safety and Pathology of Labor*, # 5, pp. 3–9, and # 6, pp. 3–11 (1930) (in Russian)
6. Bernstein, N.A.: The problem of interrelation between coordination and localization. Arch. Biol. Sci. **38**, 1–35 (1935). (in Russian)
7. Bernstein, N.A.: On the Construction of Movements. Medgiz, Moscow (1947). (in Russian)
8. Bernstein, N.A.: The Co-ordination and Regulation of Movements. Pergamon Press, Oxford (1967)
9. Bobath, B.: Adult Hemiplegia: Evaluation and Treatment. William Heinemann, London (1978)
10. Danna-Dos-Santos, A., Slomka, K., Zatsiorsky, V.M., Latash, M.L.: Muscle modes and synergies during voluntary body sway. Exp. Brain Res. **179**, 533–550 (2007)
11. d'Avella, A., Saltiel, P., Bizzi, E.: Combinations of muscle synergies in the construction of a natural motor behavior. Nat. Neurosci. **6**, 300–308 (2003)
12. DeWald, J.P., Pope, P.S., Given, J.D., Buchanan, T.S., Rymer, W.Z.: Abnormal muscle coactivation patterns during isometric torque generation at the elbow and shoulder in hemiparetic subjects. Brain **118**, 495–510 (1995)
13. Diedrichsen, J., Shadmehr, R., Ivry, R.B.: The coordination of movement: optimal feedback control and beyond. Trends Cogn. Sci. **14**, 31–39 (2010)
14. Falaki, A., Huang, X., Lewis, M.M., Latash, M.L.: Impaired synergic control of posture in Parkinson's patients without postural instability. Gait Posture **44**, 209–215 (2016)
15. Feldman, A.G.: Functional tuning of the nervous system with control of movement or maintenance of a steady posture. II Controllable parameters of the muscle. Biophysics **11**, 565–578 (1966)

16. Feldman, A.G.: Central and Reflex Mechanisms of Motor Control. Nauka, Moscow (in Russian) (1979)
17. Feldman, A.G.: Once more on the equilibrium-point hypothesis (λ-model) for motor control. J. Mot. Behav. **18**, 17–54 (1986)
18. Feldman, A.G., Orlovsky, G.N.: The influence of different descending systems on the tonic stretch reflex in the cat. Exp. Neurol. **37**, 481–494 (1972)
19. Hoehn, M., Yahr, M.: Parkinsonism: onset, progression and mortality. Neurology **17**, 427–442 (1967)
20. Jerde, T.E., Soechting, J.F., Flanders, M.: Coarticulation in fluent fingerspelling. J. Neurosci. **23**, 2383–2393 (2003)
21. Jo, H.J., Maenza, C., Good, D.C., Huang, X., Park, J., Sainburg, R.L., Latash, M.L.: Effects of unilateral stroke on multi-finger synergies and their feed-forward adjustments. Neuroscience **319**, 194–205 (2016)
22. Jo, H.J., Lucassen, E., Huang, X., Latash, M.L.: Changes in multi-digit synergies and their feed-forward adjustments in multiple sclerosis. J. Mot. Behav. **49**, 218–228 (2017)
23. Johansson, R.S., Westling, G.: Roles of glabrous skin receptors and sensorimotor memory in automatic control of precision grip when lifting rougher or more slippery objects. Exp. Brain Res. **56**, 550–564 (1984)
24. Klous, M., Danna-dos-Santos, A., Latash, M.L.: Multi-muscle synergies in a dual postural task: evidence for the principle of superposition. Exp. Brain Res. **202**, 457–471 (2010)
25. Krishnamoorthy, V., Goodman, S.R., Latash, M.L., Zatsiorsky, V.M.: Muscle synergies during shifts of the center of pressure by standing persons: Identification of muscle modes. Biol. Cybern. **89**, 152–161 (2003)
26. Latash, M.L.: Synergy. Oxford University Press, New York (2008)
27. Latash, M.L.: Towards physics of neural processes and behavior. Neurosci. Biobehav. Rev. **69**, 136–146 (2016)
28. Latash, M.L., Scholz, J.F., Danion, F., Schöner, G.: Structure of motor variability in marginally redundant multi-finger force production tasks. Exp. Brain Res. **141**, 153–165 (2001)
29. Latash, M.L., Shim, J.K., Smilga, A.V., Zatsiorsky, V.: A central back-coupling hypothesis on the organization of motor synergies: a physical metaphor and a neural model. Biol. Cybern. **92**, 186–191 (2005)
30. Latash, M.L., Huang, X.: Neural control of movement stability: lessons from studies of neurological patients. Neuroscience **301**, 39–48 (2015)
31. Latash, M.L., Zatsiorsky, V.M.: Joint stiffness: myth or reality? Hum. Mov. Sci. **12**, 653–692 (1993)
32. Latash, M.L., Zatsiorsky, V.M.: Biomechanics and Motor Control: Defining Central Concepts. Academic Press, New York, NY (2016)
33. Lewis, M.M., Lee, E.-Y., Jo, H.J., Park, J., Latash, M.L., Huang, X.: Synergy as a new and sensitive marker of basal ganglia dysfunction: a study of asymptomatic welders. Neurotoxicology **56**, 76–85 (2016)
34. Liddell, E.G.T., Sherrington, C.S.: Reflexes in response to stretch (myotatic reflexes). Proc. R. Soc. Lond., Series B **96**, 212–242 (1924)
35. Loeb, G.E.: What might the brain know about muscles, limbs and spinal circuits? Prog. Brain Res. **123**, 405–409 (1999)
36. Martin, V., Scholz, J.P., Schöner, G.: Redundancy, self-motion, and motor control. Neural Comput. **21**, 1371–1414 (2009)
37. Massion, J.: Movement, posture and equilibrium–interaction and coordination. Prog. Neurobiol. **38**, 35–56 (1992)
38. Matthews, P.B.C.: The dependence of tension upon extension in the stretch reflex of the soleus of the decerebrate cat. J. Physiol. **47**, 521–546 (1959)
39. Mattos, D., Latash, M.L., Park, E., Kuhl, J., Scholz, J.P.: Unpredictable elbow joint perturbation during reaching results in multijoint motor equivalence. J. Neurophysiol. **106**, 1424–1436 (2011)

40. Mattos, D., Kuhl, J., Scholz, J.P., Latash, M.L.: Motor equivalence (ME) during reaching: is ME observable at the muscle level? Mot. Control **17**, 145–175 (2013)
41. Mattos, D., Schöner, G., Zatsiorsky, V.M., Latash, M.L.: Motor equivalence during accurate multi-finger force production. Exp. Brain Res. **233**, 487–502 (2015)
42. Olafsdottir, H., Yoshida, N., Zatsiorsky, V.M., Latash, M.L.: Anticipatory covariation of finger forces during self-paced and reaction time force production. Neurosci. Lett. **381**, 92–96 (2005)
43. Park, J., Jo, H.J., Lewis, M.M., Huang, X., Latash, M.L.: Effects of Parkinson's disease on optimization and structure of variance in multi-finger tasks. Exp. Brain Res. **231**, 51–63 (2013)
44. Park, J., Lewis, M.M., Huang, X., Latash, M.L.: Dopaminergic modulation of motor coordination in Parkinson's disease. Parkinsonism Relat. Disord. **20**, 64–68 (2014)
45. Park, J., Wu, Y.-H., Lewis, M.M., Huang, X., Latash, M.L.: Changes in multi-finger interaction and coordination in Parkinson's disease. J. Neurophysiol. **108**, 915–924 (2012)
46. Parsa, B., O'Shea, D.J., Zatsiorsky, V.M., Latash, M.L.: On the nature of unintentional action: a study of force/moment drifts during multi-finger tasks. J. Neurophysiol. **116**, 698–708 (2016)
47. Parsa, B., Zatsiorsky, V.M., Latash, M.L.: Optimality and stability of intentional and unintentional actions: II. Motor equivalence and structure of variance. Exp. Brain Res. **235**, 457–470 (2017)
48. Prilutsky, B.I., Zatsiorsky, V.M.: Optimization-based models of muscle coordination. Exerc. Sport Sci. Rev. **30**, 32–38 (2002)
49. Pruszynski, J.A., Scott, S.H.: Optimal feedback control and the long-latency stretch response. Exp. Brain Res. **218**, 341–359 (2012)
50. Reisman, D., Scholz, J.P.: Aspects of joint coordination are preserved during pointing in persons with post-stroke hemiparesis. Brain **126**, 2510–2527 (2003)
51. Reisman, D.S., Scholz, J.P.: Workspace location influences joint coordination during reaching in post-stroke hemiparesis. Exp. Brain Res. **170**, 265–276 (2006)
52. Sainburg, R.L., Ghilardi, M.F., Poizner, H., Ghez, C.: Control of limb dynamics in normal subjects and patients without proprioception. J. Neurophysiol. **73**, 820–835 (1995)
53. Santello, M., Bianchi, M., Gabiccini, M., Ricciardi, E., Salvietti, G., Prattichizzo, D., Ernst, M., Moscatelli, A., Jörntell, H., Kappers, A.M., Kyriakopoulos, K., Albu-Schäffer, A., Castellini, C., Bicchi, A.: Hand synergies: Integration of robotics and neuroscience for understanding the control of biological and artificial hands. Phys. Life Rev. **17**, 1–23 (2016)
54. Scholz, J.P., Schöner, G., Hsu, W.L., Jeka, J.J., Horak, F., Martin, V.: Motor equivalent control of the center of mass in response to support surface perturbations. Exp. Brain Res. **180**, 163–179 (2007)
55. Scholz, J.P., Danion, F., Latash, M.L., Schöner, G.: Understanding finger coordination through analysis of the structure of force variability. Biol. Cybern. **86**, 29–39 (2002)
56. Scholz, J.P., Reisman, D., Schoner, G.: Effects of varying task constraints on solutions to joint coordination in a sit-to-stand task. Exp. Brain Res. **141**, 485–500 (2001)
57. Scholz, J.P., Schöner, G.: The uncontrolled manifold concept: identifying control variables for a functional task. Exp. Brain Res. **126**, 289–306 (1999)
58. Schöner, G.: Recent developments and problems in human movement science and their conceptual implications. Ecol. Psychol. **8**, 291–314 (1995)
59. Shim, J.K., Latash, M.L., Zatsiorsky, V.M.: The central nervous system needs time to organize task-specific covariation of finger forces. Neurosci. Lett. **353**, 72–74 (2003)
60. Shim, J.K., Olafsdottir, H., Zatsiorsky, V.M., Latash, M.L.: The emergence and disappearance of multi-digit synergies during force production tasks. Exp. Brain Res. **164**, 260–270 (2005)
61. Todorov, E.: Optimality principles in sensorimotor control. Nat. Neurosci. **7**, 907–915 (2004)
62. Todorov, E., Jordan, M.I.: Optimal feedback control as a theory of motor coordination. Nat. Neurosci. **5**, 1226–1235 (2002)
63. Todorov, E.: Efficient computation of optimal actions. Proc. Natl. Acad. Sci. **106**, 11478–11483 (2009)

Motor Compositionality and Timing: Combined Geometrical and Optimization Approaches

Tamar Flash, Matan Karklinsky, Ronit Fuchs, Alain Berthoz, Daniel Bennequin and Yaron Meirovitch

Abstract Human movements are characterized by their invariant spatiotemporal features. The kinematic features and internal movement timing were accounted for by the mixture of geometries model using a combination of Euclidean, affine and equi-affine geometries. Each geometry defines a unique parametrization along a given curve and the net tangential velocity arises from a weighted summation of the logarithms of the geometric velocities. The model was also extended to deal with geometrical singularities forcing unique constraints on the allowed geometric mixture. Human movements were shown to optimize different costs. Specifically, hand trajectories were found to maximize motion smoothness by minimizing jerk. The minimum jerk model successfully accounted for a range of human end-effector motions including unconstrained and path-constrained trajectories. The two modeling approaches involving motion optimality and the geometries' mixture model are here further combined to form a joint model whereby specific compositions of geometries can be selected to generate an optimal behavior. The optimization serves to define the timing along a path. Additionally, new notions regarding the nature

T. Flash · M. Karklinsky · R. Fuchs
Department of CS and Applied Mathematics, Weizmann Institute of Science, Rehovot, Israel
e-mail: tamar.flash@weizmann.ac.il

M. Karklinsky
e-mail: matan.karklinsky@weizmann.ac.il

R. Fuchs
e-mail: ronit.fuchs@gmail.com

A. Berthoz
College de France, Paris, France
e-mail: alain.berthoz@gmail.com

D. Bennequin
Institut de Mathématiques de Jussieu, Paris 7, Paris, France
e-mail: bennequin@math.univ-paris-diderot.fr

Y. Meirovitch (✉)
Department of Computer Science and Artificial Intelligence Laboratory, MIT,
77 Massachusetts Ave, 02139 Cambridge, MA, USA
e-mail: yaron.mr@gmail.com

© Springer International Publishing AG, part of Springer Nature 2019
G. Venture et al. (eds.), *Biomechanics of Anthropomorphic Systems*, Springer Tracts
in Advanced Robotics 124, https://doi.org/10.1007/978-3-319-93870-7_8

155

of movement primitives used for the construction of complex movements naturally arise from the consideration of the two modelling approaches. In particular, we suggest that motion primitives may consist of affine orbits; trajectories arising from the group of full-affine transformations. Affine orbits define the movement's shape. Particular mixtures of geometries achieve the smoothest possible motions, defining timing along each orbit. Finally, affine orbits can be extracted from measured human paths, enabling movement segmentation and an affine-invariant representation of hand trajectories.

1   Introduction

1.1   Organizing Principles of Human Task Space Kinematics

Many of the fundamental ideas underlying our current understanding of human movement generation arise already when examining how humans control their hand trajectories, the hand being the end-effector of the upper limb. Despite the high dimensionality and complex mechanics inherent in any human action, the movements of a multi-degrees of freedom limb such as the upper arm, can be investigated by focusing on the motions of a single disembodied point moving through space and time [7]. This approach may seem simplistic at first. However, as has already been demonstrated by many earlier studies, fundamental questions addressing different perspectives of the problem of movement generation can be addressed and even resolved at the level of the end-effector motion.

The first issue addressed here is that of overcoming or resolving the redundancy existing at the task level; any end-effector movement can be performed in many different ways. How does the motor system select distinct trajectories when the space of all possible alternative movements is so high dimensional? Interestingly, even in the case of two dimensional hand trajectories, redundancy issues arise. Moving the tip of a pen from one point to another can be performed via an infinite-dimensional set of possible paths. The temporal aspect of movement generation introduces an additional dimension that the motor system has to deal with. Not only does the movement duration have to be selected, but also does the instantaneous movement speed. The selection or planning of a particular hand speed profile creates a specific relation between path geometry and time. Thus, a closer inspection of the task of controlling two-dimensional (2D) end-effector movements reveals the richness of possible choices. Such choices are reflected in both the geometrical features and the timing pattern chosen by the motor system when performing motor actions. When inspecting higher dimensional movements from the perspectives of joint kinematics and dynamics, the basic question of selecting a specific action out of the vast set of possible ones essentially remains the same, but is even more complicated.

To select one possible movement among the very large set of possible ones, the notion of optimality serves as a key concept. We do tend to think of human movement as being optimal, but in what sense? When examining end-effector kinematics, evidence has accumulated indicating that human movements are first and foremost kinematically optimal; that the hand trajectory, referring to both the hand path and its velocity profile, maximizes smoothness. This objective was expressed by the minimization of some integrated squared n-th order time derivative of hand position [23, 38, 51, 56]. The lowest order time derivative of position is velocity, then acceleration, then jerk, snap, etc. Even for two-dimensional hand trajectories in the horizontal plane, seemingly simple optimization cost functions yield a surprisingly rich set of possible behaviors. Free reaching movements are predicted to follow straight hand paths with single-peaked bell speed profiles [1, 42]. Obstacle avoidance or simple curved movements were successfully modeled by introducing via-points, i.e., additional points through which the hand should pass [23]. Other similar approaches were applied to model more complex behaviors, including drawing movements or path tracking movements. Thus, applying additional task constraints can redefine the optimal behavior and form more and more complex behaviors. Positions that must be passed through [23], prescribed paths [51, 56], timing requirements [54] and online trajectory corrections [21, 33, 38], may all be incorporated into the minimization of some kinematic cost. Thus, the relatively simple notion that a movement is optimal in some sense can be used to generate a diverse set of movements.

Inspecting the notion of optimization as a possible motion planning principle, however, reveals some problems. Deriving an optimal solution cannot always be carried out, especially online motion planning purposes or when time is pressing, such as in the case of required online corrections [38, 57]. When the complexity of the movement task increases, several additional difficulties arise. Computationally, it becomes harder to find an optimal action. Moreover, storing in memory all possible optimal paths and trajectories does not provide a satisfactory solution since it requires a massive memory storage capacity.

Given these difficulties, another possible solution to the motion control problem arises from another underlying notion, that of compositionality [10, 11, 2, 22, 43]. According to this notion, most movements result from the composition of elementary building blocks, i.e. motion primitives. The problem, however, is how it is possible to identify such discrete primitives from apparently continuous movements? Additionally, what exactly is meant by the term motion primitive? For instance, can a large space of different behaviors be spanned based only on the use of a smaller number of motion templates?

Consider, for example, the simplest candidates, straight strokes. These strokes are the first that come to mind given the relatively straight paths and the bell-shaped velocity profiles characterizing reaching movements [1, 42]. These straight geometric paths are traversed with stereotypical bell shaped velocity profiles. However, even in the case of simple straight hand motion primitives, the number of such possible strokes is huge. The stroke's orientation, amplitude, and duration are three free parameters. Hence, the space of possible stroke primitives is large. Is each stroke represented by a separate motor plan? The similarity of the normalized speed profiles of different strokes suggests that this option is less likely than the possibility that a motion primitive exists; namely, given a generic motor template and the specific ori-

entation, amplitude, and duration of the required movement, these are sufficient for
forming specific required strokes. Furthermore, it appears that movement durations
and amplitudes are correlated. Such coupling happens when it is required to move as
rapidly and as accurately as possible, i.e., when movement duration depends on its
amplitude and on the target's width (Fitts's law; [18]). Hence, in general, the task of
reaching between given end-points requires the specification of either three or only
two parameters out of the three possible ones.

The above observations concerning straight movement primitives should also be
considered from a different perspective. Straight strokes are inherently invariant;
irrespectively of their amplitude, duration and orientation. The normalized speed
profile of a straight stroke is bell-shaped and is roughly the same across different
movement end-point locations and durations, at least in the case of 2D movements
in the horizontal plane. In three-dimensional space the paths are less straight and
do depend on the end-point locations, which has led to the suggestion that different
motion planning strategies subserve 3D versus 2D movements [8, 9, 14].

Hence, given the above arguments, we find that invariance is another fundamental
concept in trajectory formation, which goes hand in hand with the ability of the
system to generalize a motor plan designed for a specific task in order to accomplish
similar tasks. What types of invariants characterize the spatiotemporal features of
end-effector motions? The low dimensionality of task space offers a suitable ground
for carrying a geometric analysis of movements based on geometrical symmetries and
invariance theory. Note that geometrical principles for the planning and execution of
complex movements of different body segments were recently presented in a paper
by Bennequin and Berthoz [6].

The study of the action of a transformation group operating either in the plane
or in 3D space can provide insights into the geometrical principles guiding human
motor control. Affine transformations are the point-by-point correspondences send-
ing straight lines to straight-lines; the equi-affine transformations, in addition, are
respecting a unit of area in 2D (resp. volume in 3D). Humans move the hand through
2D task space with kinematics that indicate equi-affine invariance, following a motion
regularity called the two-thirds power law [34]. This law states that the movement
speed is proportional to the end-effector path curvature, raised to the power $-1/3$, thus
specifically slowing down along the curvier sections of the path [27, 46]. Isochrony,
the tendency of movement speed to be modified such that movement durations are relatively unaffected by the movement's Euclidean extent is another invariance that can
furtherly be interpreted within the realm of full-affine invariance of motor behavior
[5].

In this chapter, we discuss a few different approaches to modeling human trajectory formation. One approach involves optimization models. Another approach
involves using geometrically based descriptions. We show how these two approaches
can be combined to address both the spatial and temporal aspects of end-effector trajectory planning. Regarding optimization models, we focus on the minimum jerk
model. We thoroughly discuss geometry–based models using non-Euclidean (in the
sense of groups, not in the sense of metrics) geometries. We first briefly review the
relation between equi-affine geometry based models and the two-thirds power law.

We then proceed to review the mixture of geometries model. This model is a unifying model that suggests that movement is generated not only based on a single geometry but on a multiplicative mixture (see below) of three geometries, namely Euclidean geometry and the two non-Euclidean ones: affine and equi-affine geometries. Thus, it was suggested that movement speed (tangential velocity) emerges from a mixture of speeds, each being constant within its associated geometry. These geometric speeds are combined, dictating the net speed and timing of the movement. The model also assumes different possible combinations of the three geometrical speeds, characterized by the different weighted contributions of the three geometries to the actual movement. The relation between the mixture weight parameters and task optimality has not yet been sufficiently investigated but here we advance the possibility that both optimization and geometric mixtures may explain the observed kinematic behavior. One possibility is that the mixture of different geometries is formed to generate the smoothest possible movement. To examine whether this indeed can be the case, movements along several exemplary paths were examined, and their velocity profiles were modeled.

We compare the properties of the mixed geometry model to simpler models; in particular, a mixed geometric parameter allows moving through inflection points without having a singularity [5]. Its free parameters (the weights of the three geometries) are constrained to follow certain mathematically simple relationships at the singularity points for the velocities not to become infinitely large. We use the mixed geometry model to account for experimental data of both hand drawing and locomotion trajectories. The free weight parameters were selected among all possible weight parameters in order to achieve the best fit of the predicted velocity profiles to the experimental data as well as sufficient constancy of the weight parameters during long enough temporal intervals [5]. Based on these calculations, several observations on the possible mixtures of geometries used to generate different trajectories are discussed. Inspecting the free parameters (weights) that account for the experimental data, we found that different alternative mixtures might result in quite similar velocity profiles, which were practically indistinguishable across different mixtures.

Continuing our interest in geometric invariance, we also examine a set of plausible candidate geometric motion primitives and describe how these primitives may emerge based on the main notions of invariance and optimization. In particular, following Meirovitch [39] we propose that these primitives could geometrically correspond to the affine orbits; geometric orbits resulting from the combined actions of the Euclidean, equi-affine and full-affine transformation groups on points in the plane. The speed profile along primitive affine orbits may satisfy both the mixed geometry and the minimum jerk models, yielding motions that are both invariant and maximally smooth. We also discuss how affine orbits can be used for the segmentation of experimentally recorded end-effector trajectories.

2 Modeling Task Space Kinematics Using Optimization Principles

2.1 The Minimum Jerk Model Defining the Smoothest Trajectory

The optimality of movement based on the minimum jerk model states that a trajectory $r(t) = (x(t), y(t))$ is optimal in the sense that the cumulative squared jerk, i.e., the squared jerk cost integrated over the entire movement duration, is minimal:

$$J = \int_0^T \left(\dddot{x}^2 + \dddot{y}^2 \right) dt$$

The minimum jerk model enables to predict how the motor system operates under different task requirements such as the generation of point-to-point reaching or obstacle-avoidance movements [19, 23].

For point-to-point movements, given some boundary conditions, the two-dimensional trajectory predicted by this jerk minimization model is such that $x(t)$ and $y(t)$ are fifth order time polynomials. In the simplest case of zero speed and acceleration at the movement start-point and end-point, the resulting trajectories are straight paths with symmetric bell-shaped velocity profiles, closely resembling stereotypical human behavior.

To model curved trajectories, the movements were assumed to start at some initial point, pass through one or several additional intermediate points (termed via-points), and end at some specified end-point. For example, using the minimum jerk model, the optimization process predicts the movement that should be generated between the initial and final positions while passing through each via-point along the way at an a priori unspecified time. The solution of this minimization problem defines the movement between each pair of consecutive points as a 5th order time-dependent polynomial, with equality position constraints obeyed by the movement segments on both sides of each via-point. The model predicts the path geometry and full kinematic profile including internal timing.

For various applications, it is sometimes more helpful to examine the jerk cost after some normalization. For instance, if we assume a movement duration T and an amplitude S to be specified, then a normalized version, the unit-less normalized jerk, can be defined $J_N = \frac{T^6}{S^2} J$ as which makes it easier to compare and examine the jerk costs across different movement shapes. Other approaches to normalize the jerk cost were based, for example, on the spectral arc-length metric [3].

For a given path, path-constrained optimization deals with the problem of finding the optimal speed profile along the prescribed path [51, 56]. The predicted speed profile, the solution of the path-constrained optimization problem, is specifying the dependency of the end-effector speed on the path shape (geometry).

The relations between the predictions arising from optimization and those resulting from equi-affine invariance, which we discuss next, were examined from different perspectives [24, 27, 39, 49, 51, 60]. It is interesting to mention that the equi-affine parametrization (i.e. the two-thirds power law) corresponds to the case where the jerk vector $(\dddot{x}(t), \dddot{y}(t))$ is proportional to the velocity vector $(\dot{x}(t), \dot{y}(t))$. Possible extensions of the minimum jerk model naturally arise if one examines the time derivatives of some order k being different from $k = 3$ which relates to the time derivative of position, i.e., jerk used in the minimum jerk model. The minimum acceleration model, with $k = 2$ was also used to model human behavior during reaching tasks [4]. The minimum acceleration cost also appears to be a good candidate for describing human locomotion [41]. The minimum snap model with $k = 4$ was used as an underlying optimization cost for the control algorithm of robotic quadcopter swarms [40] as well as for object manipulation movements [16].

A more general extension comes when the cost arising from Euclidean jerk or acceleration is replaced with a cost arising from Riemannian metrics used on the configuration manifold describing the configuration of the human arm. This approach was developed by Biess et al. [9]. In their study, the geodesics of the integrated kinetic energy cost were used to predict the optimal geometric movement paths, and the velocities along these geodesics were dictated based on the minimization of the third derivative of the Euclidean distance with respect to time when moving along the resulting optimal paths.

2.2 Invariance Achieved Through Power Laws and Isochrony

As described above, point-to-point reaching movements are both spatially and temporally invariant. Invariance in movement generation applies to more than just reaching movements. Other examples are curved and scribbling movements. In particular, we consider the two-thirds power law describing how the geometry and timing of curved human movements are related. The influence of path geometry on timing is modeled by the two-thirds power law: $\omega = C\kappa^{2/3}$, relating angular speed ω to curvature κ [34]. An equivalent formulation of the two-thirds power law is:

$$v = \gamma\kappa^{-\beta}$$

relating tangential velocity (speed) v with curvature κ, with exponent whose value is: $\beta = \frac{1}{3}$, and with γ being the piecewise constant named the velocity gain factor. This law captures the phenomenon that human movement speeds slow down during more curved segments of the trajectories.

The two-thirds power law or similar power laws were found in smooth pursuit eye movements [15], full body locomotion [29, 61], leg motions [31], speech [55] visual perception of motion [36, 59] and motor imagery [32].

Other studies have shown that a generalized form of a power law holds for shapes other than ellipses and those originally tested, and that the value of the power law exponent depends on the shape of the movement path. Maximization of smoothness through the minimum jerk model or other minimum squared derivatives models successfully predicted the power law values [51]. These predictions included the power law exponent values, based on the order of the time derivative of position used by the minimum squared derivatives model and the value of the curvature modulation frequency [30].

Another approach initially used to account for the two-thirds power law was based on a geometrical approach, specifically showing that the two-thirds power law is equivalent to the movement having a constant equi-affine speed [27, 46]. Equi-affine speed designated the time derivative of equi-affine arc-length which is mathematically defined as $d\sigma/dt$, the derivative of equi-affine arc-length distance with respect to time, where the equi-affine arc-length distance is defined as: $\sigma = \int \kappa^{1/3} ds$, s being the Euclidean arc-length distance [20, 25, 27]. Differentiating both sides of the last mathematical expression with respect to time and assuming that the equi-affine speed is constant, corresponding to the velocity gain factor $\frac{d\sigma}{dt} = \gamma$, results in the two-thirds power law [20].

The importance of the equi-affine description lies not only in enabling to express the two-thirds power law in geometrical terms but also in suggesting a geometrical framework for the description of human motion. The formulation of the relation between the two-thirds power law and constant equi-affine speed enabled to formulate the idea that a possible mathematical framework for analyzing movement similarities and invariance might involve the introduction of group theory and the moving frame method [5, 20]. The use of group theory enables to consider the movement along a given trajectory by repeatedly applying some incremental transformation on the end-effector position. Similarly, using one member of a group of transformations it is possible to transform one motion into another and to compare among different differential invariants. Mainly, the generalizations of arc-lengths and curvatures according to each geometry should remain the same (invariant), when operated on by a member of this specific group of transformations (seen as a symmetry of the geometry, which is defined by the group).

Another important variable that characterizes movements is their total duration, and a significant question in studying motor control is how the brain selects the durations for different movements. In this context, it is pertinent to describe a second important behavioral characteristic of human motion, namely global isochrony. The total durations of human movements sub-linearly depend on movement amplitudes; when two figural forms, differing only in their spatial scales, are drawn, they have roughly equal durations [60, 62]. Related temporal regularities also appear in the production of goal directed movements, such as movements constrained to pass through via-points. In this case, the durations between the movement's start and via-point and between the via-point and end-point are nearly equal, a phenomenon that was termed local isochrony [60, 62] and was successfully captured by the minimum jerk model [23].

3 Mixed Geometry as a Unifying Model of Task Space Invariance

While the equi-affine description has accounted for the two-thirds power law, it does not account for global isochrony. Moreover, no theory currently exists explaining how movement duration is selected, even if it does follow the two-thirds power law. Hence, to deal with these issues and to suggest a more comprehensive theory, the findings presented here regarding invariance of motor actions were integrated into a unifying model, the mixed geometry model [5]. This theory of movement generation is based on movement invariance with respect to three families of geometric transformations; the three classical transformation groups of full-affine, equi-affine, and Euclidean transformations.

Full-affine transformations include translations, rotations, dilatations, stretching and shearing. They do not preserve angles or distances but preserve only parallelisms of lines and their incidence. Equi-affine transformations are a subgroup of affine transformations that preserve area, and Euclidean transformations (also called rigid transformations) include translations and rotations and preserve lengths and angles. The importance of the three mixed geometries arises from their relations to the observed features of human motion. The full-affine speed is of importance because it provides a theoretical prediction of the isochrony principle; full-affine transformations preserve the affine arc-length of curved segments, and if full-affine speed is constant then it preserves the movement time across different affine transformations. Hence it predicts global isochrony, namely the maintenance of global duration [5]. The equi-affine geometry provides a theoretical formulation of the two-thirds power law by stating that the equi-affine speed of a movement is constant, which is equivalent to moving according to the two-thirds power law. The constancy of Euclidean speed is natural because the Euclidean metrics of space have a physical meaning since they correspond to the accepted notion of distance. The motor system is not fully invariant to non-Euclidean full-affine and equi-affine transformations, and it is not categorically invariant to Euclidean transformations since in many tasks motion time sub-linearly scales with extent. Hence, these three geometries must be combined through the mixture model, which allows accounting for the observed phenomena across a broad variety of movements and tasks.

The mixed geometry model states that movement properties are best represented by a mixture of the three geometries, full-affine, equi-affine, and Euclidean. Given the strong dependency of movement time and local kinematics on geometry, it is assumed that within each geometry the geometrical speed is constant. Hence, movement duration is proportional to the canonical invariant parameter within that particular geometry. We then assume that the time differential arises from the mixture of the three time differentials, each associated with its own geometry, with some fixed weights, represented by a trio of weight parameters.

A graphical way to imagine this would be of three different length differentials which are combined by the motor system using some constant weights to form a combined new length differential. With a slight abuse of notation, we denote this new mixture length by z which represents a mixture of arc lengths arising naturally from the three transformation groups. For ρ the full-affine arc length, σ the equi-affine arc-length and s the Euclidean (standard) arc length, z is some mixture of their values:

$$dz = d\rho^{\beta_0} d\sigma^{\beta_1} ds^{\beta_2}$$

The combination of the $\beta's$ coefficients form a convex sum; their sum is 1 (to be compatible with the division by the time differential dt) and they are all non-negative. We will denote the trio $\beta_0, \beta_1, \beta_2$, corresponding to the full-affine, equi-affine and Euclidean weight parameters by $\bar{\beta}$, termed the mixture trio weights parameter. The mixed geometry model goes beyond the geometric description and states that the movement speed corresponding to the time derivative of the mixed geometry length parameter is proportional to dt is constant. Given that each of the arc-lengths depends on its own curvature we obtain:

$$v_0 = C_0 \kappa^{-\frac{1}{3}} |k_1|$$
$$v_1 = C_1 \kappa^{-\frac{1}{3}} \quad ; \text{ where the total sum of the three exponents is equal to 1.}$$
$$v_2 = C_2$$

where v_0 is the Euclidean velocity under constant full-affine speed, v_1 is the Euclidean velocity under constant equi-affine speed and v_2 is constant Euclidean velocity. The Euclidean and equi-affine curvatures are marked by κ and k_1, respectively, and the different C_i-s are the constant geometrical speeds, each associated with its own geometry while the β_i-s are the corresponding weights. Using the expressions above which define each of the speeds as a function of the specific geometric curvature, the mathematical expression for the motion speed according to the mixed geometry is then:

$$v = v_0^{\beta_0} v_1^{\beta_1} v_2^{\beta_2}$$

where the three non-negative exponents sum to one.

3.1   The Geometric Singularities

The two-thirds power law has one main drawback as a generative model for motion speed along arbitrary paths. Inflections points, having zero Euclidean curvature $\kappa = 0$, are not traceable using the two-thirds power law; the zero curvature yields infinite speed when passing through an inflection point. Thus, the model is limited to the generation of movements that only wind in one direction, namely, movements that may not turn back and wind to the opposite direction (e.g., from the anti-clockwise to the clockwise directions). Augmenting the mixed geometry model with singularity analysis leads to specific mixture weights, $\bar{\beta}$, which are required to guarantee finite nonzero speeds through singularities [5]. To traverse inflection points with $\kappa = 0$, the trio $\bar{\beta}$, must satisfy the relation:

$$\beta_1 = 3\beta_0.$$

Parabolic points, defined as points of zero equi-affine curvature can be traversed with any mixing parameter that has no full-affine component:

$$\beta_0 = 0.$$

This ability of the mixed geometry model to enable travelling through singularity points suggests a new interpretation of the role these points play in forming human movement. Rather than being break points of the motor plan, as suggested, for instance, by Viviani and Cenzato [58], the singularity points are best considered as some sort of via-points; points that the system must travel through with specific constraints on its parameters [5, 39], but without stopping nor re-planning. This type of via-points, however, assumes a different constraint than the one assumed at via-points by the minimum jerk model. Such constraints go hand in hand with the idea that continuity is guaranteed when moving through some intermediate points and that the segmented appearance of movements may not necessarily imply segmented control [52]. To summarise, the geometric singularities discussed here play a different role compared to cusps and movement end-points, when it comes to human motor control.

3.2 Motion Primitives Predicted by the Mixture of Geometries Model

The two candidate geometric movement primitives discussed so far were straight and parabolic segments [20, 27, 49]. Movement primitives, however, may have additional predefined geometric shapes, which might be accompanied by a kinematic rule prescribing the speed of movement along these geometric paths.

Straight movements are known to be the default mode of executing point-to-point movements. The nearly straight paths are traveled with a bell shaped speed profile, which could result from jerk minimization [23]. Thus, they serve as natural kinematic movement primitives. Interestingly, some curved movements, e.g., in target switching tasks, may be generated from the superposition of straight kinematic motion primitives [21, 28]. Hence, rather than having a concatenation of one stroke after the other, a second movement primitive could be executed while the first one has not yet been completed.

Parabolas, which are equi-affine geodesics [27], are the next set of possible geometric movement primitives. Affine transformations can be used to generate any parabolic stroke from the canonical parabolic template, $y = x^2$, and in order to compactly form a complicated path, a few parabolic strokes can be concatenated. The kinematically defined speed along a parabola reveals an interesting principle. Handzel and Flash [20, 26, 27] have shown that moving at a constant equi-affine speed is equivalent to obeying the two-thirds power law. Following this observation, Polyakov et al. [47, 48] found that the paths of trajectories that obey the two-thirds power law, minimize jerk, and are invariant under equi-affine transformations are parabolic paths.

Interestingly, analysis of monkeys' well-practiced scribbling trajectories has revealed that they are well approximated by long parabolic strokes. Unsupervised segmentation of simultaneously recorded multiple neuron activities using a Hidden Markov model yielded discrete states which when projected on the movement data gave distinct parabolic elements [48, 49]. Moreover, based on the analysis of firing rates of motor cortical neuronal activities recorded from monkeys it was found that the firing of part of the cells is better tuned to equi-affine rather than to Euclidean speed. Thus, the evidence from neurophysiological studies supported the suggestion that parabolas are promising candidates for serving as kinematic motor primitives.

3.3 Mixture of Geometries for Describing Human Behavior

The works of Bennequin et al. [5] and Fuchs [24] included a comparison of the predictions of the mixed geometry model to measured human drawing and walking trajectories, including movements along several predesignated paths. The human paths were segmented according to the kinematic fit given by the trio of mixed geometry parameters. Bennequin et al. [5] compared the human drawings and locomotion trajectories for several shapes against the kinematic predictions of the mixed geometry model. The end-effector trajectories of these movements were segmented by fitting within each segment three constant weights; β_0, β_1, and β_2.

These weights represent the mixture of geometries, i.e. the involvement of the full-affine, equi-affine and Euclidean geometries in the produced kinematics (see Introduction and Bennequin et al. [5]. The weights, that were assumed to be piecewise constant, were derived for various figural forms (cloverleaf, limaçon and lemniscate) and modalities (drawing, locomotion) and then compared according to the distribution of the constant weights (the $\bar{\beta}$ values. Figure 1 depicts the results of fitting a mixture model and the segments that result, based on the notion that within each segment we have constant β values. Figure 1 additionally depicts comparisons between predicted and measured paths and velocity profiles for both drawing (left panel) and locomotion (right panel) trajectories. Figure 2 displays the β values derived for the drawing and locomotion movements. We also present the distribution of the β values derived for the different shapes and tasks (drawing, locomotion). These are presented by the points in the triangles, which are color-coded based on the number of trials optimally fitted by the respective values. A detailed description can be found in Bennequin et al. [5].

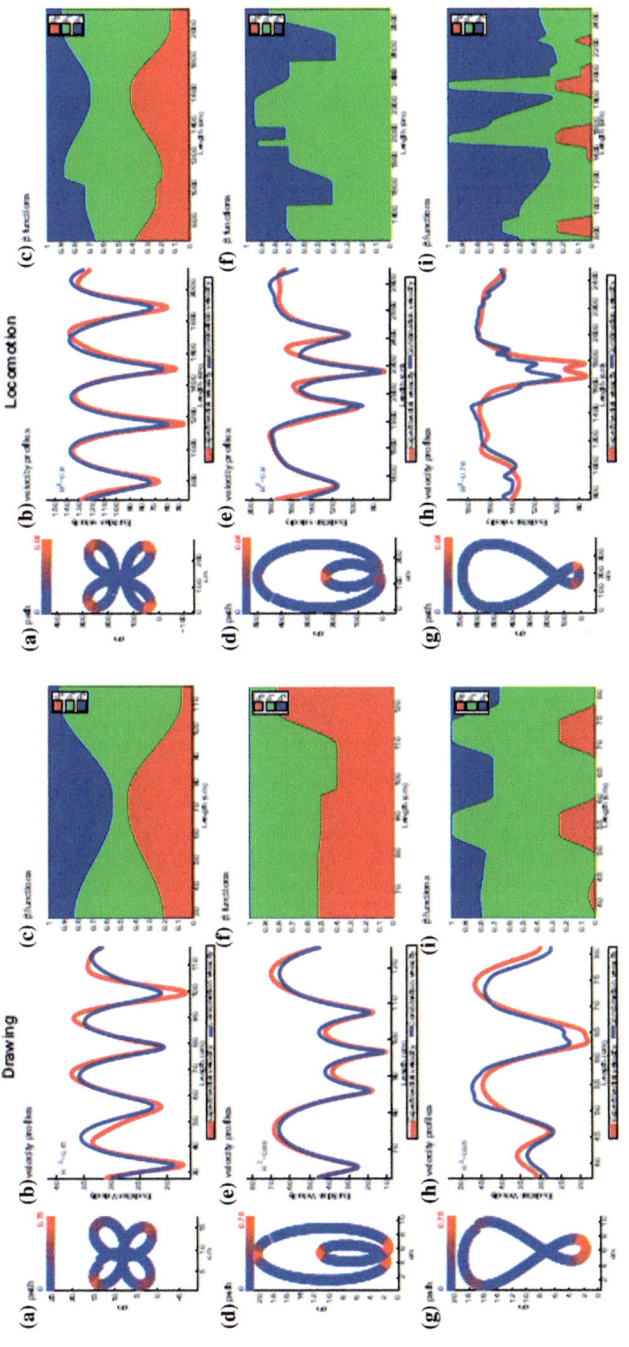

Fig. 1 Mixed geometry modeling of drawing (left panel) and locomotion (right panel) trajectories. Every row shows an example of a path of a drawing (left) and locomotion (right) trials. The shapes include a cloverleaf, an oblate limaçon and an asymmetric lemniscate. Panels (A), (D) and (G) show the paths drawn by the subject. The colors marked on the paths represent the Euclidian curvature. Blue segments have relatively low curvature (0) and red segments have a higher curvature (0.75). Color scale is shown at the top of the panel. Panels (B), (E) and (H) show the velocity profiles of the drawing (left) and locomotion (right). Red, experimental velocity profile; blue, velocity profile predicted by the model of the combination of geometries. Panels (C), (F) and (I) show values of the b functions. Red area, value of the β_1 weight; blue area, value of the β_0 weight; green area value of the β_2 weight. The values are aggregated one above the other such that their sum equals 1

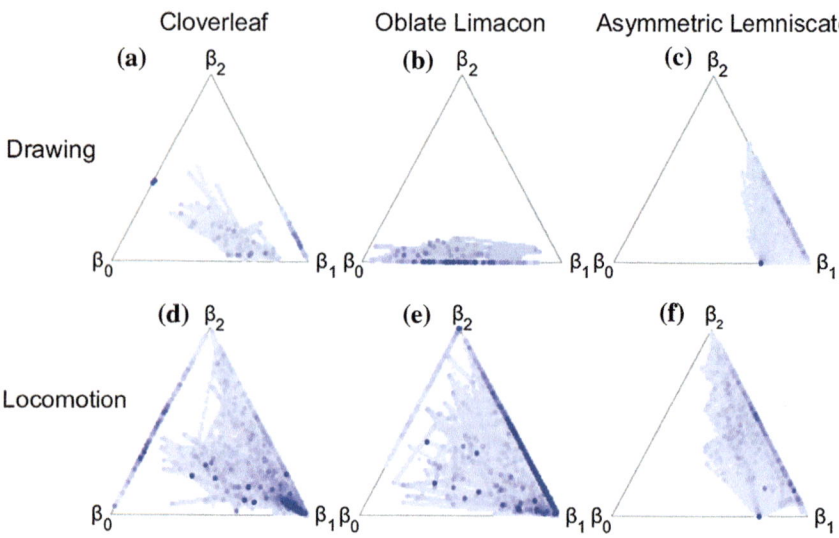

Fig. 2 Representation of the values of the three β weights during the different trials. The distribution of the β weights aggregated over all trials of the same figural form. A point within the triangle gives the values of the β_0, β_1 and β_2 weight parameters where $\beta_0 + \beta_1 + \beta_2 = 1$. The values of β_2 weight function for such a point are equal to the area delineated by the small triangle created by passing lines between this specific point and the two bottom vertices. The values of β_1 are equal to the area delineated by the small triangle created by passing lines between this specific point and the left bottom and top vertices. The values of the β_0 weight function are equal to the area delineated by the small triangle created be passing lines between this point and the right bottom and top vertices. For example, a point on the triangle's edge marked by β_1 is a point where of $\beta_1 = 1$. For a point located at the top vertex, $\beta_2 = 1$ and $\beta_0 = \beta_1 = 0$. In the center of the triangle $\beta_0 = \beta_1 = \beta_2 = \frac{1}{3}$. The color of any point within the large triangle indicates the number of times that that specific combination of β weight values was found. A white point shows a combination that did not appear in any of the trials. A dark blue point represents a combination occurring many times. *Panel (A)* contains all the trials of the drawing of cloverleaves. *Panel (B)* contains all the trials of the drawing of oblate limaçon. *Panel (C)* contains all the trials of the drawing of asymmetric lemniscate. *Panel (D)* contains all the trials of the locomotion of cloverleaves. *Panel (E)* contains all the trials of the locomotion of oblate limaçon. *Panel (F)* contains all the trials of the locomotion of asymmetric lemniscate

3.4 The Geometrical Redundancy in the Mixed Geometry Model

A reexamination of the speed profiles generated by different geometrical mixtures revealed that the mixed geometry model exhibits statistical redundancies [39]; for various paths, it was found that different values of $\bar{\beta}$ trios yield highly similar speed profiles.

For the cloverleaf template, a set of $\bar{\beta}$ trios was found to provide equally good matches ($R^2 > 0.98$) between the mixed geometry model predictions and the experimental data. All these $\bar{\beta}$ trios, which were statistically indistinguishable, obeyed

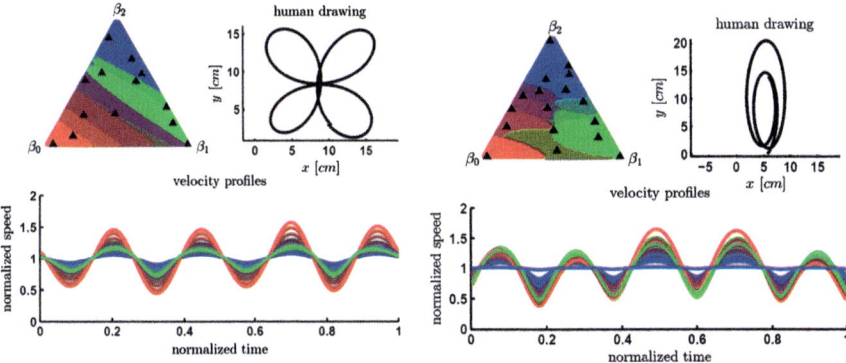

Fig. 3 Redundancy of the mixed geometry model for cloverleafs and limaçons. For each template, the mixed geometry triangle in the *top-left panel* is colored according to the statistical equivalence of parametrizations of the limaçon, the template is drawn on *the top right panel* and *the bottom panel* depicts the different speed profiles that match the different sets of speed profiles. A set of different possible mixtures following a linear relation between β_0 and β_1 yield highly similar results

linear relations between the β_0 and β_1 values, as depicted in Fig. 3. The redundancy map appearing in the upper left panel was calculated by using the following algorithm. The parameter space was quantized by obtaining a discrete set of possible β values that represent distinct, statistically distinguishable speed profiles. The speed profile corresponding to $\beta_1 = 1$ (equi-affine parametrization, or the two-third power law), was calculated and was referred to as the representative profile of the first equivalence group of parameters. Then, all $\bar{\beta}$ weight trios whose speed profiles were statistically indistinguishable from this representative profile were marked as belonging to the first group. A representative for the next equivalence group was chosen as the one giving the best agreement, in terms of R^2, with the previous representative. The process was iterated until all $\bar{\beta}$ weights were examined. Each of the groups for an analytic cloverleaf and the analytic limaçon are shown in Fig. 3, using different patches of color for different equivalent sets.

Thus, the distribution of values appearing in Fig. 2 (taken from [5]) for cloverleaf drawings can be explained by the redundancy map (Fig. 3). We suggest that the control procedure must be invariant with respect to the profiles belonging to the same equivalence class. In particular, the profiles represented in Fig. 3 are all similar from the kinematic output point of view. This suggests that humans may select a straight line in the $\bar{\beta}$ parameters space rather than a unique point. To elucidate whether the redundancy also appears for the real data, the above statistical grouping was also carried out on the actual measured paths. The same statistical tendency as detected for the analytical curves was also seen for the human data, as is shown in Fig. 4.

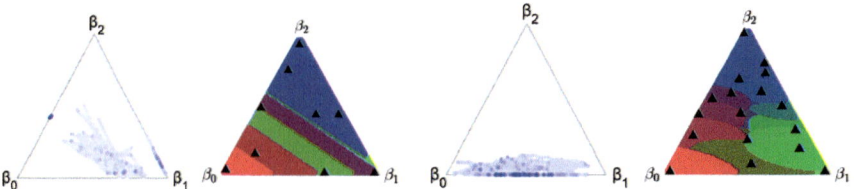

Fig. 4 The variance of the human data presented in the rightmost panel for the cloverleaf is mostly explained by one equivalence set in the second panel from right. Similar results are shown for the limaçon data (the two next panels) which suggests that the different segments in the human data employed mixed geometry weights that are statistically indistinguishable

3.5 Analysis of the Jerk Cost of Mixed Geometry Profiles

While in the above section, we showed that different geometrical mixtures can give rise to required paths; human data show that not all possible $\bar{\beta}$ mixtures within the triangle are used. Notice, however that in the above analysis we did not demand of the resulting velocity profiles to match those observed in human movements. We have shown that when the mixture of geometries model was used to account for the human data, only a subset of possible $\bar{\beta}$ trios was selected. Is it possible that the specific mixtures of geometries being generated are those that optimize behavior?

We inspected which mixtures of geometric speeds yield optimal speed profiles for each predefined movement path [24]. For each geometrical shape, we looked for the unique best geometric mixture describing a full cycle of movement, that yields minimal normalized jerk J_N or normalized acceleration, A_N. We looked for a $\bar{\beta}$ trio that minimizes these costs, but without allowing it to have different segments within a single cycle (one trio accounts for the mixture along the entire path). For this purpose, we calculated J_N for each constant $\bar{\beta}$ trio, for a dense set of $\bar{\beta}$ trios; selecting $\beta_i \in \{0, 0.01, 0.02, , ..0.99, 1\}$. The calculations were made for eight analytically described shapes: one ellipse with eccentricity of 0.97, one cloverleaf, three lemniscates with loop length ratios of 1: 1, 1: 2 and 1: 3 and three limac ons with loop length ratios 1: 3, 1: 5, and 1: 7. We examined the predictions of jerk minimization in explaining the observed parameters of the mixed geometry model. For each movement template we studied what $\bar{\beta}$ trio produces the minimal normalized jerk J_N.

For the ellipse, the J_N minimizing geometrical mixtures had $\beta_2 = 0$ which means that the contribution of the Euclidean geometry had to vanish for this trajectory. Additionally, along an ellipse, the equi-affine curvature is constant. Hence, both full-affine and equi-affine parameterizations result in a $v = \gamma \kappa^{-\frac{1}{3}}$ power law speed profile, and so is the speed profile of their mixture. Hence, analytically the value of the normalized jerk, J_N, is identical for all $\bar{\beta}$ with $\beta_2 = 0$. This case replicates the theoretical predictions of Richardson and Flash [51].

For the lemniscates which have two inflection points and four parabolic points, the set of candidate $\bar{\beta}$ trios for J_N minimization is restricted, because, as was shown above, a parabolic point must have a mixture with $\beta_0 = 0$ and an inflection point must have a mixture with $\beta_1 = 3\beta_0$ [5]. Therefore, we obtain only one feasible $\bar{\beta}$ trio for the entire path (i.e., without segmentation); constant Euclidean velocity, $\beta_2 = 1$.

For the cloverleaf, we found that the optimal geometrical mixture is a linear combination of the geometries, where $\beta_2 = 0.39 - 0.5\beta_1$ in agreement with the prediction of linear combinations from the above statistical analysis of the mixed geometry predictions by Meirovitch [39]. For the limaçon, we found that the optimal geometrical mixture is another linear relation.

These results show that the geometric mixtures yielding the minimal normalized jerk also yield a good description of the geometric mixtures that subjects use for drawing shapes, as long as there are no singularities in the template path. The latter case is likely to require a segmentation of the path into segments according to those singularities.

3.6 Human Data Analysis: Jerk Costs of Movements Arising from Different Geometrical Mixtures

Do the jerk minimizing mixtures match the mixtures selected by the human motor system? For each movement template we looked for the mixture parameter trio $\bar{\beta}$ that minimizes the jerk. The inferred mixtures of geometries derived for different trajectories are shown in Fig. 5 presented in barycentric coordinates. Each panel shows the mixture $\bar{\beta}$ of the different geometries using barycenter positive coordinates $\beta_0, \beta_1, \beta_2$ that produce the minimal normalized jerk J_N for the measured human movement paths. For all movement templates except for the lemniscates, the jerk minimizing parameterization matched human data in drawing them. For the ellipse, the two-thirds power law behavior predicted by jerk minimization is well known to be a good representation of human movement. For the cloverleaf and limaçon the J_N values of subjects' drawings again resembled those obtained from jerk minimization. For the lemniscates, the constant Euclidean speed profile differs significantly from the human speed profiles. This suggests that lemniscates are better represented using some segmentation allowing a change in the $\bar{\beta}$ parameters between consecutive segments. Together, these results show that human movements minimize jerk and that the $\bar{\beta}$ trios, inferred from jerk minimization, are quite similar to those derived from the mixed geometry model, and are showing very similar linear trends between the values of the different β parameters to those observed through the statistical redundancy analysis presented above.

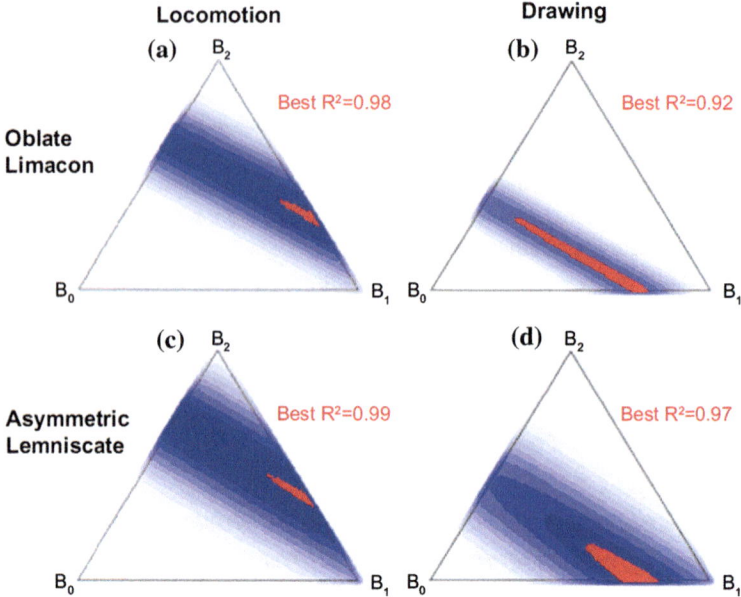

Fig. 5 The triangles show the trios of β parameters obtained from jerk minimization. The color of each point gives the value of the normalized jerk of the velocity profile created by moving along the analytic curve with the geometrical combination that the point represents. The darker a point, the lower the jerk. Red points are those with the lowest jerk

4 Affine Orbits as Geometric Motion Primitives

The suggestion that movements are stereotypical and are constructed through a sequential composition of simple building blocks is a fundamental idea in the study of motor control [22, 23, 35, 43]. The very nature of these building blocks is under debate. Kinematic motion primitives, spatio-temporal building blocks that specify an end-effector movement in time and space, are one possibility of such components. The manner in which the motor system specifies and composes kinematic motion primitives is currently being investigated.

Following Meirovitch [39], we suggest a family of prototypic geometric templates that may serve as motion primitives: affine orbits, which we use in the representation and segmentation of complex human end-effector trajectories. We first describe the properties of affine orbits, then their parameterizations, which provide maximally smooth trajectories, and finally present an algorithm for the segmentation of recorded movement data into geometric affine orbit primitives.

4.1 The Definition and Classification of Affine Orbits

Following Meirovitch [39], we now examine affine orbits, defined as the Lie 1-parameter group orbits of the affine group acting on Euclidean space, Thus each orbit corresponds to a 1-dimensional vector space of the of 2×2 sized matrices, with a single generating matrix

$$A \in gl_2(R),$$

where $A = \begin{bmatrix} a & b \\ c & d \end{bmatrix}$ is a constant matrix, termed the generator matrix.

In general, the Lie algebra of a Lie sub-group of the group of invertible matrices is its tangent space at the Identity matrix; if the sub-group is given by a set of equations, its Lie algebra is defined by taking the common zeros of the differential of these equations. The subspaces V of $gl_2(R)$ that are Lie algebras of some Lie sub-group are characterized by the fact that the bracket $XY - YX$ of any pair of elements X, Y of V also belongs to V. Then, in particular, any sub-vector space of dimension 1 is the Lie algebra of a sub-group, because in this case, X and Y are proportional and the bracket is zero. This subgroup is obtained by exponentiation.

Thus, the resulting trajectory of the affine orbit, $r(\zeta)$ is represented by:

$$r(\zeta) = \exp(A\zeta)p_0,$$

where p_0 is some fixed point in Euclidean space.

The parameter ζ is the orbit's natural arc-length parameter that is specified by the selection of the matrix A. We next examine the relation between this parameter and the geometric canonical parameter and other geometric properties of the curve. The shape of each orbit is dependent on the structure of its generating matrix A. The relation between the structure of A being the group generator and the type of orbit is the following. The Euclidean orbits consist of points, straight lines and circles. A point is the trivial orbit, which is associated with the matrix A being a matrix with 0 values for all its entries (and then its exponent is the identity element of the respective Lie group). Straight lines can be generated by any matrix A with real and identical eigenvalues. Circles are generated by any skew-symmetric matrix A.

Equi-affine orbits generalize the Euclidean ones and include the conic sections: ellipses, hyperboles and parabolas. Ellipses are generated by any matrix A for which $trace(A) = 0$ and $det(A) > 0$. Hyperboles are generated by any matrix A for which $trace(A) = 0$ and $det(A) < 0$. Parabolas are characterized by an equation defining their eigenvalues; $\alpha = 0$ for α defined as: $\alpha = det(A) - \frac{2}{9}trace^2(A)$. The parameter α is a useful shorthand, and we term it the parabolicity of the affine orbit.

Full-affine orbits are best sorted based on the value of the eigenvalues of A, denoted by λ_1, λ_2. For real eigenvalues, if the matrix is diagonalizable, either both eigenvalues are the same, and the orbit is a straight line or if the eigenvalues are real and different then the orbit can be represented by $y = x^\lambda$, in some x, y coordinate system which is achieved by an affine transformation of the canonical coordinate frame. The latter type of orbit we call here a monomial, although this does not precisely fit this function type. If the eigenvalues are real and the matrix is not diagonalizable, then both eigenvalues are equal, $\lambda_1 = \lambda_2$ and up to a similarity transformation the matrix A is upper triangular with nonzeros above the diagonal). Then the geometric form of the orbit is exceptional; $y = x \log(x)$ for some coordinate frame that results from an affine transformation of the canonical frame. Last, if the two eigenvalues are not real, then they are conjugate and the orbit is an elliptic logarithmic spiral (affine transform of the classical logarithmic spiral).

The different orbits derived in the manner described above are the ones having constant curvatures in their respective geometries; straight lines and circles are the orbits of the Euclidean geometry having constant Euclidean curvatures. Conic sections (parabolas, hyperbolas and ellipses) are the orbits of the equi-affine geometry and have constant equi-affine curvatures, which are 0, negative and positive for these three types of conic sections, respectively (see [20]). All affine orbits have a constant full-affine curvature (for a definition see [5]). The differential properties of an orbit, defined by the geometry, are always continuous functions of the canonical parameter, and on all the orbit's points, the geometric structure is the same up to a transformation by a member of the group.

Olver et al. [44] and Calabi et al. [13] have shown the usefulness of fundamental osculating curves of a given path. They noted that the point-wise geometric properties of the target curve are captured by the respective properties of the osculating one.

Therefore, in each of these geometries: Euclidean, equi-affine and affine, studying the osculating orbits of a general path provides us with the invariants describing the path for the associated geometry.

5 The Geometric Properties of Affine Orbits

5.1 Geometric Curvatures Along Affine Orbits

The affine orbits, being specific paths, enable to represent movement geometry and kinematics in a somewhat simplified form. Their geometric properties, represented by their curvatures, take thus the following form.

The Euclidean, equi-affine, and full-affine curvatures along the orbit at some point p on the orbit are represented by:

$$\kappa = \frac{|Ap \times A^2 p|}{|Ap|^3}, \quad k_1 = \frac{\alpha}{|Ap \times A^2 p|^{\frac{2}{3}}}, \quad k_0 = \mp \frac{2}{3} \frac{trace(A)}{|\alpha|^{\frac{1}{2}}},$$

where κ, k_1 and k_0 are the Euclidean, equi-affine and full-affine curvatures, respectively (the parabolicity of the affine orbit, α, was defined in the previous section).

The relation between the full-affine parameter of the orbit ρ and the parameter of the orbit ζ is $\rho = \zeta |\alpha|^{1/2}$.

An exception is the case of a parabola, for which the full-affine curvature is not defined. It can be traversed with equi-affine speed but not with full-affine speed.

5.2 Geometric Speeds Along Affine Orbits

The equi-affine speed along an affine orbit (see definition in Sect. "4.1") is

$$\dot{\sigma} = \exp\left(trace(A)\frac{\zeta}{3}\right)\dot{\zeta}\,|Ap_0 \times A^2 p_0|^{1/3}.$$

If $trace(A) = 0$ then the parameter ζ is defining a constant equi-affine speed. Otherwise, the equi-affine speed along an orbit is $d\sigma/dt = 0$. Hence, a constant equi-affine speed along the affine orbit is satisfied by the parameter.

$$\zeta = C_1 \ln|Ap_0 \times A2p_0|^{\left(-\frac{1}{3}\right)}\sigma + C_2.$$

Here C_1 is a geometric constant depending on A and C_2 is an arbitrary integration constant.

The parameterization of an affine orbit (as in Sect. "4.1") with a mixed geometry parameter z, defined for a given mixture trio $\bar{\beta}$ is:

$$dz = C(\bar{\beta})a^{\beta_1}e^{b\beta_1\zeta}\,|\exp(A\zeta)Ap_0|^{\beta_2}\alpha^{\beta_0/2}d\zeta.$$

Here $C(\bar{\beta})$ is a constant depending on the mixture trio $\bar{\beta}$, and α is the parabolicity constant defined in the previous section, $a = |Ap_0 \times A^2 p_0|^{1/3}$, and $b = \frac{1}{3}trace(A)$.

5.3 Mixed Geometry Parameterizations of Affine Orbits that Minimize Jerk

We now search for examples for how, using a constant mixture of geometries, one may generate speed profiles and trajectories that are extrema of jerk optimization. We show this for affine orbits.

Bright [12] and Polyakov [47] found analytic expressions for paths along which, when the movement has a constant equi-affine speed, it also yields a minimal jerk cost. Polyakov found that traversing a parabola with constant equi-affine speed yields a minimal jerk trajectory. Bright found a specific spiral for which constant equi-affine speed yields minimal jerk and other spirals for which constant Euclidean or full-affine speed profiles yield minimal jerk trajectories. Because Euclidean, equi-affine and full-affine parameterizations are special cases of the mixed geometry model, our results generalize these previous findings.

5.3.1 Monomials with a Mixed Geometry Parameterization that Minimizes Jerk

We examine monomials, generally defined as affine transformations of the standard Cartesian equation $Y^n = X^m$, for some constant integer exponents n and m. This definition includes as specific examples all parabolic and hyperbolic conic sections. We examine a specific set of monomials, whose generating matrix A is:

$$A = \begin{bmatrix} 1 & b \\ 0 & d \end{bmatrix},$$

where b and d are any real numbers.

We provide a set of mixed geometry parameterizations of monomials that are candidates for yielding jerk extrema. For some finite set of values of the entry d, a mixed geometry solution that minimizes the jerk cost exists. Solutions for the jerk minimizing mixed geometry parameters impose that d satisfies $d = m, n \in \{1, \dots, 5\}$. These correspond, up to affine transformations, to the free minimum jerk solutions of Flash and Hogan [23]. $Y^n = X^m$, where $n, m \in \{1, \dots, 5\}$. Each specific solution has a mixed geometry parameterization $\bar{\beta}$ that is a candidate for optimizing jerk along it. As a particular case, this derivation predicts that parabolas should be traversed with equi-affine parameterization in order to minimize jerk. In all of the mixtures derived above, there is no Euclidean contribution (so $\beta_2 = 0$) and the speed profiles are represented by a composition of equi-affine and full-affine parametrizations.

5.3.2 Non-elliptic Logarithmic Spirals, General Mixed Geometry Solutions

If the generator matrix is of the form:

$$A = \begin{bmatrix} 1 & b \\ -b & 1 \end{bmatrix},$$

where b is any real number. Then, any mixture parameter $\bar{\beta}$, (depending on the value of b, the inverse of the orbit rate-of-growth parameter) satisfying:

$$\beta_1 + \frac{2}{3}\beta_2 = \frac{1}{160}\left(117 + \sqrt{3}C_2 + \left(3858 + 36000b^2 - 120C_1 - \frac{226800}{C_1}b^2\right.\right.$$

$$\left.\left. - \frac{8760}{C_1} - \frac{96300}{C_1}b^4 + 378000\frac{\sqrt{3}}{C_2}b^2 + 374222\frac{\sqrt{3}}{C_2}\right)^{1/2}\right),$$

where C_1 and C_2 are constants depending on the parameter b in the above matrix representation of the generating matrix A (defined in [39]) and additionally $1 \geq \beta_i \geq 0$ for all i, is a candidate parametrization for jerk minimization along an affine orbit.

5.3.3 A Mixed Geometry Solution that Is a Candidate on All Non-elliptic Orbits

We now seek a specific mixture parameter that is valid for each non-elliptic orbit that has a generating matrix of the form

$$A = \begin{bmatrix} a & b \\ -b & a \end{bmatrix},$$

where a and b are any real numbers.

The specific trio $\bar{\beta}$, defined by $\beta_0 = \beta_1 = \frac{1}{2}$, is a mixed geometry parameter which guarantees that the first variation of the minimum jerk cost is zero.

5.4 Data Segmentation with Osculating Affine Orbits

We suggest that affine orbits are plausible natural building blocks for the description of trajectories of human movements. We describe here the segmentation algorithm developed by Meirovitch [39] that allows trajectory segmentation using the affine invariant local geometric properties of the trajectory. We examine a set of candidate affine orbits, truncated to form possible movement primitives, whose respective distances from the trajectory are calculated. An optimality criterion is used to select subsets of these primitives that reliably represent the parameterised trajectories [21]. Figure 6 depicts an example of this segmentation for the original and an affine transformed lemniscates.

The following description assumes a sampled trajectory, $r(n) \in R^2, n = 1, \ldots, N$.

For each data point i:

1. We calculate $\psi_i(v)$, the osculating affine orbit.

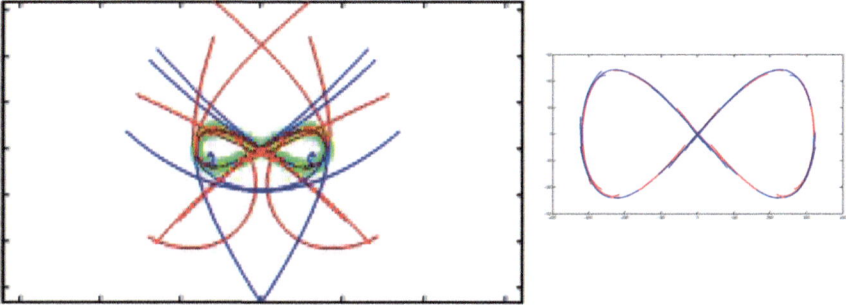

Fig. 6 In the *left panel*, the osculating affine orbits were calculated for every 20th point on one lemniscate, The osculating curves were restricted by a large Hausdorff threshold. Since this metric is not affine invariant the restricted osculating orbit of L1 and L2 differ according to their extent. Each osculation point divided the osculating curve into two branches before and after the point, referred to as "left" (blue) and "right" (red) branches. In the *right panel*, the osculating orbits were calculated and subsequently restricted using a relatively small Hausdorff distance on a scaled lemniscate

2. We then find the maximal boundaries $v_1 < v_2$ such that the one-sided Housdorff distance between the osculating orbit and the curve (taking into account only the distances of points on the orbit from the sampled trajectory) is bounded by a small number ϵ_0

$$Housdorff\left(\{\psi_i(v)\}_{v=v_1}^{v_2}, r(n)_{n=1}^N\right) < \epsilon_0.$$

3. We then project the boundaries $\psi_i(v_1)$ and $\psi_i(v_2)$ on the data points $(n1), r(n2)$.
4. Next, we store the value $S_i = (n_1, n_2)$ (overall we will repeat and collect S_i for each of the points on the sampled trajectory $r(n)$).
5. We then use dynamic programming to choose a subset of $\{S_i\}_{i=1}^N$ with segments that are compatible with each other (allowing no overlaps of the segments S_i), while maximizing the number of samples in each S_i [37].

The segmentation process is affine invariant, in the sense that the osculating orbits matching an affine transformation of a path are the affine transformations of the osculating orbits matching the original path. This is true except for a minute detail that the trimming of the orbits is based on Euclidean Hausdorff distance which is not affine invariant. Based, however, on numerical simulations, we could conclude that the affine invariance seems to hold, and the segmentation of the affine transformed lemniscate is the affine transformation of the segments of the original lemniscate (Fig. 7).

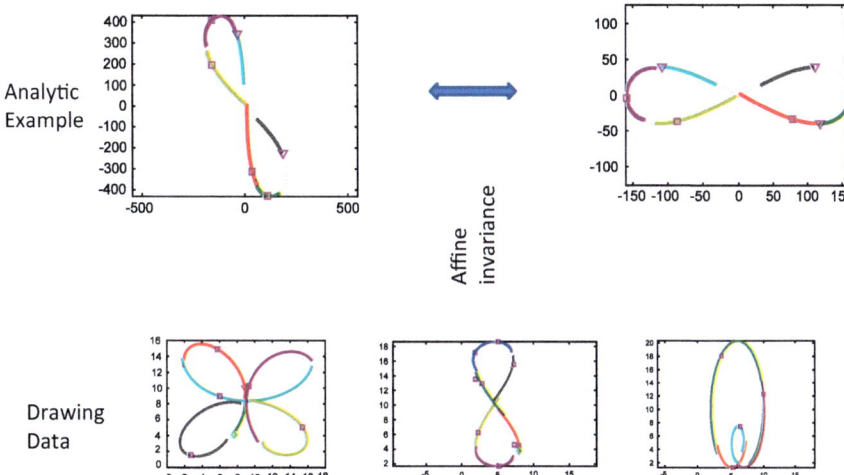

Fig. 7 An optimal segmentation method was adapted and used to select a subset of osculating segments, where for each osculation point three segments were generated according to "left", "right" and "left-right" branches of the osculating curve, where "left-right" included both the "left" and "right" sides of the osculating orbit (see Fig. 6). Triangles, diamonds and squares mark the osculation points in correspondence to whether the selected segments were "left", "right" or "left-right", respectively. The similarity between the segmentations in the two leminscates with respect to the geometry stems from the affine invariance of the osculating orbit. It should be noted that the trimming according to the Hausdorff distance is not an affine invariant, but still under the threshold of the algorithm the difference seems negligible. The colors of the segments are given for the sake of illustration

6 Discussion

In this chapter, we discussed how the concepts of invariance and optimization play different yet complementary roles in the description of how the human motor system plans movement. We examined the mixed geometry model in theory and practice, showing that for some templates a redundancy appears; entirely different mixture parameters produce highly similar speed profiles. Following the results of the mixed geometry model for drawing data, we considered the theoretical aspects of the specific selection of geometric mixtures. First, we noticed that some conditions constrain the space of possible mixtures; singularity points dictate specific mixtures. Next, we reexamined the practical implications of the variety of geometric mixtures. For specific templates, we see that not all mixtures are distinguishable from each other and that different mixtures may yield a similar behavior. We examined the idea that the mixture of geometries may be selected to account for an optimality criterion. Testing various templates reveals that humans select mixtures of geometries that minimize jerk.

We discussed a new theory of motion primitives based on the composition of the classical Euclidean, equi-affine and full-affine geometries [39]. The shapes of these primitives are orbits of 1-parameter subgroups acting on the points in the task space. The non-trivial orbits are straight lines and circles (Euclidean geometry), parabolas, ellipses and hyperbolas (equi-affine geometry), and elliptic logarithmic spirals and monomials (full-affine geometry). After examining the geometric properties and descriptions of the affine orbits, we provide examples of mixed geometry parameterizations of some of these affine orbits that may allow optimal movement along them with respect to jerk minimization.

6.1 Affine Orbits as Motion Primitives

Representing complex movements as a composition of affine orbits that serve as geometrical primitives, is plausible and useful for several reasons.

First, from a theoretical point of view, the geometrical simplicity of orbits makes them attractive candidates for serving as primitives. The symmetry properties of orbits are not only the Euclidean ones obeyed by circles and straight lines, but additionally the non-Euclidean symmetries, that proved to be highly useful in describing the visual properties of shapes in computer vision research [13, 17, 44]. The orbits transform among themselves by specific transformations. Affine mappings permute the set of affine orbits. Any two points along a given orbit can be affinely mapped one upon the other such that the orbit maps to itself. The affine orbits generalize previously suggested movement primitives; straight movement primitives and parabolic movement primitives [23, 48]. Note that a positive direct test of affine invariance reflected in the duration of hand drawings was presented in Pham and Bennequin [45].

Second, we demonstrated some simple mixed geometry parameterizations of affine orbits that may satisfy constrained jerk minimization. This is a generalization of the fact that obeying the two-thirds power law by moving with a constant equi-affine speed along parabolas, automatically minimizes the jerk of the movement [20, 27]. Thus some subsets of geometric primitives are easily assigned with kinematics that are optimal. Additionally, for each affine orbit that is a circular logarithmic spiral, there exists a special mixture of geometries that may minimize the jerk along that orbit.

Third, a movement representation using affine orbits is compact, in the sense that full-affine invariants such as full-affine curvature and arc-length are preserved under affine transformations. The same movement plan, a canonical orbit, can yield different actual paths, according to the affine mapping used in transforming the canonical orbit. Once the shape of primitives is decided upon, the manner of segmenting a movement and extracting these primitives is important. The segmentation algorithm we suggested identifies locally osculating affine orbits and temporally concatenates them as building blocks. The identified set of primitives describing a complex movement is inherently affine invariant. Not only is each primitive by itself affine invariant, but more importantly, a set of concatenated affine orbits describing a path is mapped to a set of affine orbits describing the mapped path. This occurs because osculating affine orbits are mapped to other osculating orbits by affine transformations, unlike best fitting primitives, which are not necessarily mapped to best fitting primitives under affine transformations.

6.2   The Nature of Kinematic Motion Primitives

We now speculate regarding the nature of the motion primitives used by the human motor system.

First, the relation between timing and geometry is unclear. Does the primitive entail a kinematic pattern, or is it just dictating the geometric form? In case that the primitive's description provides only the geometric form, it could be that the timing of motion is dictated at another level, and is possibly selected for the entire composite movement rather than for its primitive components.

Second, the variability of neural patterns and actual movement execution may prove to be an inherent part, dictated by the motion primitives being selected. Perhaps a noisy statistical representation is an essential part of motor execution to the extent that it makes little sense to debate regarding the mean behavior without paying attention to the statistical properties of motor noise. Recently, new statistical frameworks were developed (e.g. [50]) allowing to determine systematic patterns and differences across experimental conditions, participants and repetitions. Such methods are important, since unlike in robotic systems, the physiology of biological systems generates highly variable outputs due to inherent noise in biological sensing of the body and the environment and in the neural commands and muscles' activation patterns underlying motor execution.

Third, a question arises whether human movements are discrete or continuous, i.e. whether they are planned as a whole or by composing together several segments. The notion of a primitive by itself is suggestive of the existence of a set of discrete components that are performed one after the other, or in the case of co-articulation, each starting after the previous one has begun but not necessarily been completed.

Fourth, even if the basic primitives indeed are centrally represented, the manner according to which they are generated may make a large difference. As the human motor system is capable of learning, it is possible that new motion primitives arise when a movement that was previously generated as a concatenation of simpler primitives becomes a single new motion primitive [11, 53]. A process where new primitives emerge out of preexisting ones may also be accompanied by primitive refinement according to some optimality criteria. Thus, a set of previous primitives, first concatenated over time and then adjusted to be molded together and smooth, may form a new motion primitive. This process is interesting when examining movement kinematics but even more so when examining the underlying neural processes.

References

1. Abend, W., Bizzi, E., Morasso, P.: Human arm trajectory formation. Exp. Brain Res. **105**, 331–348 (1982)
2. Abeles, M., Diesmann, M., Flash, T., Geisel, T., Hermann, M., Teicher, M.: Compositionality in neural control: an interdisciplinary study of scribbling movements in primates. Frontiers in computational neuroscience **7**, 103 (2013)
3. Balasubramanian, S., Melendez-Calderon, A., Roby-Brami, A., Burdet, E.: On the analysis of movement smoothness. J. Neuroengineering Rehabil. **12**(1), 112 (2015)
4. Ben-Itzhak, A., Karniel, A.: Minimum acceleration criterion with constraints implies bang-bang control as an underlying principle for optimal trajectories of arm reaching movements. Neural Comput. **20**(3), 779–812 (2008)
5. Bennequin, D., Fuchs, R., Berthoz, A., Flash, T.: Movement timing and invariance arise from several geometries. PLoS Comput. Biol. **5**(7), e1000426 (2009)
6. Bennequin, D., Berthoz, A.: Several geometries for movements generations. In: Laumond, J.-P., Mansard, N., Lasserre, J.-B. (eds.) Geometric and Numerical Foundations of Movements vol. 117, pp. 13–43. Springer Tract in Advanced Robotics (2017)
7. Bernstein, N.: The Co-ordination and Regulation of Movements. Pergamon Press, Oxford (1967)
8. Biess, A., Nagurka, M., Flash, T.: Simulating discrete and rhythmic multi-joint human arm movements by optimisation of nonlinear performance indices. Biol. Cybern. **95**(1), 31–53 (2006)
9. Biess, A., Liebermann, D.G., Flash, T.: A computational model for redundant human three-dimensional pointing movements: integration of independent spatial and temporal motor plans simplifies movement dynamics. J. Neurosci. **27**(48), 13045–13064 (2007)
10. Bizzi, E., Tresch, M.C., Saltiel, P., d Avella, A.: New perspectives on spinal motor systems. Nat. Rev. Neurosci. **1**(2), 101–108 (2000)
11. Bizzi, E., Mussa-Ivaladi, F.A.: Motor learning through the combination of primitives. Philos. Trans. Royal Soc. London Ser. B-Biol. Sci. **355**, 1755–1759 (2000)
12. Bright, I.: Motion planning through optimisation. Master's thesis, Weizmann Institute of Science (2007)
13. Calabi, E., Olver, P.J., Shakiban, C., Tannenbaum, A., Haker, S.: Differential and numerically invariant signature curves applied to object recognition. Int. J. Comput. Vision **26**(2), 107–135 (1998)
14. Desmurget, M., Pélisson, D., Rossetti, Y., Prablanc, C.: From eye to hand: planning goal-directed movements. Neurosci. Biobehav. Rev. **22**(6), 761–788 (1998)
15. de' Sperati, C., Viviani, P.: The relationship between curvature and velocity in two-dimensional smooth pursuit eye movements. J. Neurosci. **17**(10), 3932–3945 (1997)
16. Dingwell, J.B., Mah, C.D., Mussa-Ivaldi, F.A.: Experimentally confirmed mathematical model for human control of a non-rigid object. J. Neurophysiol. **91**(3), 1158–1170 (2004)
17. Faugeras, O., Keriven, R.: (1996) On projective plane curve evolution. In Lecture Notes in Control and Information Sciences, pp. 66–73 (1996)
18. Fitts, P.M.: The information capacity of the human motor system in controlling the amplitude of movement. J. Exp. Psychol. **47**, 381–391 (1954)
19. Flash T.: Organizing principles underlying the formation of hand trajectories. Doctoral dissertation, Massachusetts Institute of Technology, Cambridge, MA (1983)
20. Flash, T., Handzel, A.A.: Affine differential geometry analysis of human arm movements. Biol. Cybern. **96**(6), 577–601 (2007)
21. Flash, T., Henis, E.: Arm trajectory modifications during reaching towards visual targets. J. Cogn. Neurosci. **3**(3), 220–230 (1991)
22. Flash, T., Hochner, B.: Motor primitives in vertebrates and invertebrates. Curr. Opin. Neurobiol. **15**(6), 660–666 (2005)
23. Flash, T., Hogan, N.: The coordination of arm movements—an experimentally confirmed mathematical-model. J. Neurosci. **5**(7), 1688–1703 (1985)

24. Fuchs, R.: Human motor control: geometry, invariants and optimisation, Ph.D. thesis, Department of CS and applied Mathematics, Weizmann Institute of Science, Rehovot, Isreal (2010)
25. Guggenheimer, H.W.: Differential Geometry. Dover Publications, (1977, June)
26. Handzel, A., Flash, T.: Affine differential geometry analysis of human arm trajectories. Abs. Soc. Neurosci. **22** (1996)
27. Handzel, A.A., Flash, T.: Geometric methods in the study of human motor control. Cognitive Studies **6**(3), 309–321 (1999)
28. Henis, E., Flash, T.: Mechanisms underlying the generation of averaged modified trajectories. Biol. Cybern. **72**(5), 407–419 (1995)
29. Hicheur, H., Vieilledent, S., Richardson, M.J.E., Flash, T., Berthoz, A.: Velocity and curvature in human locomotion along complex curved paths: a comparison with hand movements. Exp. Brain Res. **162**(2), 145–154 (2005)
30. Huh, D., Sejnowski, T.J.: Spectrum of power laws for curved hand movements. Proc. Natl. Acad. Sci. **112**(29), E3950–E3958 (2015)
31. Ivanenko, Y.P., Grasso, R., Macellari, V., Lacquaniti, F.: Two-thirds power law in human locomotion: role of ground contact forces. NeuroReport **13**(9), 1171–1174 (2002)
32. Karklinsky, M., Flash, T.: Timing of continuous motor imagery: the two-thirds power law originates in trajectory planning. J. Neurophysiol. **113**(7), 2490–2499 (2015)
33. Kohen, D., Karklinsky, M., Meirovitch, T., Flash T., Shmuelof, L.: The effects of shortening preparation time on the execution of intentionally curved trajectories: optimisation and geometrical analysis. Frontiers Human Neuroscience (in press)
34. Lacquaniti, F., Terzuolo, C., Viviani, P.: The law relating kinematic and figural aspects of drawing movements. Acta Physiol. (Oxf) **54**, 115–130 (1983)
35. Lashley, K.: The problem of serial order in psychology. In: Cerebral mechanisms in behavior. Wiley, New York (1951)
36. Levit-Binnun, N., Schechtman, E., Flash, T.: On the similarities between the perception and production of elliptical trajectories. Exp. Brain Res. **172**(4), 533–555 (2006)
37. Meirovitch, Y: Kinematic Analysis of Israeli Sign Language. Master thesis, The Weizmann Institute (2008)
38. Meirovitch, Y., Bennequin, D., Flash, T.: Geometrical Invariance and Smoothness Maximization for Task-Space Movement Generation. IEEE Trans. Rob. **32**(4), 837–853 (2016)
39. Meirovitch, T.: Movement decomposition and compositionality based on geometric and kinematic principles. Department of Computer Science and Applied Maths Ph. D. dissertation, Weizmann Institute of Science, Rehovot, Israel (2014)
40. Mellinger, D., Kumar, V.: Minimum snap trajectory generation and control for quadrotors. IEEE *Robotics and Automation (ICRA)*, pp. 2520–2525, 2011
41. Mombaur, K., Laumond, J.P., Yoshida, E.: An optimal control model unifying holonomic and nonholonomic walking. In: Humanoid Robots. 8th IEEE-RAS International Conference on pp. 646–653 (2008)
42. Morasso, P.: Spatial control of arm movements. Exp. Brain Res. **42**(2), 223–322 (1981)
43. Mussa-Ivaldi, F.A., Solla, S.A.: Neural primitives for motion control. IEEE J. Oceanic Eng. **29**(3), 640–650 (2004)
44. Olver, P.J., Sapiro, G., Tannenbaum, A., et al.: Differential invariant signatures and flows in computer vision: A symmetry group approach. In: Geometry driven diffusion in computer vision (1994)
45. Pham, Q.-C., Bennequin, D.: Affine invariance of human hand movements: a direct test. preprint arXiv Biology:1209.1467 (2012)
46. Pollick, Frank E., Sapiro, Guillermo: Constant affine velocity predicts the 1/3 power law pf planar motion perception and generation. Short Commun. **37**(3), 347–353 (1996)
47. Polyakov, F.: Analysis of monkey scribbles during learning in the framework of models of planar hand motion. Ph.D. thesis, The Weizmann Institute of Science (2001)
48. Polyakov, F., Drori, R., Ben-Shaul, Y., Abeles, M., Flash, T.: A compact representation of drawing movements with sequences of parabolic primitives (2009)

49. Polyakov, F., Stark, E., Drori, R., Abeles, M., Flash, T.: Parabolic movement primitives and cortical states: merging optimality with geometric invariance. Biol. Cybern. **100**(2), 159–184 (2009)
50. Raket, L.L., Grimme, B., Schoner, G., Christian, I., Markussen, B.: Separating timing, movement conditions and individual differences in the analysis of human movement. PLoS Comput. Biol. **12**(9), e1005092 (2016)
51. Richardson, J.M.E., Flash, T.: Comparing smooth arm movements with the two- thirds power law and the related segmented-control hypothesis. J. Neurosci. **22**(18), 8201–8211 (2002)
52. Schaal, S., Sternad, D.: Segmentation of endpoint trajectories does not imply segmented control. Exp. Brain Res. **124**, 118–136 (1999)
53. Sosnik, R., Hauptmann, B., Karni, A., Flash, T.: When practice leads to co-articulation: the evolution of geometrically defined movement primitives. Exp. Brain Res. **156**(4), 422–438 (2004)
54. Tanaka, H., Krakauer, J.W., Qian, N.: An optimization principle for determining movement duration. Neurophysiol. **95**(6), 3875–3886 (2006)
55. Tasko, S.M., Westbury, J.R.: Speed–curvature relations for speech-related articulatory movement. J. Phonetics **32**(1), 65–80 (2004)
56. Todorov, E., Jordan, M.I.: Smoothness maximization along a predefined path accurately predicts the speed profiles of complex arm movements. J. Neurophysiol. **80**, 696–714 (1998)
57. Todorov, E., Jordan, M.I.: Optimal feedback control as a theory of motor coordination. Nat. Neurosci. **5**(11), 1226–1235 (2002)
58. Viviani, P., Cenzato, M.: Segmentation and coupling in complex movements. J. Exp. Psychol. Hum. Percept. Perform. **11**(6), 828–845 (1985)
59. Viviani, P., Stucchi, N.: Biological movements look uniform: evidence of motor-perceptual interactions. J. Exp. Psychol. Hum. Percept. Perform. **18**(3), 603 (1992)
60. Viviani, P., Flash, T.: Minimum-jerk, two-thirds power law, and isochrony: converging approaches to movement planning. J. Exp. Psychol. Hum. Percept. Perform. **21**(1), 32–53 (1995)
61. Vieilledent, S., Kerlirzin, Y., Dalbera, S., Berthoz, A.: Relationship between velocity and curvature of a human locomotor trajectory. Neurosci. Lett. **305**(1), 65–69 (2001)
62. Viviani, P., McCollum, G.: The relation between linear extent and velocity in drawing movements. Neuroscience **10**(1), 211–218 (1983)

Review of Anthropomorphic Head Stabilisation and Verticality Estimation in Robots

Ildar Farkhatdinov, Hannah Michalska, Alain Berthoz
and Vincent Hayward

Abstract In many walking, running, flying, and swimming animals, including mammals, reptiles, and birds, the vestibular system plays a central role for verticality estimation and is often associated with a head stabilisation (in rotation) behaviour. Head stabilisation, in turn, subserves gaze stabilisation, postural control, visual-vestibular information fusion and spatial awareness via the active establishment of a quasi-inertial frame of reference. Head stabilisation helps animals to cope with the computational consequences of angular movements that complicate the reliable estimation of the vertical direction. We suggest that this strategy could also benefit free-moving robotic systems, such as locomoting humanoid robots, which are typically equipped with inertial measurements units. Free-moving robotic systems could gain the full benefits of inertial measurements if the measurement units are placed on independently orientable platforms, such as a human-like heads. We illustrate these benefits by analysing recent humanoid robots design and control approaches.

1 Introduction

Spatial awareness is crucial for free-moving robots. For instance, legged humanoid robots must be aware of their body motion and orientation with respect to gravity in order to maintain stable posture and gait. Recent technological developments in sens-

I. Farkhatdinov (✉)
School of Electronic Engineering and Computer Science,
Queen Mary University of London, London, UK
e-mail: i.farkhatdinov@qmul.ac.uk

I. Farkhatdinov
Department of Bioengineering, Imperial College London, London, UK

H. Michalska
McGill University, Montreal, QC, Canada

A. Berthoz
Collège de France, Paris, France

V. Hayward
Sorbonne Universités, Institut des Systèmes Intelligents
et de Robotique, ISIR, F-75005 Paris, France

© Springer International Publishing AG, part of Springer Nature 2019
G. Venture et al. (eds.), *Biomechanics of Anthropomorphic Systems*, Springer Tracts
in Advanced Robotics 124, https://doi.org/10.1007/978-3-319-93870-7_9

ing and control for robotic systems contributed to the appearance of new humanoid robots with human-like (anthropomorphic) behaviours. However, the performance of robots is still lagging far behind the abilities of biological systems. For example, it is clear that that a powerful and adaptive sensorimotor control system enables humans, for example, to balance on fallen trunks or run on sandy beaches, not mentioning the locomotion prowesses of felines, of mountain goats, and so on.

The purpose of this chapter is to examine the phenomenon of head stabilisation displayed by biological and robotic systems and the importance of such behavior in motion sensing and body balancing. Head stabilisation has been extensively studied by neuroscientists and recently several robotic systems have adopted this human-like control to improve inertial sensing, gaze and postural control, and interaction qualities of the robots. Functions of the vestibular system and its primary roles are first reviewed in Sect. 2. Head stabilisation in humans is next discussed in Sect. 3, which is then followed by the overview of inertial sensing and head stabilisation in robotics systems in Sects. 5 and 6.

2   Sensing Body Movements in Animals and Robots

2.1   Sense of Motion

Humans and animals have a complex multisensory system which provides the central neural system (CNS) with information about the external world. For us, like for other living creatures, it is important to understand the spatial orientation and localization of our bodies with respect to the objects of the external world. This can be accomplished by the help of diverse sensory systems: vision, audition, tactile, proprioception and vestibular. Change of visual flow in the retina would indicate that the position of the body changed with respect to an observed environment. Change of difference in acoustic flow in the ear would indicate that the body moves with respect to a sound source. Likewise, tactile and proprioceptive inputs would signal position change of the body if the latter interacts with an environment. However, independent sensory information of one modality only may not be sufficient for correct motion perception. For instance, change of visual flow can be caused by movements of eyes, sound source movement may be estimated wrongly if the head rotates, etc. Then, the problem is solved by multisensory integration [45]. The CNS can efficiently process multisensory information and provide the brain with a proper motion sensation. Visual, audition, tactile and proprioception sensory inputs, excluding vestibular, can be easily put out of action while vestibular inputs will always be there. Even in case of zero gravity vestibular organs will respond to linear accelerations and angular velocities of the body motion. It is thus reasonable to claim that the vestibular system plays a key role in perceiving self-motion. The principles which govern vestibular organs and their roles for postural control will be described next.

2.2   Vestibular System

The vestibular system detects motion of the head in space and generates reflexes that are crucial for survival, such as stabilising gaze and maintaining head and body posture. In addition, the vestibular system provides its owner with a subjective sense of movement and orientation in space. The vestibular sensory organs are located in close proximity to the cochlea, as shown in Fig. 1a. The vestibular system consists of two types of organs: the two otolith organs and the three semicircular canals. Two types of otolith organs (the saccule and utricle) sense linear acceleration which includes gravitational and translational forces. Semicircular canals sense angular velocities in three planes. Receptor signals are sent through the vestibular nerve fibres to the neural structures that are responsible for eye movements, posture and balance control. As a result, the vestibular organs constitute our sixth sense - the sense of motion which allows us to perceive and control bodily movements [16]. Vestibular processing is highly multimodal, for instance, visual/vestibular and proprioceptive/vestibular sensory inputs are dominant for gaze and postural control, but at the same time vestibular system itself plays an important role in everyday activities and contributes to a various range of functions [5, 35].

Each inner ear has three **semicircular canals** arrayed approximately at right angles to each other. The real geometry has evolved during evolution depending upon the need of each species [37]. Semicircular canals are sensitive to angular accelerations [121]. Each canal is comprised of a circular path of fluid continuity, interrupted at the ampulla by a watertight, elastic membrane called the cupula [5]. Figure 1b shows a schematic view of one semicircular canal. Each canal is filled with a fluid called endolymph. When the head rotates in the plane of a semicircular canal, inertial forces causes the endolymph in the canal to lag behind the motion of the head [28]. The motion of the endolymph applies pressure to the membrane

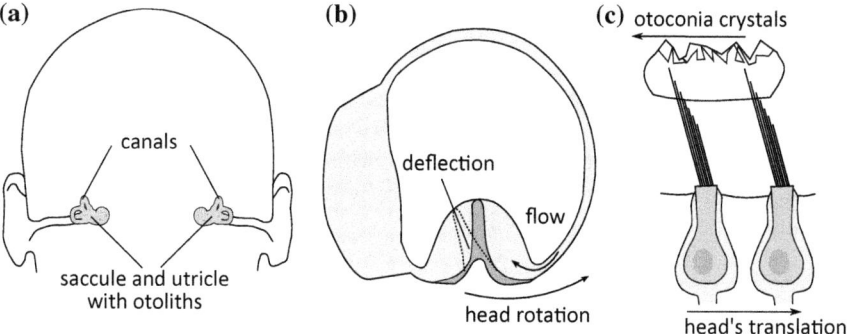

Fig. 1 The vestibular organs located in each inner ear (**a**). Main organs are three almost orthogonal semicircular canals (superior, posterior and lateral) and two otolith organs (utricle and saccule). Semicircular canals measure angular velocity of the head (**b**). Otolith organs measure linear gravitoinertial acceleration (**c**)

of the cupula and its deflection causes shearing stress in the hair cells [28]. The corresponding electrical signals are generated and transmitted through neurons in a way similar to the way it is done in the otoliths. Although the semicircular canals respond to angular acceleration, the neural output from the sensory cells represents the velocity of rotation. This suggests that the input signal is integrated due to the mechanics of the canals, mainly because of the increase in viscous properties of the fluid in a tight canal [36, 76]. The vestibular system transmits the components of the angular velocity of the head to the CNS which forms a three-dimensional angular velocity vector describing the body motion. More advanced experimental studies with human subjects supported the torsion-pendulum representation of the canal dynamics [68, 75, 107]. In [139], it was suggested to model semicircular canals with the heavily-damped second order system which behaved as an angular-velocity meter.

The otolith organ comprised utricle and saccule, both sensitive to linear acceleration of the head in linear motion and static tilt in the vertical planes. The utricle and the saccule are arranged to respond to the motion in three dimensions. When the head is upright, the saccule is vertical and it measures linear accelerations in the sagittal plane, specifically movements up and down. The utricle is horizontally oriented and measures accelerations in interaural transverse (horizontal) plane (anterio-posterior and medio-lateral accelerations) [70]. Both, the utricle and the saccule, contain sheets of hair cells a sensory epithelium (macula). An otolithic membrane (otoconia) composed of calcium carbonate crystals sits atop of hair cells. Figure 1c shows the simplified view of otolithic hair cells during an accelerated motion to the right. In response to linear acceleration, the crystals are left behind due to their inertia. Linear acceleration of the head causes otoconia's motion which in turn causes shear forces acting on the hair cells. In addition during head tilt they can measure the changes of head angle with respect to the gravity vector. Complex molecular level mechanic-electro-chemical mechanism of interaction between hair cells results in the generation of electrical signals which are sent to the neural structures for further processing [28]. The displacement of the large saccular otolith of a ruff was measured in [39] and the dynamic identification suggested that the mechanics of the otolith could be described by a critically damped second-order system with a resonant frequency of 50 Hz. A recent model of the geometry of the utriculus and sacculus has suggested that the curved shape of the striola (which contains the dynamic sensory cells) are in fact the projection of a curved virtual striola giving the ensemble of the two sensors the capacity to measure directly the 3D acceleration vector of the movements for the head in space [41].

2.3 Sensory Data Integration and Processing

Sensory outputs from otoliths and semicircular canals are processed in neural systems on different levels. Some information is processed in low-level neural networks, while some are directed to CNS where integration with other sensory modalities may occur.

L. R. Young with his group proposed an optimal estimator model in [21]. A concept of the internal model was introduced which comprised the dynamic model information about the sensory organs and head-neck system. This internal model was considered to be known for the CNS. L. R. Young used an optimal estimator (Kalman filter) to estimate the human's orientation based on the outputs from visual, vestibular, proprioceptive and tactile sensory systems. Assumptions about sensor dynamics and noise statistic from internal model were used to correct the estimated states which represented spatial orientation. Estimated states, called as perceptions, contained angular orientation of the head, its angular velocity, inertial translation and inertial velocity.

In similarity with [21], an internal model based approach was proposed by D.M. Merfeld and colleagues in [99]. Their sensory integration model was based on simulating the neural processing of gravito-inertial cues. In [101, 151] experiments were performed to analyse the vestibular-ocular response during tilt and rotation of the head and the body. The presented experimental findings were consistent with the hypothesis that the nervous system resolves the ambiguous measurements of gravitoinertial forces into neural estimates of gravity and linear acceleration. A human model for this vestibular signals processing was presented in [100]. A linear system was used to model the semicircular canals' dynamics, and a simple gain (identity matrix) was used to model the otoliths. It was suggested that the CNS is able to perform the operation of gravitational vector transformation (rotation) from the world frame to the head's frame.

More recently, J. Laurens and J. Droulez built a Bayesian processing model of self-motion perception in [90]. It was proposed that the brain processes these signals in a statically optimal fashion, reproducing the rules of Bayesian inference. It was also proposed that the Bayesian based processing uses the statistics of natural head movements. The outputs of semicircular canals and head's angular velocity were assumed to be subjects to Gaussian noise. By using particle filtering, a three dimensional model of vestibular signal processing model was developed based on optimal estimation. The model was successfully tested by computational experiments, as well, it was proved to be efficient in modelling the vestibulo-ocular reflex.

Among the vestibular information processing models mentioned above, the concepts of internal model and estimator or observer are crucial for the system description. Detailed reviews of existing vestibular information processing models can be found in [96, 126]. Interestingly, in robotic systems, the concept of the internal model has been known for decades, since the early works of E. Kalman in linear filtering [82] and D. Luenberger in state estimation for linear systems [95].

2.4 Roles of the Vestibular System

Humans rely on the multiplicity of sensory inputs and sophisticated anticipatory mechanisms to solve the control problems subserving standing, walking, running, jumping, dancing, and so on. Vestibular inputs play a central role in all these tasks, which are achieved through a combination of postural movements and forces and torques exerted against the environment. Better understanding of the functioning of the vestibular system may have important implications in design and control of robotic systems.

Vision. In humans, the head-located vestibular system is known to participate in a number of functions that include gaze stabilisation through the vestibulo-ocular reflex [17, 18, 40, 57]. The vestibulo-ocular reflex stabilises the gaze to ensure clear and still vision. It is a reflex of eye movement that stabilises the image projected on the retina during head movements. Produced eye movements are in the direction opposite to head movements. The reflex has both rotational and translational components which are driven by semicircular canals and otoliths, respectively.

Self-motion perception. The vestibular system is a key sensory organ that enables perception of body motion [19]. A human with a healthy vestibular system can always tell if the body is moving even if other senses such as vision, audition are absent. Vestibular system provides us with the ability to distinguish between self-generated motion and the external one. It has been shown that vestibular only information is sufficient to reconstruct current and past passive body locations [77, 78].

Balance. The vestibular system plays a dominant role in the coordination of postural reflexes, such as vestibulocollic reflex. The vestibulocollic reflex is responsible for maintaining head and body posture. This reflex stabilises the head with respect to inertial space. It produces commands to move the head in the direction opposite to the direction of the actual velocity of the head [10, 47]. Another important role of the vestibular system is vestibulospinal reflex which coordinates head and neck movement with respect to the trunk of the body. The goal of the reflex is maintaining the head in an upright position [3]. Together, vestibulocollic and vestibulospinal reflexes are responsible for self-balancing control [2, 98, 127, 145, 146].

Perception of verticality. The vestibular system is the principal sensory system which is able to perceive perceive gravitational forces. When the body is motionless, otoliths measure the gravitational acceleration vector, whose components render the sense of absolute verticality [20, 149]. Knowledge of gravitational verticality is essential for balancing and posture control, as it allows to determine spatial orientation of the body [30].

Frame of reference. The vestibular system, like embodied inertial sensors, provides the CNS with a head-centred frame of reference. It may be suggested that low-level balance and posture control is realised in this frame of reference. Spatial transformations from head-fixed and world inertial frame can be performed based on vestibular system measurements. Therefore, this embodied frame of reference is directly related to the world inertial frame and enables the neural system to perform stable posture control independently from other sensory inputs, such as tactile or proprioception information on ground inclination. In this way, ground-independent posture and balance control can be implemented.

3   Head Stabilisation Behaviour in Animals and Humans

Various human motion experimental studies showed that humans stabilise their heads while performing locomoting, balancing or other postural tasks. It has been initially proposed by T. Pozzo and A. Berthoz that humans stabilise their heads in rotation for different locomotor tasks, such as free walking, walking in place, running in place and hopping [115]. In the experiments with ten healthy subjects humans stabilised their heads in such a way as the maximum angular amplitude of Frankfort's plane (plane of horizontal semicircular canals) did not exceed 20°. It was suggested that such stabilisation is probably due to a cooperation between both the measurement of head rotations by the semi-circular canals and the measure of translations by the utriculus and sacculus (otolith organs). The plane of the stabilisation is determined by the task; it can vary and be controlled by gaze. Further experiments showed that total darkness does not significantly influence the stabilisation of the head, which stresses the importance of this behaviour in the coordination of the multiple degrees of freedom of the body during gait [116].

Later in [117], it was shown that head stabilisation occurred also in the frontal plane during the maintenance of monopodal and bipodal equilibrium on unstable rocking platforms. The head remained stable relative to the vertical, despite large translations in the frontal plane. Head angular stabilisation close to vertical orientation was essential for effective postural control during those complex balancing tasks.

Some recent behavioural studies by A. Berthoz and colleagues showed that head orientation is anticipated during locomotion relatively to walking direction [64, 73, 113]. This may suggest that head orientation and gaze stabilisation is important for motion planning during locomotion, and both visual and vestibular cues can be processed by the CNS in a better way if the head is stabilised and oriented towards the walking direction. Additional studies suggested that the motion of the head, together with gaze control, is closely related to optimal postural control during locomotion [6].

In [84], a control mechanism model for head stabilisation was described. Angular velocities of the head and trunk in space were recorded in seated subjects during external perturbations of the trunk. The passive mechanics of the head was changed by adding additional mass in different experimental trials. It was shown that head stabilisation in a horizontal plane in yaw orientation did not differ much with respect to the changed inertia of the head. For pitch motion, the response of the head-neck stabilisation controller was changed when additional mass was added to the head. However, in all the cases subjects were stabilising their heads while their trunk was perturbed.

In [25], balancing on a moving platform with subjects with and without the bilateral vestibular loss was studied. Results showed that subjects with vestibular loss were unable to perform that task properly, and their trunk and head were not stabilised in space, while healthy subjects were stabilising their heads regardless to the motion of the platform.

Similar to humans, head stabilisation was observed in many other animals. For instance, head stabilisation was reported for cats [61]. Similar head stabilisation was found in monkeys during locomotion [148]. The head was stabilised and absolute values of pitch and roll head movements did not exceed 7°. The similar result was obtained for running monkeys [42]. For horses, a weaker head stabilisation effect was reported [43].

Head stabilisation was observed for birds, as well. In [65], head stabilisation in chickens during jumping and walking on surfaces of different slopes. Results suggested that the head was stabilised during locomotion but the angle of stabilisation increased with increasing downwards slope of the walking surface and decreased with increasing upwards slope. Head stabilisation was reported for other birds such as herons, as well [83]. In most of the cases, different types of herons stabilised their heads' spatial orientation and location while their body was disturbed by harmonic oscillations. Head bobbing during locomotion is a very well-known behaviour for birds like pigeons, egrets, whooping cranes [34, 38, 55, 56, 136]. Head-bobbing behaviour results in stabilisation of the head to the surroundings for specific phases of the locomotion cycle. This bobbing is coupled with the locomotion and is mainly under the control of visual system inputs [55, 106].

In the studies mentioned above, it was shown that orientation of the head is naturally stabilised during locomotion, balancing. These studies emphasized the importance of head stabilisation as a part of the general postural control in humans and animals. In some cases, head stabilisation may be related to the task performed by a human, but naturally, the head stabilisation may be the result of vestibular-ocular interactions, as well [69]. However, in all of these cases vestibular information is important for the head-neck control system, and sometimes it is the only source of information available [23]. One of the key roles of head stabilisation is related to the establishment of a stable reference frame in which spatial perception, multiple sensory integration, and postural control are performed. The stable frame of reference based on vestibular information provides the brain with a mobile reference frame, which, in cooperation with vision and gaze allows a 'top-down' control of locomotion [8, 15].

4 Sensing Motion in Robots with Inertial Sensors

4.1 Inertial Sensors

In robotic systems, inertial sensors play a role similar to the vestibular system in humans and animals. In this subsection, we give a short introduction to inertial sensors and their applications in robotic systems. Modern inertial sensors are electromechanical transducers which measure translation and rotation of their bodies. Recent technological developments provided robotics with various types of inertial sensors [12, 144]. The most common inertial sensors are accelerometers, gyrometers and inclinometers. Very often they are combined together and then they are called inertial measurement units (IMU).

An accelerometer measures linear acceleration which is a result of linear forces and gravitational forces applied. Modern accelerometers are usually implemented in MEMS based electronic circuits and can have sensitivity in one, two or three axes. Some examples of different technologies used in accelerometer design can be found in [4, 26, 87, 108, 118]. In a typical accelerometer, a rigid body is constrained to linear translation inside the frame of the sensor. The body is attached to the frame with a spring of a known stiffness. Assuming linear stiffness, low friction and small displacements acceleration of the sensor's body can be calculated from the body's displacement. Importantly, accelerometers react to both translational and gravitational accelerations. In absence of translational motion, they can thus be used as tilt meters (inclinometers). However, this causes a significant problem for acceleration sensing in robotics which is often called gravito-inertial ambiguity. It is impossible to distinguish between translational and gravitational components of acceleration when a robotic body is being accelerated in translation and rotation at the same time.

An inclinometer is a sensor used to measure the absolute angular orientation of a body with respect to gravitational acceleration. The physical principle of sensing is the same as for the accelerometer, but an inclinometer's primary function is to measure angular orientation with respect to the gravitational acceleration vector. A very simple type of inclinometer can be implemented with a pendulum which tends to keep its verticality, and in steady state, it is aligned with the gravitational acceleration vector. The measurement is an angle between the pendulum and the frame of the sensor. This angle is usually measured by magnetic, optical or electromechanical angular encoders attached to the joint between the sensor frame and the pendulum. Usually, a moving body in an inclinometer is damped to achieve convergence and steady measurements. In the cases when the frame of the sensor is accelerated, the measurement from the inclinometer is no longer tilt-only related, and the sensor behaves in the same way as the tilted accelerometer as it responses both to translational and gravitational acceleration. Modern inclinometers use different electromechanical, magnetic, hydraulic, optical and other physical effects. Some design examples can be found in [92, 97, 102, 150].

A gyrometer (or a gyroscope) is a sensor for measuring the angular velocity. Modern gyrometers, similar to accelerometers are implemented in electronic circuits based on MEMS technology [66, 105]. A common MEMS based gyrometer measures angular velocity by means of Coriolis acceleration [60].

4.2   Verticality Estimation Methods

Verticality estimation is a crucial task for the stability of mobile robotic systems. In particular humanoid robots and other walking machines require accurate knowledge of verticality to perform balancing and postural tasks in the gravitational field. In this subsection, we look into some basic methods of verticality estimation methods used in robotic systems. Here we limit our review to approaches which use inertial measurements only. Theoretically, simple time integration of gyrometers output will provide us with the angular orientation of a body (robot) in space. However, initial conditions have to be known, which is not the case for many real-life applications. Another problem with rate sensors is bias, which changes with time, temperature and other conditions. An accelerometer can be used as a tilt sensor to measure the vertical orientation of the robot in the cases when the linear acceleration of the robot is zero or negligible. But most of robotic systems such as humanoid robots are moving continuously with variable velocities and are the subjects to unpredictable mechanical impacts. Generally speaking, inertial sensors are noisy because they pick-up vibrations that become added to the low-frequency components of interest of the acceleration and velocity signals. Gyroscopic measurements also suffer from bias and are highly sensitive to dynamic errors. To combat these problems, different approaches to the design state observers and sensor fusion methods have been proposed to improve inertial measurements [13].

In [4], a nonlinear regression model to improve the accuracy of a low-g MEMS accelerometer was proposed. It was assumed that accelerometer was not translated and its measurements provided the tilt information only. Similarly, in [119] accelerometer was used as a tilt sensor. Accurate tilt sensing was achieved with a linear kinematic model which included scale factor, bias and misalignment. It was assumed that the body does not perform any translational motion which makes this approach limited to rotations only. Gyrometer and accelerometer measurements were used together with a Kalman filter for sensors alignment and calibration errors compensation in [67]. Similar to previous cases only rotations were considered.

The vertical orientation of a flying robot was estimated in [9]. The measurement system included an inclinometer and a rate gyro. The data coming from the sensors was fused through a complementary filter, which compensated the slow dynamics of the inclinometer. It was assumed that the robot did not perform fast accelerated motions.

A sensor fusion approach for verticality estimation in mobile robots which are translated and rotated simultaneously was proposed in [138]. To resolve the ambiguity of translational and gravitational components in measurements the odometry of the mobile robot was used. Angular measurements from the robot's wheels, joints and knowledge of its kinematics were used to provide the estimator with additional information about the robot's orientation.

In [120] drift-free attitude estimation for accelerated rigid bodies was described. The attitude estimation problem for an accelerated rigid body using gyros and accelerometers was solved with a switching algorithm which employed Kalman filters. Switching was performed based on the values of estimated linear acceleration of the body. The main idea of the algorithm was to consider the translational accelerations as disturbances that were measured partially by the accelerometer. When the acceleration of the body was too high, the accelerometer measurements were considered completely unreliable to provide verticality measurements switching algorithm forced the estimator to rely on the rate gyros by setting the accelerometer noise covariance matrix to infinity.

Additional inclination sensor was used in [91]. A state estimation technique was developed for sensing inclination angles using a low-bandwidth tilt sensor along with an inaccurate rate gyro and a low-cost accelerometer. The model of rate gyro included an inherent bias along with sensor noise. The tilt sensor was modelled as a pendulum and was characterized by its own slow dynamics. These sensor dynamics was combined with the gyrometer model to achieve high-bandwidth measurements using an optimal linear state estimator. Acceleration of the system in the inertial frame was measured by additional accelerometer and was considered as a known input for the estimator. However, in many robotic systems knowledge of the acceleration vector expressed world inertial frame is unavailable or requires additional global measurements.

In [46] acceleration of an unmanned aerial vehicle expressed in the world frame was estimated by an airflow sensor. This estimation together with the raw measurement from accelerometer was used to calculate the gravitational component of the acceleration. But the performance of the air flow sensor and may not be suitable for human-like walking robotic systems.

A multiple accelerometer based sensory system was used for attitude estimation of a rigid body in [135]. A set of accelerometers was attached to various locations of a rigid body and an optimal linear estimation algorithm was developed for determining pitch and roll of the body fixed at a pivot. The estimated tilt angles values were filtered based on additional angular velocity measurements.

As shown above, the attitude estimation problem for robotic systems is a complex one and often requires nonlinear state estimation techniques to solve it [33]. One of the key problems in verticality estimation based on inertial sensors only is the ambiguity in acceleration or inclinometer measurements. Additional techniques are required to separate the gravitational component of acceleration from translational one. In many cases, extra sensors may facilitate the solution. For example, global positioning system (GPS) was used in [59]; magnetic field sensors were utilized in [54, 103], additional bearing information was required in [11]; landmark measurements were used in [141], Earth horizon sensor was utilized in [63], active vision system was employed in [22]. In humanoid robots, it was shown that inertial sensing can be efficiently used to estimate the robot's attitude when combined with body joint and links relative position sensors and legs contact information [14, 123] or based on additional information from an external range sensor [48]. Like robotic systems, living creatures are solving the same problem of estimation gravitational verticality using a limited set of sensors.

5 Anthropomorphic Head Stabilisation in Human-like Robots

5.1 Location of Inertial Sensors in Humanoid Robots

In living creatures, past and present, the vestibular system is located in the head which is explained by their evolutionary development and specific roles of vestibular organs. In robotic systems, a common design wisdom wants that the inertial sensors to be located close to the centre of mass of a humanoid robot. Most of the humanoid robots have their IMU located in the main body close to their centre of mass. Locating an IMU close to the centre of mass provides information about the motion of the centre of mass, of rotations about it, and allows the application of simple models (inverted pendulum) for locomotion and balancing control tasks. In Table 1, we summarised our survey results on the location of inertial sensors in some of the humanoid robots used in research.

As we can see from the table, most of the robots have their IMU, or, we may call it, artificial vestibular system, in the main body: between hip joints, torso, trunk or pelvis. Only some robots have their IMU located in the head. In robots like Cog, [24], and iCub, [137], the IMU is located in their heads. However, the original design of these robots did not include locomotion functionality, since they were mainly used for human-robot interaction and computational neuroscience studies. The ARMAR-III robot, [7], has its inertial sensors in the head, but the lower body of the robot is based on a wheeled platform, which makes it very different from legged biological systems. To our knowledge, very few physical humanoid robots use inertial sensors in the head for posture control during locomotion and balancing. Another example, the CB humanoid, [31, 71], uses two IMUs: one in the head and another one in the torso. Information from both of them is used to coordinate eyes, head and torso movements [132]. In [86] and [49], robotic heads were equipped with IMUs which were used for head stabilisation during locomotion. In [86], direct measurements of the head's angular orientation were fed back to linear control of the head. However, the performance of this head stabilisation system was quite low, because there were no filtering or estimation methods used for sensory data processing. In [49], a feedback learning algorithm was employed to stabilise the head orientation independently from the trunk motion. A neural network was used to learn the unknown head dynamics. More recent robotic systems used IMU located in the head for cooperative head-neck movements and gaze control, which enabled the humanoid robot to achieve human-like gaze and head control behaviour.

5.2   Head Stabilisation in Robots

In robotic systems, inertial sensors provide control systems with important information on actual robot's state such as its body's acceleration, velocity, orientation. This information is important for maintaining stable motion while performing a required task. Accelerometers are often utilized as tilt sensors and gyroscopes are used to measure the change in the robot's orientation. Next, we describe some of the robotic applications in which head stabilisation plays an important role to achieve better performance and human-like behaviour.

Balancing and walking. Classical humanoid robot locomotion strategy is based on zero moment point (ZMP) control originally proposed by M. Vukobratovic [143]. In this case, knowledge about the robot kinematic configuration and mass distribution was sufficient for stable locomotion. Modern humanoid robotics requires robots to move faster and perform more dynamic tasks which require the use of additional

Table 1 Location of the inertial sensors in the humanoid robots

Robot's name	IMU location	References	Year
Cog	Head	[24]	1999
ASIMO	Upper body	[74, 124]	1999
ATLAS (DASL-1)	Pelvis	[79]	2002
H7	Upper body	[81]	2004
KHR- 3 HUBO	Torso	[112]	2005
FHR- 1 FIBO	Hip	[147]	2006
KHR- 2	Torso	[85]	2006
CB	Head and torso	[31, 71]	2007
iCub	Head	[111, 137]	2007, 2012
MAHRU- R	Pelvis	[29]	2008
ARMAR- III	Head	[7]	2008
Lola	Upper body	[27, 93]	2009
Nao	Torso	[62]	2009
AAU- BOT1	Trunk	[133]	2009
Sarcos Primus	Hip	[131]	2010
KOBIAN	Head	[86]	2012
WABIAN	Head	[49]	2012
ROMEO	Head	[114]	2014
COMAN	Waist	[1]	2014
MOMARO	Head	[125]	2016

sensory information, such as inertial measurements to enhance balance control of the robot.

For example, in [133], development of AAU-BOT1 humanoid robot is described. In this robot, an IMU was placed in the trunk and was used for balance control during locomotion. The orientation of the robot measured by accelerometers and gyroscopes and foot-ground force reactions measured by force sensors were considered to be the dominant information for walking controller.

Tilt and angular velocity of humanoid's torso were used for balance and walking control in [85]. Tilt was measured by accelerometers attached to the torso of the robot. Direct measurements were used by torso roll and pitch controller and predicted motion controller to avoid tilt over situations.

Inertial measurement unit was used for humanoid balance control in [31, 71]. Three-axis accelerometer and gyro were attached close to the centre of mass in the trunk of the robot. Angular velocity information was used for the centre of mass measurement and calculation of the desired ground interaction force direction for maintaining a stable posture.

Balance control of humanoid robot was described in [29]. The robot's body was equipped with an IMU for measuring trunk's angular velocity and orientation in space. This information was used for the robot's actual posture calculation together with the measurements from joint sensors.

Biologically inspired postural and reaching control for the humanoid robot was presented in [134]. A humanoid robot was equipped with accelerometer and gyrometer in the head. Normal and tangential contact forces between the robot and the platform were measured by force sensors. Unlike in most of the humanoid research literature, in [134], the platform's (ground's) non zero inclination was considered to be unknown. Unknown ground tilt and disturbance forces were estimated based on measurements from accelerometer and gyrometer. Estimation was implemented by Kalman filtering. It was claimed that inertial sensors and their signals processing could be called as the artificial vestibular system of the humanoid robot [134]. In [114] a sophisticated version of IMU and eye control integration was presented for a humanoid robot. The system was based on coordinated human-like control of head-neck and gaze system which enabled to achieve natural saccadic movements in the head's frame of reference.

Gaze and head stabilisation. In many recently developed human-like robots, the vestibulo-ocular reflex is realised for gaze stabilisation. Common human-like robot's head is equipped with two video cameras. Stereo information from these cameras is used for robot's localization and visual perception of the environment. For instance, it may be very important for planning grasping tasks. This visual flow should be stabilised by implementation gaze stabilisation like it is done in biological systems. In addition, this gaze stabilisation behaviour makes a humanoid robot's more natural and realistic which is important in human-robot interaction scenarios.

One of the first humanoid ocular motor control systems was described in [128, 129]. A biomimetic gaze stabilisation based measurements from three gyros in the head of the robot was implemented. Feedback-error learning algorithm with neural networks was implied to achieve human-like vestibulo-ocular and optokinetic reflexes. Similarly, in [110], the humanoid robot's head was equipped with an IMU and video cameras. Measurements from the IMU and analysis of visual flow from cameras were used for achieving efficient visual stabilisation. Inertial sensors provided short latency measurements of rotations and translations of the robot's head. Visual flow information provided a delayed estimate of the motion across the image plane. A self-tuning neural network was used to learn to integrate visual and inertial information and to generate proper ocular-motor control signals.

Vision only based gaze stabilisation was presented in [58]. Adaptive frequency oscillators were used to learn the frequency, phase and amplitude of the optical flow from the robot's cameras during locomotion. The developed vision-based control was used for gaze stabilisation during periodic locomotion and for visual object tracking.

In [31], a three-axis gyrometer and an accelerometer in the head of the humanoid were used for controlling the gaze and implementing visual attention system. In [94], biomimetic eye-neck coordination control was proposed and used for visual target tracking. Proprioceptive feedback from neck joints and vestibular signals from inertial sensors in the head were processed together. The control system was tested with the iCub humanoid robot.

6   Benefits of Head Stabilisation in Robots

6.1   Towards a Robotic Vestibular System

In the previous section, we have presented some example of robots with IMUs located in their heads. We can call such IMUs as a *robotic vestibular system*. In fact, in humanoid robotics research, the term *robotic vestibular system* or *artificial vestibular system* was already in use. However, strictly speaking not in all of the cases inertial sensors may be interpreted as the artificial vestibular system because of the ways the sensors are implemented or used. In [142], one of the first attempts to design artificial vestibular system was presented. It was suggested to integrate bi-axial accelerometer and uni-axial gyrometer in a single inertial sensing unit. Later, in [109] a three-axial artificial vestibular system was described. It included a sensing unit with three orthogonal planes. Each plane was equipped with MEMS based accelerometer and gyrometer. Design approaches described in [109, 142] are similar to the one used in classical sensor engineering when a set of individual inertial sensors were integrated together in a single IMU. An attempt to realise a biomimetic angular rate sensor based on the biomechanical model of the semicircular canals was presented in [32]. However, no further experimental results were presented. A more general point of view to the artificial vestibular system as a part of a multi-sensory system for a robotic rat was presented in [104]. A more thorough example of an artificial vestibular system implementation can be found in [114] where mutual integration of IMU sensing and head control for natural gaze control for was demonstrated for a humanoid robot.

6.2   Advantages of Head Stabilisation in Human-like Robots

A stabilised robotic head would presumably benefit from the same advantages as those of natural heads, hence we suggest that humanoid robots should also adopt a similar head stabilisation behaviour. Head stabilisation improves sensing robustness of visual and vestibular organs located in the head. It is important for the visual sensory system and gaze stabilisation, as well as it enables the establishment of a stable frame of reference for motion planning within the robots' body.

Among numerous other potential advantages, it is important to mention that a horizontally stabilised head facilitates the estimation of the gravitational vertical during locomotion. Knowledge of the direction of the gravity vector, that is of the gravitational vertical, is essential to achieve stance and locomotion as the gravitational vertical may be poorly estimated from visual cues or based on the kinematic relationship of the robot to the ground. However, when accelerometers cannot distinguish between translational and gravitational components of acceleration when translated and tilted at the same time. We suggest that a stabilised robotic head may improve the quality of inertial measurements of the sensors located in this head. First of all, we suppose that the head stabilised independently from the motion of the trunk will be less affected by external forces and disturbances which may occur during locomotion [51, 52]. Second, if the head is stabilised in upright vertical position, then at least one of the axis of inertial sensors will be aligned with gravitational acceleration. This means that a simpler estimation method can be used to process the measurements from inertial sensors, such as accelerometers. In certain situations, it may be assumed that accelerometer located in the upright stabilised head will measure independently gravitational and translational components of acceleration. However, stabilisation of a robotic head with respect to the gravitational verticality requires knowledge of this verticality, and in our case, the only available information related to gravitational verticality can be obtained from inertial sensors, in particular from an accelerometer. This means that the control loop has to be closed by the feedback of the measured or estimated vertical orientation of the head. As a result, we obtain an observer-based closed loop control system. The task of the observer is to estimate the vertical orientation of the head based on inertial sensor measurements without the explicit knowledge of translational acceleration of the head. Hence, our head observation-stabilisation system can be considered as self-sufficient or in other words ideothetic.

Idiothetic sensing, or sensing entirely based on states measured with reference to one's own body, is of course not special to robots. In aerospace engineering, flying and rocketing vehicles also use IMU. Long ago, it was noticed that guidance was greatly simplified if the inertial sensors were placed on stabilised platforms [44]. In such systems, the application of fundamental mechanical principles and stabilising control provided engineers with possibilities to establish the gravity referenced Earth's inertial frame without the need for other external references. Ideothetic inertial sensing can be useful to any type of mobile robotic systems [13].

Recently, a concept of top-down control was proposed for humanoid robotics. It was suggested the posture control during locomotion is governed in the head's frame of reference. As the head is stabilised, a stable frame of reference is achieved. Inertial measurements from the vestibular system would be expressed in this frame of reference and would be used for gaze control. Gaze direction anticipates on head orientation which in turn anticipates on the body segments during locomotion [15, 80, 88] and this behaviour can be used to control steering in assistive walking robots [53]. Following the top-down control organisation, a humanoid robot can be controlled in its stabilised head's frame of reference as it was done in [72, 130]. In [130], the humanoid robot was teleoperated and its desired walking direction and posture were given to the controller in head's frame. This example of robot control with top-down organization differs from the classical humanoid locomotion control when posture configuration was defined in the world (ground) reference frame. In most of the humanoid walking applications ground is assumed to be flat and rigid, and its inclination has to be known. However, the concept of top-down control organization suggests that the only information which is required is joint (proprioceptive) information of the robot's body with respect to a stabilised head's frame of reference. Head stabilisation also contributes to whole body dynamic stabilisation during locomotion in humanoid robots, as it was demonstrated in [89]. Interestingly, head stabilisation contributed to the stabilisation while the head's mass accounted only for 7% of the total mass of the body.

6.3   Current State of Research and Challenges

In this final subsection, we would like to re-iterate the importance of head stabilisation for robots through demonstrating some successful implementations. Research on human-like head stabilisation and control has been intensified over last years. In [114] a set of experiments for human-like gaze control for a humanoid robot was presented. Efficiently coordinated head and eye movements were achieved through the development of anthropomorphic algorithms based on the head's IMU measurements and position sensing of head-neck and head-eye electric motors. Experiments with ROMEO humanoid robot demonstrated that eyes-only, head-only and combined head-eye coordinated movements corresponded to the natural behaviour observed in humans, which in turn contributed to realistic and intuitive human-robot interaction.

The problem of compensating for disturbances induced in a humanoid robot's cameras due to self-generated body movements was approached in [122]. The head's angular velocity measured by an embedded IMU was used to compensate the head's and the eyes' movements, which improved the quality of the robot's optical flow.

An internal model-based adaptive gaze stabilisation for a humanoid robot was proposed in [140]. The internal model was based on the coordination of vestibulo-collic and vestibulo-ocular reflexes, and it includes a self-learning and adaptation capabilities. As a result, the experiments with a simulated robotic platform demonstrated that the model was able to stabilise the robot's head independently of the torso movements and at the same time to control gaze for tracking visual targets.

Three head stabilisation control approaches for humanoid robots were described and implemented in [50]. The approaches were two types of inverse kinematic based control and bioinspired adaptive control based on feedback error learning control. All controllers used the robot's head IMU measurements and head-neck joint kinematics information. Additionally to head stabilisation behaviour, vestibulo-ocular and optokinetic reflexes were implemented for the gaze control. Control based on feedback error learning adjusted the mapping of sensory errors into control errors. It showed better performance in head stabilisation behaviour compared to a simple feedback control with inverse kinematics model, demonstrating that nonlinear adaptive techniques are more suitable for anthropomorphic robot control.

Whereas significant progress has been achieved in transferring anthropomorphic behaviours to humanoid robot control, substantial research and development challenges remain. Firstly, resolution of gravito-inertial ambiguity for inertial sensor measurements is problematic for robots with ideothetic sensing and very often employment of external sensors with a fixed frame of reference is required. One more challenging issue, also related to gravito-inertial ambiguity is the existence of several competing gaze-head stabilisation approaches, however, due to technical limitation none of them is mimicking the simplicity of the biological system and achieve sufficient performance. Further development of gaze stabilisation algorithms will also require integration of visual flow based movement compensation, so that accurate visual tracking is achieved while head and gaze are stabilised. Addressing these challenges will enable development of more dynamic, stable and reliable robotic systems, and will be beneficial not only to humanoid robots but also to other types of mobile systems, such as jumping and flying robots.

Acknowledgements I. Farkhatdinov was supported by a fellowship from Ecole Doctorale, Sciences Mécaniques, Acoustique, Electronique et Robotique de Paris (UPMC). Additional funding was provided by the European Research Council, Advanced Grant PATCH, agreement No. 247300 and EU FP7 BALANCE (ICT-601003).

References

1. Ajoudani, A., Lee, J., Rocchi, A., Ferrati, M., Hoffman, E.M., Settimi, A., Caldwell, D.G., Bicchi, A., Tsagarakis, N.G.: A manipulation framework for compliant humanoid COMAN: Application to a valve turning task. In: 2014 IEEE-RAS International Conference on Humanoid Robots, pp. 664–670 (2014)
2. Allum, J.H.J., Adkin, A.L., Carpenter, M.G., Held-Ziolkowska, M., Honegger, F., Pierchala, K.: Trunk sway measures of postural stability during clinical balance tests: effects of a unilateral vestibular deficit. Gait Posture **14**(3), 227–237 (2001)
3. Allum, J.H.J., Honegger, F., Pfaltz, C.R.: Afferent Control of Posture and Locomotion, vol. 80 of Progress in Brain Research. Elsevier (1989)
4. Ang, W.T., Khosla, P.K., Riviere, C.N.: Nonlinear regression model of a low-g MEMS accelerometer. IEEE Sens. J. **7**(1), 81–88 (2007)
5. Angelaki, D.E., Cullen, K.E.: Vestibular system: the many facets of a multimodal sense. Annu. Rev. Neurosci. **31**, 125–150 (2008)
6. Arechavaleta, G., Laumond, J.-P., Hicheur, H., Berthoz, A.: An optimality principle governing human walking. IEEE Trans. Robot. **24**(1), 5–14 (2008)
7. Asfour, T., Azad, P., Vahrenkamp, N., Regenstein, K., Bierbaum, A., Welke, K., Schröder, J., Dillmann, R.: Toward humanoid manipulation in human-centred environments. Robot. Autonom. Syst. **56**(1), 54–65 (2008)
8. Authie, C., Hilt, P., N'Guyen, S., Berthoz, A., Bennequin, D.: Differences in gaze anticipation for locomotion with and without vision. Front. Hum. Neurosci. (2015) (June 9)
9. Baerveldt, A.-J., Klang, R.: A low-cost and low-weight attitude estimation system for an autonomous helicopter. In: Proceedings of IEEE International Conference on Intelligent Engineering Systems, pp. 391–395. IEEE (1997)
10. Baker, J., Goldberg, J., Peterson, B.: Spatial and temporal response properties of the vestibulocollic reflex in decerebrate cats. J. Neurophysiol. **54**(3), 735–756 (1985)
11. Baldwin, G., Mahony, R., Trumpf, J.: A nonlinear observer for 6 dof pose estimation from inertial and bearing measurements. In: 2009 IEEE International Conference on Robotics and Automation, pp. 2237–2242. IEEE (2009)
12. Barbour, N., Schmidt, G.: Inertial sensor technology trends. IEEE Sens. J. **1**(4), 332–339 (2001)
13. Barshan, B., Durrant-Whyte, H.: Inertial navigation systems for mobile robots. IEEE Trans. Robot. Autom. **11**(3), 328–342 (1995)
14. Benallegue, M., Lamiraux, F.: Humanoid flexibility deformation can be efficiently estimated using only inertial measurement units and contact information. In: 2014 IEEE-RAS International Conference on Humanoid Robots, pp. 246–251 (2014)

15. Bernardin, D., Kadone, H., Bennequin, D., Sugar, T., Zaoui, M., Berthoz, A.: Gaze anticipation during human locomotion. Exp. Brain Res. **223**(1), 65–78 (2012)
16. Berthoz, A.: The Brain's Sense of Movement. Harvard University Press, Cambridge (2000)
17. Berthoz, A., Droulez, J., Vidal, P.P., Yoshida, K.: Neural correlates of horizontal vestibulo-ocular reflex cancellation during rapid eye movements in the cat. J. Physiol. **419**(1), 717–751 (1989)
18. Berthoz, A., Jones, M.G., Begue, A.: Differential visual adaptation of vertical canal-dependent vestibulo-ocular reflexes. Exp. Brain Res. **44**(1) (1981)
19. Berthoz, A., Pavard, B., Young, L.: Perception of linear horizontal self-motion induced by peripheral vision (linear vection) basic characteristics and visual-vestibular interactions. Exp. Brain Res. **23**(5) (1975)
20. Bisdorff, A.R., Wolsley, C.J., Anastasopoulos, D., Bronstein, A.M., Gresty, M.A.: The perception of body verticality (subjective postural vertical) in peripheral and central vestibular disorders. Brain **119**(5), 1523–1534 (1996)
21. Borah, J., Young, L.R., Curry, R.E.: Optimal estimator model for human spatial orientation. Ann. N.Y. Acad. Sci. **545**, 51–73 (1988)
22. Bras, S., Cunha, R., Vasconcelos, J.F., Silvestre, C., Oliveira, P.: A nonlinear attitude observer based on active vision and inertial measurements. IEEE Trans. Robot. **27**(4), 664–677 (2011)
23. Bronstein, A.M.: Evidence for a vestibular input contributing to dynamic head stabilization in man. Acta Otolaryngol. **105**(1–2), 1–6 (1998)
24. Brooks, R.A., Breazeal, C.: The cog project: building a humanoid robot. In: Computation for Metaphors, Analogy, and Agents. Lecture Notes in Computer Science, vol. 1562, pp. 52–87 (1999)
25. Buchanan, J.J., Horak, F.B.: Vestibular loss disrupts control of head and trunk on a sinusoidally moving platform. J. Vestib. Res. Equilib. Orientation **11**(6), 371–89 (2002)
26. Bums, E., Homing, R., Herb, W., Zook, J., Guckel, H.: Resonant microibeam accelerometers. In: Proceedings of the International Solid-State Sensors and Actuators Conference—TRANSDUCERS '95, vol. 2, pp. 659–662. IEEE (1995)
27. Buschmann, T., Lohmeier, S., Ulbrich, H.: Humanoid robot lola: design and walking control. J. Physiol. Paris **103**(3–5), 141–148 (2009)
28. Carey, J.P., Santina, C.C.D.: Principles of applied vestibular physiology. In: Cummings Otolaryngology—Head and Neck Surgery, Chapter 163. Mosby (2005)
29. Chang, Y.-H., Oh, Y., Kim, D., Hong, S.: Balance control in whole body coordination framework for biped humanoid robot MAHRU-R. In: RO-MAN 2008—The 17th IEEE International Symposium on Robot and Human Interactive Communication, pp. 401–406. IEEE (2008)
30. Chase, W.G., Clark, H.H.: Semantics in the perception of verticality. Br. J. Psychol. **62**(3), 311–326 (1971)
31. Cheng, G., Hyon, S.-H., Morimoto, J., Ude, A., Hale, J.G., Colvin, G., Scroggin, W., Jacobsen, S.C.: Cb: a humanoid research platform for exploring neuroscience. Adv. Robot. **21**(10), 1097–1114 (2007)
32. Ciaravella, G., Laschi, C., Dario, P.: Biomechanical modeling of semicircular canals for fabricating a biomimetic vestibular system. IEEE Eng. Med. Biol. Soc. **1**, 1758–1761 (2006)
33. Crassidis, J.L., Markley, F.L., Cheng, Y.: Survey of nonlinear attitude estimation methods. J. Guidance Control Dyn. **30**(1), 12–28 (2007)
34. Cronin, T.W., Kinloch, M.R., Olsen, G.H.: Head-bobbing behavior in walking whooping cranes (*grus americana*) and sandhill cranes (*grus canadensis*). J. Ornithol. **148**(S2), 563–569 (2007)
35. Cullen, K.E.: The vestibular system: multimodal integration and encoding of self-motion for motor control. Trends Neurosci. **35**(3), 185–96 (2012)
36. Curthoys, I.S., Markham, C.H., Curthoys, E.J.: Semicircular duct and ampulla dimensions in cat, guinea pig and man. J. Morphol. **151**(1), 17–34 (1977)
37. David, R., Stoessel, A., Berthoz, A., Spoor, F., Bennequin, D.: Assessing morphology and function of the semicircular duct system: introducing new in-situ visualization and software toolbox. Sci. Rep. **6**, 32772 (2016)

38. Davies, M.N., Green, P.R.: Head-bobbing during walking, running and flying: relative motion perception in the pigeon. J. Exp. Biol. **138**(1), 71–91 (1988)
39. De Vries, H.: The mechanics of the labyrinth otoliths. Acta Otolaryngol. **38**(3), 262–73 (1951)
40. Dieterich, M., Brandt, T.: Vestibulo-ocular reflex. Curr. Opin. Neurol. **8**(1), 83–8 (1995)
41. Dimiccoli, M., Girard, B., Berthoz, A., Bennequin, D.: A functional explanation of otolith geometry. J. Comput. Neurosci. **35**(2) (2013)
42. Dunbar, D.C.: Stabilization and mobility of the head and trunk in wild monkeys during terrestrial and flat-surface walks and gallops. J. Exp. Biol. **207**(6), 1027–1042 (2004)
43. Dunbar, D.C., Macpherson, J.M., Simmons, R.W., Zarcades, A.: Stabilization and mobility of the head, neck and trunk in horses during overground locomotion: comparisons with humans and other primates. J. Exp. Biol. **211**(Pt 24), 3889–907 (2008)
44. Duncan, R.C., Gunnersen, A.L.F.S.J.: Inertial guidance, navigation, and control systems. J. Spacecraft Rockets **1**(6), 577–587 (1964)
45. Ernst, M.O., Bülthoff, H.H.: Merging the senses into a robust percept. Trends Cogn. Sci. **8**(4), 162–9 (2004)
46. Euston, M., Coote, P., Mahony, R., Hamel, T.: A complementary filter for attitude estimation of a fixed-wing UAV. In: 2008 IEEE/RSJ International Conference on Intelligent Robots and Systems, pp. 340–345. IEEE (2008)
47. Ezure, K., Sasaki, S., Uchino, Y., Wilson, V.J.: Frequency-response analysis of vestibular-induced neck reflex in cat. ii. functional significance of cervical afferents and polysynaptic descending pathways. J. Neurophysiol. **41**(2), 459–471 (1978)
48. Fallón, M.F., Antone, M., Roy, N., Teller, S.: Drift-free humanoid state estimation fusing kinematic, inertial and lidar sensing. In: 2014 IEEE-RAS International Conference on Humanoid Robots, pp. 112–119 (2014)
49. Falotico, E., Cauli, N., Hashimoto, K., Kryczka, P., Takanishi, A., Dario, P., Berthoz, A., Laschi, C.: Head stabilization based on a feedback error learning in a humanoid robot. In: 2012 IEEE RO-MAN: The 21st IEEE International Symposium on Robot and Human Interactive Communication, pp. 449–454. IEEE (2012)
50. Falotico, E., Cauli, N., Kryczka, P., Hashimoto, K., Berthoz, A., Takanishi, A., Dario, P., Laschi, C.: Head stabilization in a humanoid robot: models and implementations. Autonom. Robots **41**(2), 349–365 (2017)
51. Farkhatdinov, I., Hayward, V., Berthoz, A.: On the benefits of head stabilization with a view to control balance and locomotion in humanoids. In: 11th IEEE-RAS International Conference on Humanoid Robots 2011, Bled, Slovenia, pp. 147–152. IEEE (2011)
52. Farkhatdinov, I., Michalska, H., Berthoz, A., Hayward, V.: Modeling verticality estimation during locomotion. In: Romansy 19, Robot Design, Dynamics and Control, vol. 544 of CISM International Centre for Mechanical Sciences, Paris, France, pp. 359–366. Springer (2012)
53. Farkhatdinov, I., Roehri, N., Burdet, E.: Anticipatory detection of turning in humans for intuitive control of robotic mobility assistance. Bioinspiration Biomimetics **12**(5) (2017)
54. Fourati, H., Manamanni, N., Afilal, L., Handrich, Y.: A nonlinear filtering approach for the attitude and dynamic body acceleration estimation based on inertial and magnetic sensors: bio-logging application. IEEE Sens. J. **11**(1), 233–244 (2011)
55. Frost, B.J.: The optokinetic basis of head-bobbing in the pigeon. J. Exp. Biol. **74**, 187–195 (1978)
56. Fujita, M.: Head bobbing and the body movement of little egrets (*Egretta garzetta*) during walking. J. Comp. Physiol. A Neuroethology Sens. Neural Behav. Physiol. **189**(1), 53–58 (2003)
57. Jones, G.M., Berthoz, A., Segal, B.: Adaptive modification of the vestibulo-ocular reflex by mental effort in darkness. Exp. Brain Res. **56**(1), 149–153 (1984)
58. Gay, S., Ijspeert, A., Santos Victor, J.: Predictive gaze stabilization during periodic locomotion based on adaptive frequency oscillators. In: 2012 IEEE International Conference on Robotics and Automation, pp. 271–278. IEEE (2012)
59. Gebre-Egziabher, D., Hayward, R., Powell, J.: A low-cost GPS/inertial attitude heading reference system (AHRS) for general aviation applications. In: IEEE 1998 Position Location and Navigation Symposium (Cat. No. 98CH36153), pp. 518–525. IEEE (1998)

60. Geen, J., Krakauer, D.: Rate-sensing gyroscope. Technical report, ADI Micromachined Products Division (2003)
61. Goldberg, J., Peterson, B.W.: Reflex and mechanical contributions to head stabilization in alert cats. J. Neurophysiol. **56**(3), 857–875 (1986)
62. Gouaillier, D., Hugel, V., Blazevic, P., Kilner, C., Monceaux, J., Lafourcade, P., Marnier, B., Serre, J., Maisonnier, B.: Mechatronic design of NAO humanoid. In: 2009 IEEE International Conference on Robotics and Automation, pp. 769–774. IEEE (2009)
63. Grassi, M.: Attitude determination and control for a small remote sensing satellite. Acta Astronaut. **40**(9), 675–681 (1997)
64. Grasso, R., Prévost, P., Ivanenko, Y.P., Berthoz, A.: Eye-head coordination for the steering of locomotion in humans: an anticipatory synergy. Neurosci. Lett. **253**(2), 115–118 (1998)
65. Green, P.R.: Head orientation and trajectory of locomotion during jumping and walking in domestic chicks. Brain Behav. Evol. **51**(1), 48–58 (1998)
66. Greiff, P., Antkowiak, B., Campbell, J., Petrovich, A.: Vibrating wheel micromechanical gyro. In: Proceedings of Position, Location and Navigation Symposium—PLANS '96, pp. 31–37. IEEE (1996)
67. Grewal, M., Henderson, V., Miyasako, R.: Application of kalman filtering to the calibration and alignment of inertial navigation systems. IEEE Trans. Autom. Control **36**(1), 3–13 (1991)
68. Groen, J.J.: Cupulometry. Laryngoscope **67**(9), 894–905 (1957)
69. Guitton, D., Kearney, R., Wereley, N., Peterson, B.: Visual, vestibular and voluntary contributions to human head stabilization. Exp. Brain Res. **64**(1) (1986)
70. Hain, T., Ramaswamy, T., Hilmann, M.: Anatomy and physiology of vestibular system. In: Herdman, S.J. (ed.) Vestibular Rehabilitation, Chapter 1 (2007)
71. Hale, J., Cheng, G.: Full-body compliant human-humanoid interaction: balancing in the presence of unknown external forces. IEEE Trans. Robot. **23**(5), 884–898 (2007)
72. Hashimoto, K., Kang, H., Nakamura, M., Falotico, E., Lim, H., Takanishi, A., Laschi, C., Dario, P., Berthoz, A.: Realization of biped walking on soft ground with stabilization control based on gait analysis. In: Proceedings of the 2012 IEEE/RSJ International Conference on Intelligent Robots and Systems (2014)
73. Hicheur, H., Vieilledent, S., Berthoz, A.: Head motion in humans alternating between straight and curved walking path: combination of stabilizing and anticipatory orienting mechanisms. Neurosci. Lett. **383**(1–2), 87–92 (2005)
74. Hirose, M., Ogawa, K.: Honda humanoid robots development. Philos. Trans. Ser. A Math. Phys. Eng. Sci. **365**(1850), 11–9 (2007)
75. Hulk, J., Jongkees, L.B.W.: The turning test with small regulable stimuli. J. Laryngol. Otol. **62**(02), 70–75 (1948)
76. Igarashi, M., O-Uchi, T., Alford, B.R.: Volumetric and dimensional measurements of vestibular structures in the squirrel monkey. Acta Otolaryngol. **91**(5–6), 437–44 (1981)
77. Israel, I., Grasso, R., Georges-Francois, P., Tsuzuku, T., Berthoz, A.: Spatial memory and path integration studied by self-driven passive linear displacement. i. Basic properties. J. Neurophysiol. **77**(6), 3180–3192 (1997)
78. Israël, I., Rivaud, S., Gaymard, B., Berthoz, A., Pierrot-Deseilligny, C.: Cortical control of vestibular-guided saccades in man. Brain **118**(5), 1169–1183 (1995)
79. Jun, Y., Ellenburg, R., Oh, P.: From concept to realization: designing miniature humanoids for running. J. Systemics Cybern. Inform. (2010)
80. Kadone, H., Bernardin, D., Bennequin, D., Berthoz, A.: Gaze anticipation during human locomotion-top-down organization that may invert the concept of locomotion in humanoid robots. In: 19th International Symposium in Robot and Human Interactive Communication, pp. 552–557 (2010)
81. Kagami, S., Mochimaru, M., Ehara, Y., Miyata, N., Nishiwaki, K., Kanade, T., Inoue, H.: Measurement and comparison of humanoid H7 walking with human being. Robot. Autonom. Syst. **48**(4), 177–187 (2004)
82. Kalman, R.E.: A new approach to linear filtering and prediction problems 1. Trans. ASME J. Basic Eng. **82**(Series D), 35–45 (1960)

83. Katzir, G., Schechtman, E., Carmi, N., Weihs, D.: Head stabilization in herons. J. Comp. Physiol. A Sens. Neural. Behav. Physiol. **187**(6), 423–432 (2001)
84. Keshner, E.A., Hain, T.C., Chen, K.J.: Predicting control mechanisms for human head stabilization by altering the passive mechanics. J. Vestib. Res. Equilib. Orientation **9**(6), 423–34 (1999)
85. Kim, J.-Y., Park, I.-W., Oh, J.-H.: Experimental realization of dynamic walking of the biped humanoid robot KHR-2 using zero moment point feedback and inertial measurement. Adv. Robot. **20**(6), 707–736 (2006)
86. Kryczka, P., Falotico, E., Hashimoto, K., Lim, H., Takanishi, A., Laschi, C., Dario, P., Berthoz, A.: Implementation of a human model for head stabilization on a humanoid platform. In: 2012 4th IEEE RAS & EMBS International Conference on Biomedical Robotics and Biomechatronics (BioRob), pp. 675–680. IEEE (2012)
87. Lapadatu, D., Habibi, S., Reppen, B., Salomonsen, G., Kvisteroy, T.: Dual-axes capacitive inclinometer/low-g accelerometer for automotive applications. In: 14th IEEE International Conference on Micro Electro Mechanical Systems, pp. 34–37. IEEE (2001)
88. Laumond, J.-P., Arechavaleta, G., Truong, T.-V.-A., Hicheur, H., Pham, Q.-C., Berthoz, A.: The words of the human locomotion. In: Kaneko, M., Nakamura, Y. (eds.) Robotics Research, vol. 66 of Springer Tracts in Advanced Robotics, pp. 35–47. Springer (2011)
89. Laumond, J.-P., Benallegue, M., Carpentier, J., Berthoz, A.: The yoyo-man. Int. J. Robot. Res. 0278364917693292 (2017)
90. Laurens, J., Droulez, J.: Bayesian processing of vestibular information. Biol. Cybern. **96**(4), 389–404 (2007)
91. Leavitt, J., Sideris, A., Bobrow, J.: High bandwidth tilt measurement using low-cost sensors. IEEE/ASME Trans. Mechatron. **11**(3), 320–327 (2006)
92. Lin, C.-H., Kuo, S.-M.: High-performance inclinometer with wide-angle measurement capability without damping effect. In: 2007 IEEE 20th International Conference on Micro Electro Mechanical Systems (MEMS), pp. 585–588. IEEE (2007)
93. Lohmeier, S., Buschmann, T., Ulbrich, H.: System design and control of anthropomorphic walking robot LOLA. IEEE/ASME Trans. Mechatron. **14**(6), 658–666 (2009)
94. Lopes, M., Bernardino, A., Santos-Victor, J., Rosander, K., von Hofsten, C.: Biomimetic eyeneck coordination. In: 2009 IEEE 8th International Conference on Development and Learning. IEEE (2009)
95. Luenberger, D.G.: Observing the state of a linear system. IEEE Trans. Mil. Electron. **8**(2), 74–80 (1964)
96. MacNeilage, P.R., Ganesan, N., Angelaki, D.E.: Computational approaches to spatial orientation: from transfer functions to dynamic bayesian inference. J. Neurophysiol. **100**(6), 2981–96 (2008)
97. Manaf, A.B.A., Nakamura, K., Onishi, J., Matsumoto, Y.: One-side-electrode-type fluid-based inclinometer combined with cmos circuitry. In: 2007 IEEE Sensors, pp. 844–847. IEEE (2007)
98. Manchester, D., Woollacott, M., Zederbauer-Hylton, N., Marin, O.: Visual, vestibular and somatosensory contributions to balance control in the older adult. J. Gerontol. **44**(4), M118–M127 (1989)
99. Merfeld, D.M., Zupan, L., Peterka, R.J.: Humans use internal models to estimate gravity and linear acceleration. Nature **398**(6728), 615–8 (1999)
100. Merfeld, D.M., Zupan, L.H.: Neural processing of gravitoinertial cues in humans. iii. Modeling tilt and translation responses. J. Neurophysiol. **87**(2), 819–833 (2002)
101. Merfeld, D.M., Zupan, L.H., Gifford, C.A.: Neural processing of gravito-inertial cues in humans. ii. Influence of the semicircular canals during eccentric rotation. J. Neurophysiol. **85**(4), 1648–1660 (2001)
102. Mescheder, U., Majer, S.: Micromechanical inclinometer. Sens. Actuators A Phys. **60**(1–3), 134–138 (1997)
103. Metni, N., Pflimlin, J.-M., Hamel, T., Souères, P.: Attitude and gyro bias estimation for a VTOL UAV. Control Eng. Pract. **14**(12), 1511–1520 (2006)

104. Meyer, J.-A., Guillot, A., Girard, B., Khamassi, M., Pirim, P., Berthoz, A.: The psikharpax project: towards building an artificial rat. Robot. Autonom. Syst. **50**(4), 211–223 (2005)
105. Mottier, P., Pouteau, P.: Solid state optical gyrometer integrated on silicon. Electron. Lett. **33**(23) (1997)
106. Necker, R.: Head-bobbing of walking birds. J. Comp. Physiol. A Neuroethology Sens. Neural Behav. Physiol. **193**(12), 1177–1183 (2007)
107. Niven, J., Hixson, W.: Frequency response of the human semicircular canals: I. Steady-state ocular nystagmus response to high-level sinusoidal angular rotations. Technical report, NASA Naval School of Aviation Medicine (1961)
108. Novack, M.J.: Design and fabrication of a thin film micromachined accelerometer. Ph.D. thesis, Massachusetts Institute of Technology (1992)
109. Paian, F., Laschi, C., Miwa, H., Guglielmelli, E., Dario, P., Takanishi, A.: Design and development of a biologically-inspired artificial vestibular system for robot heads. In: 2004 IEEE/RSJ International Conference on Intelligent Robots and Systems (IROS), vol. 2, pp. 1317–1322. IEEE (2004)
110. Panerai, F., Metta, G., Sandini, G.: Learning visual stabilization reflexes in robots with moving eyes. Neurocomputing **48**(1–4), 323–337 (2002)
111. Paramiggiani, A., Maggiali, M., Natale, L., Nori, F., Schmitz, A., Tsagarakis, N., Victor, J.S., Becchi, F., Sandini, G., Metta, G.: The design of the iCub humanoid robot. Int. J. Humanoid Rob. 1250027 (2012)
112. Park, I.-W., Kim, J.-Y., Lee, J., Oh, J.-H.: Mechanical design of humanoid robot platform KHR-3 (KAIST humanoid robot - 3: HUBO). In: 5th IEEE-RAS International Conference on Humanoid Robots, 2005, pp. 321–326. IEEE (2005)
113. Pascal, P., Ivanenko, Y., Grasso, R., Berthoz, A.: Spatial invariance in anticipatory orienting behaviour during human navigation. Neurosci. Lett. **339**(3), 243–247 (2003)
114. Pateromichelakis, N., Mazel, A., Hache, M., Koumpogiannis, T., Gelin, R., Maisonnier, B., Berthoz, A.: Head-eyes system and gaze analysis of the humanoid robot Romeo. In: 2014 IEEE/RSJ International Conference on Intelligent Robots and Systems (IROS 2014), pp. 1374–1379. IEEE (2014)
115. Pozzo, T., Berthoz, A., Lefort, L.: Head stabilisation during various locomotor tasks in humans. Exp. Brain Res. **82**(1), 97–106 (1990)
116. Pozzo, T., Berthoz, A., Lefort, L., Vitte, E.: Head stabilization during various locomotory tasks in humans ii. Patients with bilateral vestibular deficits. Exp. Brain Res. **85**, 208–217 (1991)
117. Pozzo, T., Levik, Y., Berthoz, A.: Head and trunk movements in the frontal plane during complex dynamic equilibrium tasks in humans. Exp. Brain Res. **106**(2), 327–338 (1995)
118. Puers, R., Reyntjens, S.: Design and processing experiments of a new miniaturized capacitive triaxial accelerometer. Sens. Actuators A Phys. **68**(1–3), 324–328 (1998)
119. Qian, J., Fang, B., Yang, W., Luan, X., Nan, H.: Accurate tilt sensing with linear model. IEEE Sens. J. **11**(10), 2301–2309 (2011)
120. Rehbinder, H., Hu, X.: Drift-free attitude estimation for accelerated rigid bodies. Automatica **40**(4), 653–659 (2004)
121. Reisine, H., Simpson, J.I., Henn, V.: A geometric analysis of semicircular canals and induced activity in their peripheral afferents in the rhesus monkey. Ann. N.Y. Acad. Sci. **545**, 10–20 (1988)
122. Roncone, A., Pattacini, U., Metta, G., Natale, L.: Gaze stabilization for humanoid robots: a comprehensive framework. In: 2014 14th IEEE-RAS International Conference on Humanoid Robots (Humanoids), pp. 259–264. IEEE (2014)
123. Rotella, N., Bloesch, M., Righetti, L., Schaal, S.: State estimation for a humanoid robot. In: 2014 IEEE/RSJ International Conference on Intelligent Robots and Systems, pp. 952–958 (2014)
124. Sakagami, Y., Watanabe, R., Aoyama, C., Matsunaga, S., Higaki, N., Fujimura, K.: The intelligent ASIMO: system overview and integration. In: IEEE/RSJ International Conference on Intelligent Robots and System, vol. 3, pp. 2478–2483. IEEE (2002)

125. Schwarz, M., Rodehutskors, T., Schreiber, M., Behnke, S.: Hybrid driving-stepping locomotion with the wheeled-legged robot momaro. In: 2016 IEEE International Conference on
Robotics and Automation (ICRA), pp. 5589–5595 (2016)
126. Selva, P., Oman, C.M.: Relationships between Observer and Kalman filter models for human
dynamic spatial orientation. J. Vestib. Res. Equilib. Orientation **22**(2), 69–80 (2012)
127. Shepard, N.T., Telian, S.A., Smith-Wheelock, M., Raj, A.: Vestibular and balance rehabilitation therapy. Ann. Otol. Rhinol. Laryngol. **102**(3 Pt 1), 198–205 (1993)
128. Shibata, T., Schaal, S.: Biomimetic gaze stabilization based on feedback-error-learning with
nonparametric regression networks. Neural Netw. **14**(2), 201–216 (2001)
129. Shibata, T., Vijayakumar, S.: Humanoid oculomotor control based on concepts of computational neuroscience. In: IEEE-RAS International Conference on Humanoid Robots, Japan
(2001)
130. Sreenivasa, M.N., Soueres, P., Laumond, J.-P., Berthoz, A.: Steering a humanoid robot by its
head. In: 2009 IEEE/RSJ International Conference on Intelligent Robots and Systems, pp.
4451–4456. IEEE (2009)
131. Stephens, B.J., Atkeson, C.G.: Dynamic balance force control for compliant humanoid robots.
In: 2010 IEEE/RSJ International Conference on Intelligent Robots and Systems, pp. 1248–
1255. IEEE (2010)
132. Sugimoto, N., Morimoto, J., Hyon, S.-H., Kawato, M.: The eMOSAIC model for humanoid
robot control. Neural Netw. Official J. Int. Neural Netw. Soc. **29-30**(null), 8–19 (2012)
133. Svendsen, M.S., Helbo, J., Hansen, M.R., Popovic, D.B., Stoustrup, J., Pedersen, M.M.: Aau-
bot1: a platform for studying dynamic, life-like walking. Appl. Bionics Biomech. **6**(3–4),
285–299 (2009)
134. Tahboub, K.A.: Biologically-inspired postural and reaching control of a multi-segment
humanoid robot. Int. J. Biomechatron. Biomed. Robot. **1**(3), 175–190 (2011)
135. Trimpe, S., D'Andrea, R.: Accelerometer-based tilt estimation of a rigid body with only rotational degrees of freedom. In: 2010 IEEE International Conference on Robotics and Automation, pp. 2630–2636. IEEE (2010)
136. Troje, N., Frost, B.: Head-bobbing in pigeons: how stable is the hold phase? J. Exp. Biol.
203(5), 935–940 (2000)
137. Tsagarakis, N.G., Metta, G., Sandini, G., Vernon, D., Beira, R., Becchi, F., Righetti, L., Santos-
Victor, J., Ijspeert, A.J., Carrozza, M.C., Caldwell, D.G.: icub: the design and realization of
an open humanoid platform for cognitive and neuroscience research. Adv. Robot. **21**(10),
1151–1175 (2007)
138. Vaganay, J., Aldon, M., Fournier, A.: Mobile robot attitude estimation by fusion of inertial
data. In: Proceedings IEEE International Conference on Robotics and Automation, pp. 277–
282. IEEE Computer Society Press (1993)
139. Van Buskirk, W.C., Watts, R.G., Liu, Y.K.: The fluid mechanics of the semicircular canals. J.
Fluid Mech. **78**(01), 87–98 (1976)
140. Vannucci, L., Tolu, S., Falotico, E., Dario, P., Lund, H.H., Laschi, C.: Adaptive gaze stabilization through cerebellar internal models in a humanoid robot. In: 2016 6th IEEE International
Conference on Biomedical Robotics and Biomechatronics (BioRob), pp. 25–30. IEEE (2016)
141. Vasconcelos, J., Cunha, R., Silvestre, C., Oliveira, P.: A nonlinear position and attitude
observer on SE(3) using landmark measurements. Syst. Control Lett. **59**(3–4), 155–166 (2010)
142. Veltink, P.H., Luinge, H.J., Kooi, B.J., Baten, C.T.M., Slycke, P.: The artificial vestibular
system—design of a tri-axial inertial sensor system and its application in the study of human
movement. In: Proceedings of the International Society for Postural and Gait Research, number 1 (2001)
143. Vukobratovic, M., Juricic, D.: Contribution to the synthesis of biped gait. IEEE Trans. Biomed.
Eng. BME-**16**(1), 1–6 (1969)
144. Welch, G., Foxlin, E.: Motion tracking: no silver bullet, but a respectable arsenal. IEEE
Comput. Graph. Appl. **22**(6), 24–38 (2002)
145. Winter, D.: Human balance and posture control during standing and walking. Gait Posture
3(4), 193–214 (1995)

146. Winter, D.A., Patla, A.E., Prince, F., Ishac, M., Gielo-Perczak, K.: Stiffness control of balance in quiet standing. J. Neurophysiol. **80**(3), 1211–1221 (1998)
147. Wongsuwarn, H., Laowattana, D.: Experimental study for a FIBO humanoid robot. In: IEEE Conference on Robotics, Automation and Mechatronics (2006)
148. Xiang, Y., Yakushin, S.B., Kunin, M., Raphan, T., Cohen, B.: Head stabilization by vestibu-locollic reflexes during quadrupedal locomotion in monkey. J. Neurophysiol. **100**(2), 763–80 (2008)
149. Yelnik, A.P.: Perception of verticality after recent cerebral hemispheric stroke. Stroke **33**(9), 2247–2253 (2002)
150. Yotter, R., Baxter, R., Ohno, S., Hawley, S., Wilson, D.: On a micromachined fluidic incli-nometer. In: TRANSDUCERS '03. 12th International Conference on Solid-State Sensors, Actuators and Microsystems, vol. 2, pp. 1279–1282. IEEE (2003)
151. Zupan, L.H., Peterka, R.J., Merfeld, D.M.: Neural processing of gravito-inertial cues in humans. i. Influence of the semicircular canals following post-rotatory tilt. J. Neurophys-iol. **84**(4), 2001–2015 (2000)

The Physics and Control of Balancing on a Point in the Plane

Roy Featherstone

Abstract This chapter presents a new model of the physical process of balancing in a vertical plane, in which the essential parameters of a robot's balancing behaviour are distilled into just two numbers, regardless of the complexity of the robot. It also presents a balance control system based on this model, and a simple method for leaning in anticipation of future movements. The result is a control system that requires relatively little computation, yet allows a robot to make large, fast movements without falling over. Furthermore, by seeking to control the balance model only, instead of trying to control the whole robot, one achieves a separation between the potential complexity of the robot's complete equations of motion and the simplicity of its balancing behaviour. The chapter concludes with a simulation study to illustrate the kind of performance that can be achieved, and a brief discussion on how the theory can be extended to 3D.

1 Introduction

'Balancing' is a term that has acquired two meanings in the robotics community. According to one meaning, illustrated in Fig. 1a, balancing is the activity of maintaining a robot in an unstable upright posture. This kind of balancing is needed if the robot is standing on a point, a line, or an area that is very narrow in at least one direction; and it requires a type of balance controller that works by manipulating the moment of gravity about the support. A robot in this situation must be continually in motion in order to avoid falling over.

According to the other meaning, illustrated in Fig. 1b, balancing is the activity of maintaining a robot in a stable upright posture. This type of balancing is used when the robot has a substantial area (usually a polygon) of support. A robot in this situation can balance simply by standing still. However, if movement is required then it is necessary to make sure that the robot does not reach its tipping point. This is

R. Featherstone (✉)
Department of Advanced Robotics, Istituto Italiano di Tecnologia,
via Morego 30, 16163 Genova, Italy
e-mail: roy.featherstone@iit.it

© Springer International Publishing AG, part of Springer Nature 2019
G. Venture et al. (eds.), *Biomechanics of Anthropomorphic Systems*, Springer Tracts
in Advanced Robotics 124, https://doi.org/10.1007/978-3-319-93870-7_10

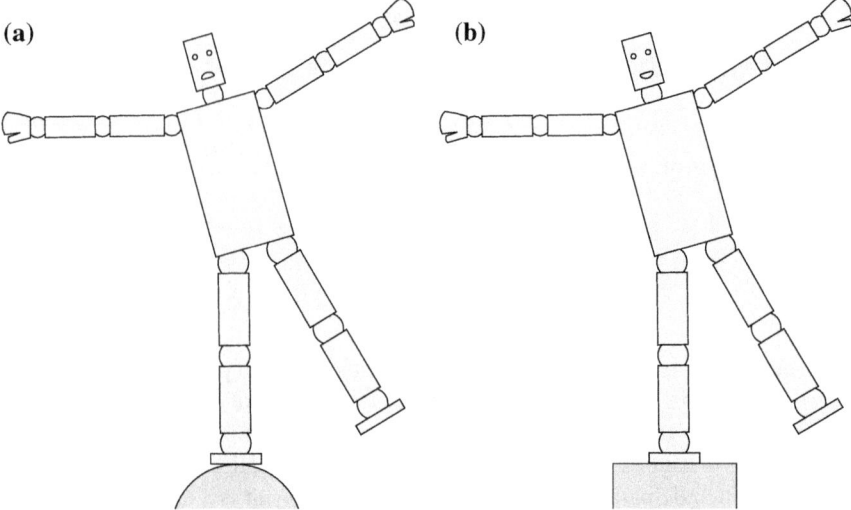

Fig. 1 Two types of balancing in robotics

usually accomplished by a motion planner that looks for a movement that satisfies
a given motion command while keeping the robot's zero moment point inside its
support polygon. Once the planner has found such a motion, it can then be executed
by a simple motion controller.

This chapter is concerned with the kind of balancing shown in Fig. 1a. In particular,
it examines the problem of a planar robot that must balance itself on a single point
in a vertical plane while simultaneously executing motion commands. The objective
of this study is to develop a model of the physics of balancing in the plane that (1)
facilitates the development of high-performance balance control systems, and (2)
serves as a step towards a similar model of balancing in 3D. The meaning of 'high
performance' in this context is that the robot should be able to track accurately a
command to make large, fast movements without losing its balance.

The reason for choosing to study this kind of balancing is that it is physically more
difficult than to balance on a large area of support. Thus, although a competent legged
robot needs both skills, it is the former that imposes the most stringent requirements
on the robot's hardware and software: the sensors must be good enough, the joints
must be fast enough, and the control system must be sufficiently effective, or else
the robot will fall.

The main contributions of this work are: (1) a new model of the physics of balancing on a point, in which the essential features of the robot's balancing behaviour
have been reduced to just two numbers; (2) a simple balance control system that
acts directly on this model; and (3) a simple method of making the robot lean in
anticipation of future movements. The resulting performance substantially exceeds
that of previously published balance controllers.

From a theoretical point of view, the main difference between the approach taken here and that in the existing literature is the idea of controlling only the robot's balancing behaviour, not the whole robot. In existing works, the development begins with the complete, closed-form symbolic equations of motion of the robot—see, for example, Eqs. 1 and 2 in [15], Eqs. 1 and 2 in [12], Eqs. 1–3 in [17], Eqs. 1–4 in [16] and Eqs. 1 and 2 in [4]. These equations are then manipulated in various ways in order to arrive at a formula for a feedback control law. This approach works well when the robot is simple, but it does not scale well because the size and complexity of the equations of motion grow very quickly with the number of joints in the robot. By seeking to control only the robot's balancing behaviour, as expressed in the new model, one achieves a decoupling between the simplicity of balance control and the complexity of the robot, so that the former remains simple no matter how complex the latter becomes.

The rest of this chapter is organized as follows. Section 2 develops the new model of balancing for the special case of a planar double pendulum; then Sect. 3 describes both the balance controller and the method of leaning in anticipation for this special case. Section 4 extends these results to the case of a general planar robot in which the balance controller is only one component in a larger control system; and Sect. 5 presents some simulation results that illustrate the kind of performance that can be achieved. Finally, Sect. 6 discusses briefly how to extend the theory from 2D to 3D. Readers can find a more detailed treatment of these topics in [9].

2 The New Model

Figure 2 shows an inverted double pendulum representing a simple robot balancing on a single point. It consists of two bodies, representing a leg and a torso, and two revolute joints, representing the contact with the ground and an actuated hip joint.

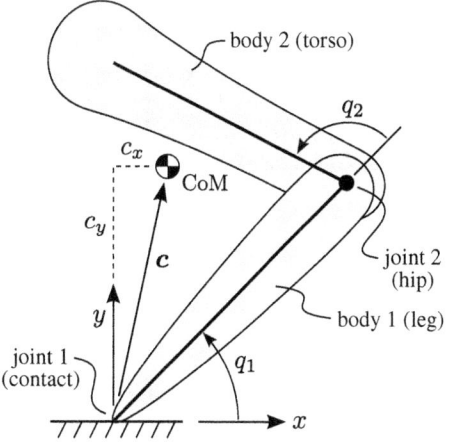

Fig. 2 An inverted double pendulum representing a simple planar robot balancing on a point

By modelling the contact with a revolute joint, we are assuming that the robot never loses contact with the ground, and that any movement of the contact point (e.g. due to rolling, slipping or local deformation) is negligible in comparison with the dimensions of the robot. If either the ground or the bottom of the leg is soft then the contact will occupy a small area instead of a single point, but any such contact can be approximated by a point contact located at the centre of pressure.

The robot has two degrees of motion freedom, of which only one is actuated. Its current state is given by the variables q_1, q_2, $\dot{q}_1$ and $\dot{q}_2$, which are the joint position and velocity variables; and the location of its centre of mass (CoM) relative to the support point is given by the vector c, having coordinates c_x and c_y expressed in a coordinate system with its origin at the contact point. The equation of motion of the robot is

$$\begin{bmatrix} H_{11} & H_{12} \\ H_{21} & H_{22} \end{bmatrix} \begin{bmatrix} \ddot{q}_1 \\ \ddot{q}_2 \end{bmatrix} + \begin{bmatrix} C_1 \\ C_2 \end{bmatrix} = \begin{bmatrix} 0 \\ \tau_2 \end{bmatrix},$$ (1)

where H_{ij} are the elements of the joint-space inertia matrix, C_i contain the Coriolis, centrifugal and gravitational terms, $\ddot{q}_i$ are the joint acceleration variables, and τ_2 is the torque at joint 2. Algebraic expressions for H_{ij} and C_i are not required, but we do need to know their numeric values, which can be computed efficiently using standard dynamics algorithms [6, 10].

The conditions for the robot to be in a balanced position are $c_x = 0$ and $\dot{q}_1 = \dot{q}_2 = 0$, and the balance controller's only means of attaining such a state is through motion of joint 2. However, the robot is also subject to the position command signal $q_c(t)$, which is an input to the balance controller specifying the commanded value of q_2 as a function of time. Clearly, it is not possible for joint 2 to track the command signal exactly if it also has to perform the movements needed to maintain the robot's balance; so some degree of tracking error is inevitable.

We now proceed to the analysis. Any mechanism that balances on a single point has the following special property, which is central to the activity of balancing: the only force that can exert a moment about the support point is gravity. If we define L to be the total angular momentum of the robot about the support point then we find that

$$\dot{L} = -mgc_x,$$ (2)

where m is the mass of the robot and g is the magnitude of gravitational acceleration (a positive number). This equation implies that

$$\ddot{L} = -mg\dot{c}_x$$ (3)

and

$$\dddot{L} = -mg\ddot{c}_x.$$ (4)

We also have

$$L = p_1 = H_{11}\dot{q}_1 + H_{12}\dot{q}_2,$$ (5)

which follows from a special property of joint-space momentum that is proved in the appendix of [9]: if p_i is the momentum variable of joint i then, by definition, $p_i = \sum_j H_{ij} \dot{q}_j$; but if joint i does not participate in any kinematic loop then p_i is also the component in the direction of motion of joint i of the total momentum of the set of bodies supported by joint i. As joint 1 is rotational, and it supports the whole robot, it follows that p_1 is the total angular momentum of the robot about the support point, hence $p_1 = L$.

Observe that $\dot{L}$ is simply a constant multiple of c_x, and that L and $\ddot{L}$ are both linear functions of the robot's velocity, implying that the condition $L = \ddot{L} = 0$ is equivalent to $\dot{q}_1 = \dot{q}_2 = 0$ (assuming linear independence). So the three conditions for balance can be written as

$$L = \dot{L} = \ddot{L} = 0 . \tag{6}$$

The significance of this result is that if a control system can make L converge to zero then it will have the side-effect of making $\dot{L}$ and $\ddot{L}$ also converge to zero, thereby satisfying all three conditions. Thus, any controller that successfully drives L to zero will cause the robot to balance, but will not necessarily bring q_2 to the commanded angle. (This is not a new result—see, for example, [13, §4].)

We now introduce a fictitious extra joint between joint 1 and the base, which is a prismatic joint acting in the x direction. To preserve the numbering of the existing joints, the extra joint is called joint 0. This joint never moves, and therefore never has any effect on the dynamics of the robot. Its purpose is to increase the number of coefficients in the equation of motion, which now reads

$$\begin{bmatrix} H_{00} & H_{01} & H_{02} \\ H_{10} & H_{11} & H_{12} \\ H_{20} & H_{21} & H_{22} \end{bmatrix} \begin{bmatrix} 0 \\ \ddot{q}_1 \\ \ddot{q}_2 \end{bmatrix} + \begin{bmatrix} C_0 \\ C_1 \\ C_2 \end{bmatrix} = \begin{bmatrix} \tau_0 \\ 0 \\ \tau_2 \end{bmatrix} . \tag{7}$$

The position and velocity variables of joint 0 are always zero, and τ_0 takes whatever value is necessary to ensure that $\ddot{q}_0 = 0$. The reason for adding this joint is that the special property of joint-space momentum, which we used earlier to deduce that $p_1 = L$, also implies that p_0 is the linear momentum of the whole robot in the x direction. So $p_0 = m\dot{c}_x$. With the extra coefficients in Eq. 7 we can write

$$p_0 = H_{01} \dot{q}_1 + H_{02} \dot{q}_2 = m\dot{c}_x = -\ddot{L}/g , \tag{8}$$

so that we now have a pair of linear equations relating L and $\ddot{L}$ to the two joint velocities:

$$\begin{bmatrix} L \\ \ddot{L} \end{bmatrix} = \begin{bmatrix} H_{11} & H_{12} \\ -gH_{01} & -gH_{02} \end{bmatrix} \begin{bmatrix} \dot{q}_1 \\ \dot{q}_2 \end{bmatrix} . \tag{9}$$

Solving this equation for $\dot{q}_2$ gives

$$\dot{q}_2 = Y_1 L + Y_2 \ddot{L} , \tag{10}$$

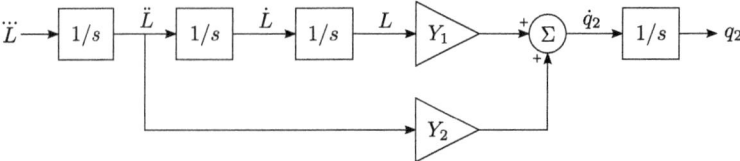

Fig. 3 New plant model for balancing

where

$$Y_1 = \frac{H_{01}}{D}, \qquad Y_2 = \frac{H_{11}}{gD} \tag{11}$$

and

$$D = H_{12}H_{01} - H_{11}H_{02} . \tag{12}$$

Clearly, this only works if $D \neq 0$. The physical significance of $D = 0$ is explained below. From a control point of view, a problem also arises if $Y_1 = 0$, and this too is discussed below. Note that Y_1 and Y_2 are not constants, but depend on the values of the joint position variables.

Equation 10 is the key to a new model of balancing, which is shown in Fig. 3 in the form of a block diagram.[1] This model serves as the *plant* that is to be controlled by the balance controller in Sect. 3. This means that the job of the controller is to calculate suitable values for the plant's input ($\dddot{L}$), given only the values of the plant's output (q_2), the command signal, the plant's state variables (q_2, L, $\dot{L}$ and $\ddot{L}$) and its parameters (Y_1 and Y_2).

By virtue of its construction, this plant has the special property that if a control system seeks to control the robot via this plant, then the resulting motion is such that q_2 follows the command signal in a way that maintains the robot's balance. To see how this is possible, consider the case of a constant value for q_c, meaning that joint 2 is being commanded to move to a specified constant angle. If the control system is stable then q_2 will converge to the commanded value, implying that $\dot{q}_2$ will converge to zero; but the only way that $\dot{q}_2$ can converge to zero, assuming that the system is stable, is if L, $\dot{L}$ and $\ddot{L}$ also converge to zero, thereby satisfying the conditions for balance in Eq. 6. Eventually, the plant reaches a steady state in which $q_2 = q_c$ and $L = \dot{L} = \ddot{L} = 0$. Furthermore, if a disturbance should occur then the control system immediately responds in a way that is calculated to bring the plant back towards its steady state, thereby restoring the robot's balance.

If the command signal varies then the situation is more complicated, and there do exist commands that can make the robot fall over. In particular, there exist commands that will cause the robot to deliberately tip over in a measured way—a behaviour that is necessary if the robot is to lean in anticipation of future movements. Nevertheless, it is possible to design a control system that will let the robot perform large, fast

---

[1]The lines in this diagram represent signals, and the blocks represent operations on those signals such as integration ($1/s$), summation, and multiplication by a scalar.

movements without falling over—so large, in fact, that the practical limit on what the robot can do is set by the physical limits of the robot itself (joint motion and speed limits, actuator torque limits, etc.) rather than the performance limits of the control system. We shall develop such a control system in Sect. 3.

Note that Y_1 and Y_2 are calculated directly from the elements of the joint-space inertia matrix in Eq. 7, which can be obtained using any standard method for calculating the joint-space inertia matrix of a robot. So no special code is needed to calculate these quantities.

Physical Meaning of Y_1 and Y_2

The two gains Y_1 and Y_2 are related in a simple way to two physical properties of the mechanism: the natural time constant of toppling and the linear velocity gain [7, 8]. The former quantifies the rate at which the robot begins to fall in the absence of movement of the actuated joint; and the latter measures the degree to which motion of the actuated joint influences the motion of the CoM.

If there is no movement in the actuated joint then the robot behaves as if it were a single rigid body, and its motion is governed by the equation of motion of a simple pendulum:

$$I\ddot{\theta} = mgc(\cos(\theta_0) - \cos(\theta)) \tag{13}$$

where I is the rotational inertia of the robot about the support point, $c = |\mathbf{c}|$ is the distance between the CoM and the support point, $\theta = \tan^{-1}(c_y/c_x)$ is the angle of the CoM from the x axis, and the term $mgc\cos(\theta_0)$ is a hypothetical constant torque acting at the support point, which serves to make θ_0 an equilibrium point of the pendulum. Linearizing this equation about θ_0, and defining $\phi = \theta - \theta_0$, results in the following equation:

$$I\ddot{\phi} = mgc_y\phi, \tag{14}$$

which has solutions of the form

$$\phi = Ae^{t/T_c} + Be^{-t/T_c} \tag{15}$$

where A and B are constants depending on the initial conditions, and T_c is the natural time constant of the pendulum, given by

$$T_c^2 = \frac{I}{mgc_y}. \tag{16}$$

If $c_y > 0$ then T_c is real and Eq. 15 contains both a rising and a decaying exponential. This is characteristic of an unstable equilibrium. If $c_y < 0$ then T_c is imaginary and Eq. 15 is a combination of sines and cosines, which is characteristic of a stable equilibrium. But if $c_y = 0$ then we are at the boundary between stable and unstable equilibrium and T_c is unbounded. As we are considering the problem of a robot balancing on a supporting surface, it is reasonable to assume $c_y > 0$.

From the definition of the joint-space inertia matrix [6, §6.2] we have $H_{01} = s_0^T I_0^c s_1$ and $H_{11} = s_1^T I_1^c s_1$, where $s_0 = [0\ 1\ 0]^T$, $s_1 = [1\ 0\ 0]^T$ and

$$I_0^c = I_1^c = \begin{bmatrix} I & -mc_y & mc_x \\ -mc_y & m & 0 \\ mc_x & 0 & m \end{bmatrix} \tag{17}$$

(planar vectors and matrices—see [6, §2.16]). It therefore follows that $H_{01} = -mc_y$ and $H_{11} = I$, implying that

$$T_c^2 = \frac{-H_{11}}{gH_{01}}. \tag{18}$$

On comparing this with Eq. 11 it can be seen that

$$T_c^2 = \frac{-Y_2}{Y_1}. \tag{19}$$

The linear velocity gain of a robot mechanism, G_v, as defined in [7, 8], is the ratio of a change in the horizontal velocity of the CoM to the change in velocity of the joint (or combination of joints) that is being used to manipulate the CoM. For the robot in Fig. 2 the velocity gain is

$$G_v = \frac{\Delta\dot{c}_x}{\Delta\dot{q}_2}, \tag{20}$$

where both velocity changes are caused by an impulse about joint 2. The value of G_v can be worked out via the impulsive equation of motion derived from Eq. 7:

$$\begin{bmatrix} \iota_0 \\ 0 \\ \iota_2 \end{bmatrix} = \begin{bmatrix} H_{00} & H_{01} & H_{02} \\ H_{10} & H_{11} & H_{12} \\ H_{20} & H_{21} & H_{22} \end{bmatrix} \begin{bmatrix} 0 \\ \Delta\dot{q}_1 \\ \Delta\dot{q}_2 \end{bmatrix}, \tag{21}$$

where ι_2 is an arbitrary nonzero impulse. Solving this equation for ι_0 gives

$$\begin{aligned} \iota_0 &= H_{01}\Delta\dot{q}_1 + H_{02}\Delta\dot{q}_2 \\ &= \left(H_{02} - \frac{H_{01}H_{12}}{H_{11}}\right)\Delta\dot{q}_2 = \frac{-D}{H_{11}}\Delta\dot{q}_2. \end{aligned} \tag{22}$$

But ι_0 is the ground-reaction impulse in the x direction, which is the step change in horizontal momentum of the whole robot; so we also have $\iota_0 = m\Delta\dot{c}_x$, and the velocity gain is therefore

$$G_v = \frac{\Delta\dot{c}_x}{\Delta\dot{q}_2} = \frac{\iota_0}{m\Delta\dot{q}_2} = \frac{-D}{mH_{11}}. \tag{23}$$

The two plant gains can now be written in terms of T_c and G_v as follows:

$$Y_1 = \frac{1}{mgT_c^2 G_v}, \quad Y_2 = \frac{-1}{mgG_v}, \tag{24}$$

and another interesting formula for Y_1 is

$$Y_1 = \frac{c_y}{IG_v}. \tag{25}$$

We are now in a position to explain the physical significance of the conditions $D \neq 0$, which is required by the plant model, and $Y_1 \neq 0$, which is required by the control law in the next section. $D \neq 0$ is equivalent to $G_v \neq 0$, and it is the condition for joint 2 to have an effect on the horizontal motion of the CoM. If $D = 0$ in some particular configuration then it is physically impossible for the robot to balance itself in that configuration. $Y_1 = 0$ occurs when $c_y = 0$, which is on the boundary between unstable and stable equilibrium. As we can safely assume $c_y > 0$ in the present context, the possibility of $Y_1 = 0$ can be ignored. A similar analysis appears in [2, 4].

3   The Balance Controller

The new plant model is interesting because it presents us with a particularly simple model of the physical process of balancing. However, its practical usefulness lies in the high quality of balance control that can be achieved if one designs the balance controller to control this plant instead of trying to control the robot directly. So let us now design a controller for the plant in Fig. 3.

3.1   Control Law

The first step is to choose a control law, which is a formula expressing the control system's output $(\dddot{L})$ as a function of its input (q_c) and feedback signals from the plant. In the simplest case, this formula is just a weighted sum of the input and feedback signals. Given the nature of the plant in Fig. 3, the obvious strategy to try is full state feedback, meaning that the feedback signals are the four state variables of the plant. So let us consider the following control law:

$$\dddot{L} = k_{dd}\ddot{L} + k_d\dot{L} + k_L L + k_q(q_2 - u), \tag{26}$$

where k_{dd}, k_d, k_L and k_q are feedback gains, and the input signal u is the output of a filter

$$u = q_c + \alpha_1 \dot{q}_c + \alpha_2 \ddot{q}_c \tag{27}$$

that modifies the original command signal by combining it with scalar multiples of its first two time derivatives.[2] The purpose of this filter will become clear later on; and note that we will modify this filter in Sect. 3.3.

Having chosen a control law, the next step is to analyse the dynamics of the closed-loop system (plant plus control law) in order to guide the selection of feedback gains. When the plant in Fig. 3 is subjected to the control law in Eq. 26, the resulting closed-loop equation of motion is

$$
\begin{bmatrix} \dddot{L} \\ \ddot{L} \\ \dot{L} \\ \dot{q}_2 \end{bmatrix} = \begin{bmatrix} k_{dd} & k_d & k_L & k_q \\ 1 & 0 & 0 & 0 \\ 0 & 1 & 0 & 0 \\ Y_2 & 0 & Y_1 & 0 \end{bmatrix} \begin{bmatrix} \ddot{L} \\ \dot{L} \\ L \\ q_2 \end{bmatrix} - \begin{bmatrix} k_q u \\ 0 \\ 0 \\ 0 \end{bmatrix},
\tag{28}
$$

and the characteristic equation of the coefficient matrix is

$$
\lambda^4 - k_{dd}\lambda^3 - (k_d + k_q Y_2)\lambda^2 - k_L\lambda - k_q Y_1 = 0.
\tag{29}
$$

At this point we introduce an approximation. The next step is to linearize the dynamics about the current configuration. However, instead of an exact linearization, we shall use an approximate linearization that is obtained by assuming $\partial Y_i/\partial \dot{L} = \partial Y_i/\partial q_2 = 0$. This is equivalent to assuming that Y_1 and Y_2 are constant in a small neighbourhood surrounding the current configuration. Having made this approximation, the roots of Eq. 29 become the poles[3] of the linearized system.

The simplest way to choose the feedback gains is by pole placement, which is a technique in which you first decide where you want the poles to be, and then work out the gains. If λ_1, λ_2, λ_3 and λ_4 are the desired values of the poles, then the polynomial that has these four numbers as its roots is

$$
\lambda^4 + a_3\lambda^3 + a_2\lambda^2 + a_1\lambda + a_0,
\tag{30}
$$

where

$$
\begin{aligned}
a_0 &= \lambda_1\lambda_2\lambda_3\lambda_4 \\
a_1 &= -\lambda_1\lambda_2\lambda_3 - \lambda_1\lambda_2\lambda_4 - \lambda_1\lambda_3\lambda_4 - \lambda_2\lambda_3\lambda_4 \\
a_2 &= \lambda_1\lambda_2 + \lambda_1\lambda_3 + \lambda_1\lambda_4 + \lambda_2\lambda_3 + \lambda_2\lambda_4 + \lambda_3\lambda_4 \\
a_3 &= -\lambda_1 - \lambda_2 - \lambda_3 - \lambda_4.
\end{aligned}
\tag{31}
$$

---

[2] As q_c is a computer-generated synthetic signal, it is reasonable to expect that these derivatives are available, or at least easily computable. However, as q_c is only required to be piece-wise differentiable, there may be moments when one or both derivatives are undefined. At these moments, the values of the signals $\dot{q}_c$ and $\ddot{q}_c$ should be defined to be the values of the derivatives immediately before or after (it doesn't matter which) the moment in question.

[3] Poles are attributes of a linear system that determine important aspects of its behaviour, such as whether or not it is stable, and how quickly it responds to an input. Poles are either real or complex numbers, the latter always appearing in conjugate pairs.

The gains are then obtained by matching the coefficients in Eqs. 29 and 30, resulting in

$$
\begin{aligned}
k_{dd} &= -a_3 & k_d &= -a_2 + a_0 Y_2/Y_1 \\
k_L &= -a_1 & k_q &= -a_0/Y_1 .
\end{aligned}
\tag{32}
$$

A suitable choice of poles is discussed in Sect. 3.2.

For the linearized system to be stable, it is necessary that every pole is negative, or has a negative real part. However, it can be seen from Eq. 29 that if $Y_1 = 0$ then $\lambda = 0$ is always a root of the characteristic equation regardless of the choice of gains. So we require $Y_1 \neq 0$ as a condition of stability.

Finally, the value computed by Eq. 26 is $\ddot{L}$, but the output of the control system has to be either a torque command or an acceleration command for joint 2; that is, either τ_2 or $\ddot{q}_2$. These quantities are computed as follows. First, from Eq. 4 we have $\ddot{L} = -mg\ddot{c}_x$; but $m\ddot{c}_x$ is the x component of the ground reaction force acting on the robot, which is τ_0. So $\ddot{L} = -g\tau_0$. Substituting this into Eq. 7 and rearranging to put all of the unknowns into a single vector produces the equation

$$
\begin{bmatrix}
0 & H_{01} & H_{02} \\
0 & H_{11} & H_{12} \\
-1 & H_{21} & H_{22}
\end{bmatrix}
\begin{bmatrix}
\tau_2 \\
\ddot{q}_1 \\
\ddot{q}_2
\end{bmatrix}
=
\begin{bmatrix}
-\ddot{L}/g - C_0 \\
-C_1 \\
-C_2
\end{bmatrix},
\tag{33}
$$

which can be solved for both τ_2 and $\ddot{q}_2$.

3.2 Transfer Function

A transfer function is a function, or differential equation, that expresses the output of a system as a function of its input. For linear systems, we have the following general result: a linear system that is described in state-space form by the equations

$$
\begin{aligned}
\dot{x}(t) &= Ax(t) + Bu(t) \\
y(t) &= Cx(t)
\end{aligned}
\tag{34}
$$

in which x, u and y are vectors of state, input and output variables, respectively, has a transfer function that can be expressed in the frequency domain as

$$
y(s) = C(1s - A)^{-1}Bu(s)
\tag{35}
$$

Anderson and Moore [1], where s is the Laplace-transform variable,[4] $\mathbf{1}$ denotes an identity matrix of the appropriate size, and $y(s)$ and $u(s)$ are the Laplace transforms of

---

[4]In the frequency domain, s stands for a complex frequency; but in the mapping from the time domain to the frequency domain, s corresponds to the differentiation operator. Thus, $x(t)$ maps to $x(s)$, but $\dot{x}(t)$ maps to $sx(s)$.

$y(t)$ and $u(t)$. Applying this formula to the linearized system, as described above, produces the following formula for the transfer function from the input u to the output q_2:

$$q_2(s) = \frac{a_0(1 - T_c^2 s^2)}{s^4 + a_3 s^3 + a_2 s^2 + a_1 s + a_0} u(s), \qquad (36)$$

assuming that the gains have been set according to Eq. 32. The complete transfer function from q_c to q_2 is then

$$q_2(s) = \frac{a_0(1 - T_c^2 s^2)(1 + \alpha_1 s + \alpha_2 s^2)}{s^4 + a_3 s^3 + a_2 s^2 + a_1 s + a_0} q_c(s), \qquad (37)$$

where the contribution of the filter in Eq. 27 can now be seen.

Transfer functions like these are the Laplace transforms of the impulse response of a linear system, and they have three features of interest: poles, zeros and a 'DC' gain. The last of these is the value of the transfer function at $s = 0$, and it expresses the value that the ratio of output ($q_2(t)$) to input ($q_c(t)$ or $u(t)$) eventually reaches if the input remains constant for a sufficiently long time and the system is stable. In both equations this value is 1, implying that $q_2(t)$ eventually converges to the commanded value ($q_c(t)$ or $u(t)$) if the latter is held constant for long enough.

The poles are the values of s that are the roots of the polynomial on the denominator (i.e., the four numbers $\lambda_1, \ldots, \lambda_4$ whose values we haven't chosen yet), and they determine the nature of the wave forms appearing in the impulse response. In particular, the real part of a pole determines whether its corresponding wave form grows or decays exponentially. For the system to be stable, we require that all of these wave forms decay, which is equivalent to requiring that every pole has a strictly negative real part. The zeros, on the other hand, are the values of s that are the roots of the polynomial on the numerator. They identify frequencies at which the gain of the system is zero. In a stable system, if a pole and a zero coincide then they cancel each other out.

The transfer function in Eq. 37, shows us that the closed-loop system has four poles at frequencies chosen by the control-system designer, two zeros at frequencies determined by the physical properties of the robot, and up to two more zeros chosen by the designer. These last two can be used to cancel two of the poles, thereby simplifying the transfer function.

The two zeros determined by the mechanism lie at $\pm 1/T_c$. The zero at $-1/T_c$ can be eliminated by setting one of the poles equal to $-1/T_c$. This is a good idea, partly because it simplifies the behaviour of the system, and partly because it reduces the coupling between the conflicting activities of maintaining the robot's balance and following the command signal. (The eigenvector associated with this pole involves only L, $\dot{L}$ and $\ddot{L}$, so it can be regarded as being devoted exclusively to balancing.)

However, the zero at $1/T_c$ is more problematic. This is the zero that is responsible for a phenomenon called non-minimum-phase behaviour, which is a characteristic of many physical systems involving balance. The presence of this zero in the transfer function makes the closed-loop system a non-minimum-phase system, and it degrades

Fig. 4 Typical
non-minimum-phase
behaviour: each change in
the command signal
provokes an initial response
in the opposite direction

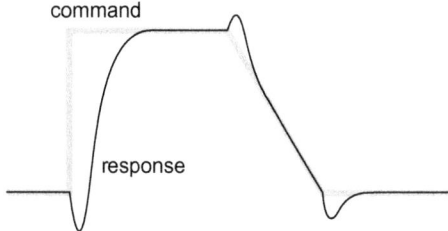

the tracking response in the way illustrated in Fig. 4. Put simply, each time the command signal changes, the robot has to alter its state of balance before it can respond, and, in order to do this, the robot has to move briefly in the wrong direction. The magnitudes of these excursions in the wrong direction can be large, especially if the command signal specifies large, fast movements and/or the designer is aiming for high performance by setting the gains high.

The physical origin of non-minimum-phase behaviour can be understood by referring back to Fig. 2 and imagining the robot to be in a similar configuration to the one shown, but with $c_x = 0$ so that the robot is balanced. Suppose that the robot receives a command to straighten up, meaning that q_2 is being commanded to decrease to zero. If the robot immediately starts to straighten then the movement causes the leg to push on the ground with a force that has a positive component in the x direction. This causes the ground to push back with a ground-reaction force that has a negative component in the x direction, which causes the CoM to accelerate to the left. The end result is that the robot loses its balance and falls to the left. The only way to prevent this from happening is if the robot first pushes its CoM a little to the right, so that gravity can exert a negative moment about the support point and begin tipping the robot over to the right. Once this has started to happen, the robot can begin to straighten itself because the straightening movement now acts to correct a balance error to the right instead of causing a balance error to the left. But the only way to push the CoM to the right is by increasing q_2. Thus, q_2 must first increase before it can follow the command to decrease.

The conventional wisdom is that the zero responsible for non-minimum-phase behaviour cannot be eliminated because the corresponding pole is unstable. However, there is a way to do it, which is the subject of the next section.

3.3 Leaning in Anticipation

If one watches how the robot behaves in response to command signals, the impression is that the robot is continually being 'surprised' by changes in the command, and is continually having to make large excursions in order to adjust its state of balance as quickly as possible to suit the new command. This is not how humans behave. We nearly always know what movements we intend to make in the immediate future (i.e. the next 1 or 2 seconds), and our current movements are tweaked in ways that facilitate the movements that we intend to make next [14].

Consider, for example, the task of pulling open a heavy door. We know that this will require a large horizontal force, and so we begin to lean backwards slightly in advance of exerting the force, in order to avoid being pulled off balance. This is a good strategy, and a good balance controller ought to be able to replicate it. However, to do so requires that the command signal be modified so that it contains information about the future. This is feasible because a robot's high-level controller, which is involved in activities such as planning, typically does know what movements it intends to make in the immediate future.

So we have two problems to solve: how to eliminate the zero at $1/T_c$, and how to incorporate information about the future into the command signal. The solution turns out to be very simple: starting sufficiently far in the future, pass the command signal through a first-order low-pass filter with a pole at $-1/T_c$, but running *backwards in time* to the current instant. This solves the first problem because a pole at $-1/T_c$ in reverse time becomes a pole at $1/T_c$ in forward time, so the filtered signal already incorporates the correct pole to eliminate the zero at $1/T_c$. And the second problem is also solved because the output of the filter is a function of both the current value of q_c and expected future values. Technically, this makes it an acausal filter.

Let q_f denote the filtered command signal, and let us redefine u as follows:

$$u = q_f + \alpha_1 \dot{q}_f + \alpha_2 \ddot{q}_f \tag{38}$$

which replaces Eq. 27. The transfer function from q_c to q_f is $1/(1 - T_c s)$, so the complete transfer function from q_c to q_2 is now

$$q_2(s) = \frac{a_0(1 + T_c s)(1 + \alpha_1 s + \alpha_2 s^2)}{s^4 + a_3 s^3 + a_2 s^2 + a_1 s + a_0} q_c(s), \tag{39}$$

which is no longer a non-minimum-phase system. Furthermore, by setting one of the poles to $-1/T_c$, as mentioned above, the last of the mechanism-dependent zeros is removed, and the transfer function simplifies to three poles and up to two zeros, all of them freely selectable by the control system designer. If the plant were linear, then it would really be true that all dependence on the behaviour of the robot has been eliminated from the transfer function (but not from the total behaviour of the robot, which depends also on the values of L, $\dot{L}$ and $\ddot{L}$). However, the plant is nonlinear, so the robot's dynamics will always have some effect on the transfer function, especially when the robot is making large, fast movements.

Ideally, the acausal filter should start at a point so far into the future as to be indistinguishable from infinity. However, looking ahead by just $3T_c$ is already enough to get 95% of the desired effect, and looking ahead by $4T_c$ gets 98%. More generally, looking ahead by αT_c gets $1 - e^{-\alpha}$ of the desired effect.

4 Extension to General Planar Robots

The theory presented in Sects. 2 and 3 can be extended in a straightforward way to the case of a general planar robot balancing on a single point. The robot may have any number of joints, of any type, and the mechanism may contain any number of branches and kinematic loops. The only restrictions that remain are the planarity assumption (all motion takes place in or parallel to a single plane) and the assumption that the robot makes contact with the ground at a single point (or centre of pressure) whose location varies so little that it may be assumed to be fixed in space.

In progressing from a double pendulum to a robot with two or more actuated motion freedoms, two aspects of the balance control problem change:

1. there is now a choice of motions to use for balancing; and
2. balancing in the plane requires only one motion freedom, so the robot now has motion freedoms that are not needed for balancing, and which can be devoted exclusively to other tasks.

Item 2 implies that the robot should be controlled by two control systems working in parallel: a balance controller that looks after the one motion freedom used for balancing, and a motion controller that looks after all of the others.

To begin the analysis, we first replace the double pendulum with a general planar mechanism, retaining only the fictitious prismatic joint and the passive revolute joint that models the contact with the ground. The rest of the mechanism is assumed to be fully actuated. As the mechanism may contain kinematic loops, not all of the joint variables will be independent. So we introduce a vector of independent generalized coordinates, $y = [y_0 \ y_1 \ y_2 \ y_3^\mathsf{T}]^\mathsf{T}$, which replaces the vector of joint variables, q. In this vector, $y_0 = q_0$ and $y_1 = q_1$, while y_2 is the coordinate used for balancing, and y_3 is a vector containing the rest of the generalized coordinates.

Except for y_0 and y_1, it is not necessary for any of the elements of y to map to particular joints. So y could be a vector of task variables, and the robot could be controlled in a task space in which y_2 is the one variable that is used to balance the robot while also carrying out a motion task.

The movement expressed by y_2 can be any one actuated joint motion, or any desired combination of actuated joint motions, provided only that the chosen movement has a nonzero effect on c_x; but for good balancing performance one should choose a movement that has a large effect. The mapping from y_2 to joint motion can be constant, or it can be allowed to vary. Abrupt changes are possible whenever the robot is stationary; but changes made while the robot is in motion will cause disturbances, and should therefore be made gradually, if at all.

The equation of motion of this robot is

$$
\begin{bmatrix}
H_{00} & H_{01} & H_{02} & \boldsymbol{H}_{03} \\
H_{10} & H_{11} & H_{12} & \boldsymbol{H}_{13} \\
H_{20} & H_{21} & H_{22} & \boldsymbol{H}_{23} \\
\boldsymbol{H}_{30} & \boldsymbol{H}_{31} & \boldsymbol{H}_{32} & \boldsymbol{H}_{33}
\end{bmatrix}
\begin{bmatrix}
0 \\
\ddot{q}_1 \\
\ddot{y}_2 \\
\ddot{\boldsymbol{y}}_3
\end{bmatrix}
+
\begin{bmatrix}
C_0 \\
C_1 \\
C_2 \\
\boldsymbol{C}_3
\end{bmatrix}
=
\begin{bmatrix}
\tau_0 \\
0 \\
w_2 \\
\boldsymbol{w}_3
\end{bmatrix},
\tag{40}
$$

in which w_2 and $\mathbf{w}_3$ are the generalized forces corresponding to y_2 and $\mathbf{y}_3$, and H_{ij} are the elements and submatrices of a generalized inertia matrix. This equation replaces Eq. 7. Equations 2–4 and 6 remain valid, but Eq. 5 becomes

$$L = H_{11}\dot{q}_1 + H_{12}\dot{y}_2 + \mathbf{H}_{13}\dot{\mathbf{y}}_3 . \tag{41}$$

Likewise, Eq. 8 becomes

$$-\ddot{L}/g = H_{01}\dot{q}_1 + H_{02}\dot{y}_2 + \mathbf{H}_{03}\dot{\mathbf{y}}_3 , \tag{42}$$

and so Eq. 9 becomes

$$\begin{bmatrix} L \\ \ddot{L} \end{bmatrix} = \begin{bmatrix} H_{11} & H_{12} \\ -gH_{01} & -gH_{02} \end{bmatrix} \begin{bmatrix} \dot{q}_1 \\ \dot{y}_2 \end{bmatrix} + \begin{bmatrix} \mathbf{H}_{13} \\ -g\mathbf{H}_{03} \end{bmatrix} \dot{\mathbf{y}}_3 . \tag{43}$$

Solving this equation for $\dot{y}_2$ gives

$$\dot{y}_2 = Y_1 L + Y_2 \ddot{L} - Y_3 \dot{\mathbf{y}}_3 , \tag{44}$$

where Y_1 and Y_2 are as given in Eq. 11, and

$$Y_3 = \frac{E}{D} \tag{45}$$

where

$$E = \mathbf{H}_{13}H_{01} - H_{11}\mathbf{H}_{03} . \tag{46}$$

Y_3 can be regarded as a mapping from the velocity vector $\dot{\mathbf{y}}_3$ to the balance disturbance it causes, which is just a scalar signal.

These equations lead us to a modified plant model, which is shown in Fig. 5. This model is identical to the one in Fig. 3 except for a new input which carries the balance disturbance signal. According to this model, the physics of balancing is essentially the same for all planar robots, no matter how complex, implying that the task of balance control is also essentially the same. By designing the balance controller to act on this plant, instead of directly on the robot, we achieve a separation between the

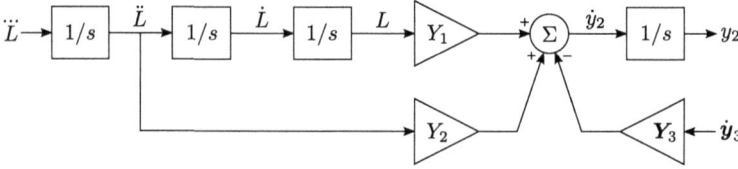

Fig. 5 Modified plant model for a general planar robot

potentially unlimited complexity of the robot's dynamics and the relative simplicity of planar balancing. We also get a controller that works for any planar robot.

The complete control system now consists of a balance controller, which is responsible for y_2, and a motion controller, which is responsible for y_3. The motion controller receives as input a command signal y_{3c}, and produces as output the value of $\ddot{y}_3$ calculated according to a suitable control law. The balance controller receives both the control signal y_{2f}, which replaces q_f, and the signal $\dot{y}_{3f}$ which is explained below. The output is still $\dddot{L}$.

The design of the balance controller is largely unaffected by the extra input in Fig. 5, and follows almost the same steps as in Sect. 3. In particular, Eq. 29 is unaffected, and the gains are still as given in Eq. 32. However, there is now a need to compensate for the disturbances caused by $\dot{y}_3$. Furthermore, it is desirable that the robot should lean in anticipation of these disturbances. This can be accomplished with the modified control law

$$\dddot{L} = k_{dd}\ddot{L} + k_d\dot{L} + k_L(L - \frac{Y_3\dot{y}_{3f}}{Y_1}) + k_q(y_2 - u),$$ (47)

which replaces Eq. 26, together with

$$u = y_{2f} + \alpha_1\dot{y}_{2f} + \alpha_2\ddot{y}_{2f}$$ (48)

which replaces Eq. 38, where y_{2f}, $\dot{y}_{2f}$, $\ddot{y}_{2f}$ and $\dot{y}_{3f}$ are the acausally filtered values of the corresponding command signals and their derivatives. As the actual disturbance depends on the actual velocity, rather than the commanded velocity, it follows that there is an assumption of accurate tracking of $\dot{y}_{3c}$ built into this control law. The rationale behind Eq. 47 is that cancellation of the incoming signal $Y_3\dot{y}_3$ requires L to be offset by an amount $Y_3\dot{y}_3/Y_1$.

Finally, the generalized forces must be calculated and mapped to the actuated joints. The first step is to solve

$$\begin{bmatrix} 0 & 0 & H_{01} & H_{02} \\ 0 & 0 & H_{11} & H_{12} \\ -1 & 0 & H_{21} & H_{22} \\ 0 & -1 & H_{31} & H_{32} \end{bmatrix} \begin{bmatrix} w_2 \\ w_3 \\ \ddot{q}_1 \\ \ddot{y}_2 \end{bmatrix} = \begin{bmatrix} -\dddot{L}/g - C_0 - H_{03}\ddot{y}_3 \\ -C_1 - H_{13}\ddot{y}_3 \\ -C_2 - H_{23}\ddot{y}_3 \\ -C_3 - H_{33}\ddot{y}_3 \end{bmatrix},$$ (49)

which is the generalization of Eq. 33. This is the point where the outputs of the motion and balance controllers ($\ddot{y}_3$ and $\dddot{L}$) are combined. The final step is to solve

$$G^T\tau_a = \begin{bmatrix} w_2 \\ w_3 \end{bmatrix},$$ (50)

where τ_a is the vector of force variables at the actuated joints, and G is the matrix that maps $[\dot{y}_2 \ \dot{y}_3^T]^T$ to the vector of actuated joint velocities (i.e., a submatrix of the Jacobian from $\dot{y}$ to $\dot{q}$). If the mechanism has the same number of actuators as actuated motion freedoms then G is square and Eq. 50 has a unique solution; but if the mechanism is redundantly actuated then G is rectangular and Eq. 50 has infinitely many solutions. In this case it is necessary to choose one particular solution in accordance with a user-defined policy on the distribution of forces among the actuators.

5 Performance in Simulation

This section illustrates the performance of the balance controller in simulation, and compares it with some existing balance controllers. A more detailed simulation study can be found in [9], which investigates how well the controller performs in the presence of modelling error, sensor noise, actuator saturation, and so on. At the time of writing, the controller has not yet been tested on a real robot, although a project is under way to do so.

The simulation study uses the triple pendulum shown in Fig. 6. The link lengths of this robot are 0.2, 0.25 and 0.35 m; and the link masses are 0.7, 0.5 and 0.3 kg, respectively. The masses are treated as point masses at the far end of each link. This is the same robot as used in the second set of simulation experiments in [9]. A triple pendulum is the simplest robot that can illustrate the general case presented in Sect. 4, and this particular robot was chosen because it has a good ability to balance, as measured by its velocity gain [7, 8].

The simulation study involves making the robot follow several linear and sinusoidal motion commands while retaining its balance. To illustrate the use of generalized coordinates that are different from the joint angles, the variables y_2 and y_3 are defined as follows:

$$y_2 = q_2 + q_3, \qquad y_3 = q_2 - q_3. \tag{51}$$

This means that the sum of the two actuated joint angles is the variable used to balance the robot and follow a motion command, while the difference between the

Fig. 6 The planar triple pendulum used in the simulation study

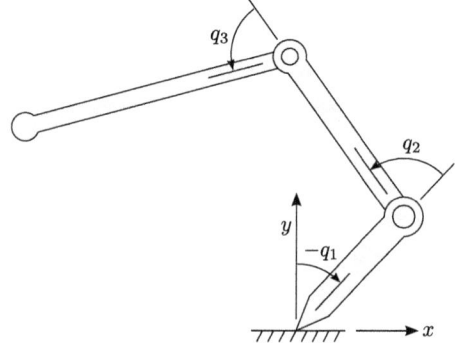

two angles is controlled by a separate motion controller. The matrix G in Eq. 50 is then the 2×2 matrix below:

$$\begin{aligned} y_2 &= q_2 + q_3 \\ y_3 &= q_2 - q_3 \end{aligned} \quad \Rightarrow \quad \begin{bmatrix} \dot{q}_2 \\ \dot{q}_3 \end{bmatrix} = \begin{bmatrix} 0.5 & 0.5 \\ 0.5 & -0.5 \end{bmatrix} \begin{bmatrix} \dot{y}_2 \\ \dot{y}_3 \end{bmatrix} = G \begin{bmatrix} \dot{y}_2 \\ \dot{y}_3 \end{bmatrix}. \tag{52}$$

The complete control system consists of the balance controller of Sect. 4, which controls y_2, and a proportional-plus-derivative (PD) position controller with exact inverse dynamics (i.e., a computed torque controller), which controls y_3. Allowing the position controller to use inverse dynamics is just a quick and simple way of ensuring that its tracking accuracy is essentially perfect everywhere, except where there are step changes in commanded velocity, so that the graphs can show the behaviour of the balance controller more clearly.

The balance controller is based on the control law in Eqs. 47 and 48. Three poles are placed at -20 rad/s and the fourth at $-1/T_c$; and the coefficients in Eq. 48 are chosen to provide two zeros at -20 rad/s, which cancel two of the poles. The transfer function is then $1/(1 + 0.05s)$. The position controller's gains can be chosen independently of those of the balance controller, but on this occasion they were chosen so as to place its poles also at -20 rad/s. The acausal filter was implemented analytically using an approximation that makes use of spot values of T_c at the beginnings and ends of each motion segment. The error introduced by this approximation is small because T_c varies in a relatively narrow range from 0.197 to 0.233 s as the robot moves.[5]

The simulation results are shown in Fig. 7. Figure 7a plots the command signals and the robot's responses, both expressed in generalized coordinates, and Fig. 7b shows the corresponding movements of the joints. The commands consist of a slow ramp, a fast ramp and a sine wave for y_3 while y_2 is held at zero, followed by a sequence of four ramps that exercise y_2 and y_3 in various combinations. The correspondence between y_i and q_i is clearly visible in these graphs. For example, when y_3 ramps from 0 to 3 rad between 0.5 and 2.5 s, it can be seen that q_2 ramps from 0 to 1.5 rad, and q_3 ramps from 0 to -1.5 rad, exactly in accordance with Eq. 51.

As expected, y_3 tracks y_{3c} accurately everywhere except where there is a step change in commanded velocity. The tracking errors are most obvious when y_{3c} comes to a halt at the end of a ramp or sine wave. But the behaviour of y_2 is a little different. Most of the time, this variable tracks y_{2c} with a delay that is very close to the 50 ms delay implied by the transfer function $1/(1 + 0.05s)$, meaning that the balance controller is behaving almost exactly as commanded. However, a small amount of tracking error can also be seen during the fast ramp and the sine wave. These movements show y_2 acting to maintain the robot's balance during the two movements that cause the largest balance disturbance. In fact, y_2 is doing this all of the time, but the movements are too small to show up in the graph.

---

[5]Other properties of the robot, like its rotational inertia about the support point, vary in a much larger range. A relatively small range of values for T_c appears to be a common feature of many robot mechanisms.

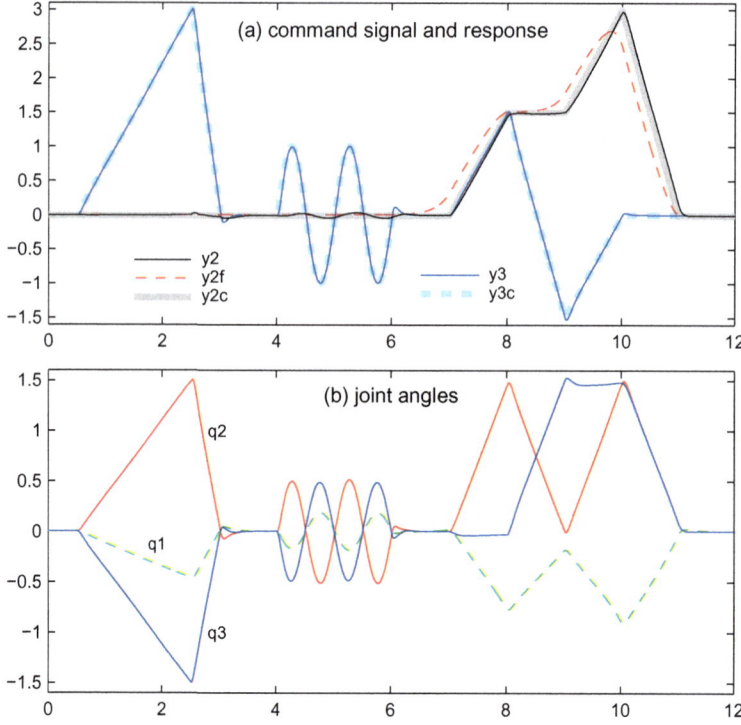

Fig. 7 Simulation results for a triple pendulum balancing while following a motion sequence (times in seconds, angles in radians)

The reason why y_2 has to move so little in order to maintain the robot's balance is because the balance controller has advance warning of the disturbance, via the signal y_{3f}, and can lean in anticipation. The action of leaning in anticipation shows up in the value of q_1 in Fig. 7b; but the amount of leaning during this particular simulation is too small to see in the graph. Clearer examples of leaning in anticipation, and of the much larger balancing movements that would be necessary in its absence, can be found in [9].

Figure 7a also plots the signal y_{2f}, which is the signal that the balance controller is actually trying to follow. (See Eq. 48.) This signal leads y_{2c} by approximately T_c (i.e., around 0.2 s); and it has a noticeably different shape to y_{2c}, due to the action of the acausal low-pass filter. Yet y_2 ends up accurately following y_{2c} in accordance with the chosen transfer function. This shows that the acausal filter is accurately compensating for the non-minimum-phase zero in the closed-loop transfer function (i.e., the zero at $1/T_c$ in the generalized-coordinate version of Eq. 37), in spite of the fact that some approximations were made in calculating y_{2f} (and also y_{3f}), as mentioned above. Further evidence that the acausal filter is doing its job can be seen in the absence of non-minimum-phase behaviour (as illustrated in Fig. 4) in the motion of y_2.

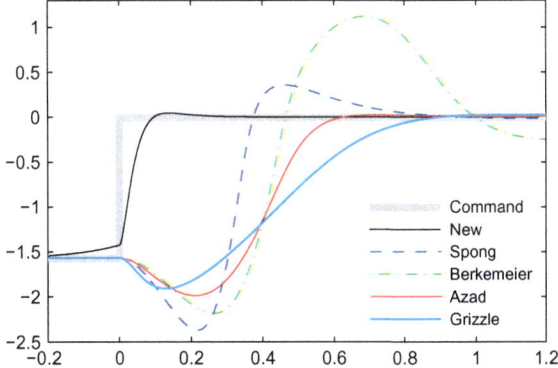

Fig. 8 Comparison of the step response of the new balance controller with those of several existing balance controllers

In summary, the balance controller makes the robot behave in a manner very close to that predicted by the theory, resulting in accurate tracking of large, fast movements without loss of balance.

To support the claim that the new controller is faster than existing balance controllers, Fig. 8 plots the step response of the new controller together with those of the balance controllers described in [4, 5, 11, 15]. This figure is a reproduction of Fig. 12 in [9], which in turn is largely based on Fig. 13 in [4]; and these two articles explain the steps that were taken to try to make the comparison fair.

The robot used in this comparison is a planar double pendulum having link lengths of 0.4 and 0.6 m, link masses of 0.49 and 0.11 kg, centres of mass at 0.1714 and 0.4364 m along each link, and rotational inertias about the CoM of 0.0036 and 0.0043 kgm^2, respectively. It is another example of a robot designed to be good at balancing as measured by its velocity gain. Other details, such as the choice of controller gains, are explained in [4, 9].

Figure 8 clearly shows that the new controller responds much more quickly than any of the others, and that the main reason is the absence of non-minimum-phase behaviour. It also shows an imperfection in the new controller: q_2 begins to creep forward ahead of the step. This effect is due to the nonlinearities in the plant model that were ignored in designing and implementing the controller and the acausal filter. Ignoring fewer nonlinearities would reduce this effect. The reason why this effect is visible in Fig. 8 but not in Fig. 7, where it also exists but is too small to see, is because the command signal in Fig. 8 is a step (i.e., a discontinuity in commanded position), whereas the command signal in Fig. 7 contains only ramps and sine waves.

6 Extension to 3D

Balancing in the plane is a useful skill, but balancing in 3D is more useful. The work presented in this chapter is really intended to help lead the way towards the goal of a high-performance balance controller in 3D. This section outlines briefly the current state of progress and the problems that remain to be solved.

The simplest way to implement a 3D balance controller, given that one already has a planar balance controller, is to make the robot balance simultaneously in two vertical planes at right angles. According to this approach, the plant model in Fig. 5 (or Fig. 3) must be modified as follows. First, the scalar state variables L, $\dot{L}$, $\ddot{L}$ and y_2 must be replaced with 2D vectors having one component for each of the two balance planes. This implies that the signals between the blocks are now 2D vector signals, and also that the input $\dddot{L}$ is a 2D vector. Next, the two gains Y_1 and Y_2 must be replaced with 2×2 matrices, and Y_3 must be replaced with a two-row matrix, so that the signal $Y_3 \dot{y}_{3f}$ is a 2D vector. (The expression $Y_3 \dot{y}_{3f}/Y_1$ in Eq. 47 becomes $Y_1^{-1} Y_3 \dot{y}_{3f}$.)

A potential difficulty now arises in the design of the balance controller. As a result of the changes in the plant model, the scalar elements in the coefficient matrix in Eq. 28 have all expanded to 2×2 matrices, making the matrix as a whole 8×8, and, more importantly, making the characteristic equation potentially very complicated. This problem goes away if the matrices Y_1 and Y_2 are both diagonal, or at least triangular, but they will have this form only if the robot mechanism has some special properties. So if the control system assumes this form then it no longer applies to general robots.

This scheme has been tried in simulation, and can be made to work under certain circumstances, but it is too restricted in what it can do. For this reason, the author and his colleagues are trying a different approach in which the task of balancing in 3D is broken down into two sub-tasks: balancing in the plane, and keeping the plane vertical.

This idea is inspired by an effort to create a highly acrobatic hopping and balancing robot, called Skippy, a simplified version of which is shown in Fig. 9. The major parts of this robot are a leg, a torso and a crossbar, all connected by actuated revolute joints. The actuator at the bend joint is powerful enough to make the robot hop, but also serves to balance the robot in the bend plane (which is the robot's saggital plane) in

Fig. 9 A simple 3D hopping and balancing robot with a well-defined saggital plane (the bend plane)

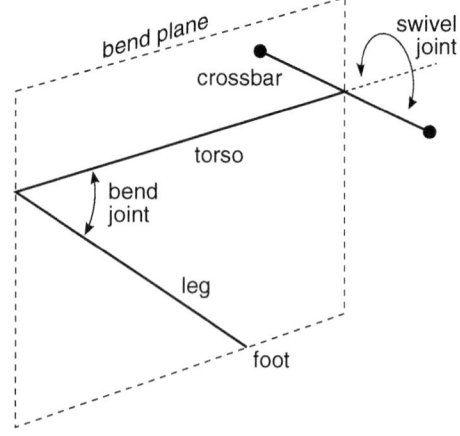

exactly the same way as the actuated joint 2 in Fig. 2. The crossbar then serves to keep the bend plane vertical, and to rotate it about a vertical line through the support point so that the robot can face in any direction.

We call these two actions 'bend control' and 'swivel control', respectively, and the complete control system is a *bend-swivel controller* [2, 3]. Clearly, the bend controller is just the balance controller described in this chapter, but the swivel controller is a little different because it has to control the robot's heading as well as keeping the bend plane vertical.

Neither of these schemes takes into account the effect of gyroscopic forces, which arise in 3D but not 2D, and neither works properly in the presence of significant gyroscopic forces. This appears to be the main theoretical obstacle to the development of a 3D balance controller that performs as well as the 2D balance controller described here, and work is under way to tackle this problem.

7   Conclusion

This chapter has presented a new way to implement balance control for a planar robot balancing on a single point. The method makes use of a new model of the physical process of balancing that has the special property that the complexity of the model does not increase with the complexity of the robot. Balance control is then implemented by designing a control system to control this model, treating it as the control system's plant, instead of trying to design a balance control system directly from the equations of motion of the robot.

The merits of this approach are: (1) the simplicity of the model, in which the essence of any planar robot's balancing behaviour is distilled down to just four state variables and two parameters; (2) the high performance of the resulting control system, as measured by its ability to make large, fast motions without losing its balance; (3) the simplicity and effectiveness of the technique of leaning in anticipation by means of an acausal low-pass filter; and (4) the decoupling of the problem of balance control from the potentially high complexity of the robot's equations of motion.

The material presented here is part of an effort that aims to produce a high-performance balance controller in 3D for a highly acrobatic hopping and balancing machine, and some of the problems that remain to be solved in pursuit of this aim were mentioned briefly. For further reading, consult [9].

References

1. Anderson, B.D.O., Moore, J.B.: Linear Optimal Control. Prentice-Hall, Englewood Cliffs, New Jersey (1971)
2. Azad, M.: Balancing and hopping motion control algorithms for an under-actuated robot. Ph.D. Thesis, The Australian National University, School of Engineering (2014)

3. Azad, M., Featherstone, R.: Balancing control algorithm for a 3d under-actuated robot. In: Proceedings of the IEEE/RSJ International Conference on Intelligent Robots & Systems, Chicago, IL, Sept. 14–18, pp. 3233–3238 (2014)
4. Azad, M., Featherstone, R.: Angular momentum based balance controller for an under-actuated planar robot. Autonom. Robots **40**(1), 93–107 (2016)
5. Berkemeier, M.D., Fearing, R.S.: Tracking fast inverted trajectories of the underactuated acrobot. IEEE Trans. Robot. Autom. **15**(4), 740–750 (1999)
6. Featherstone, R.: Rigid Body Dynamics Algorithms. Springer, New York (2008)
7. Featherstone, R.: Quantitative measures of a robot's ability to balance. In: Robotics Science & Systems 2015, Rome, 13–17 July (2015). https://doi.org/10.15607/RSS.2015.XI.026
8. Featherstone, R.: Quantitative measures of a robot's physical ability to balance. Int. J. Robot. Res. **35**(14), 1681–1696 (2016)
9. Featherstone, R.: A simple model of balancing in the plane and a simple preview balance controller. Int. J. Robot. Res. **36**(13–14), 1489–1507 (2017)
10. Featherstone, R., Orin, D.E.: Dynamics. In: Siciliano, B., Khatib, O. (eds.) Springer Handbook of Robotics, pp. 35–65. Springer, Berlin (2008)
11. Grizzle, J.W., Moog, C.H., Chevallereau, C.: Nonlinear control of mechanical systems with an unactuated cyclic variable. IEEE Trans. Autom. Control **50**(5), 559–576 (2005). https://doi.org/10.1109/TAC.2005.847057
12. Hauser, J., Murray, R.M.: Nonlinear controllers for non-integrable systems: the acrobot example. In: Proceedings of the American Control Conference, 3–5 Nov, pp. 669–671 (1990)
13. Miyashita, N., Kishikawa, M., Yamakita, M.: 3D motion control of 2 links (5 DOF) underactuated manipulator named AcroBOX. In: Proceedings of the American Control Conference, Minneapolis, MN, June 14–16, pp. 5614–5619 (2006)
14. Rabbani, A.H., van de Panne, M., Kry, P.G.: Anticipatory balance control. In: Proceedings of the 7th International Conference on Motion in Games, Los Angeles, CA, 6–8, pp. 71–76 (2014)
15. Spong, M.W.: The swing up control problem for the acrobot. IEEE Control Syst. Mag. **15**(1), 49–55 (1995)
16. Xin, X., Kaneda, M.: Swing-up control for a 3-DOF gymnastic robot with passive first joint: design and analysis. IEEE Trans. Robot. **23**(6), 1277–1285 (2007)
17. Yonemura, T., Yamakita, M.: Swing up control of acrobot based on switched output functions. In: Proceedings of the SICE 2004, Sapporo, Japan, 4–6 Aug, pp. 1909–1914 (2004)

Design and Control of a Passive Noise Rejecting Variable Stiffness Actuator

Luca Fiorio, Francesco Romano, Alberto Parmiggiani,
Bastien Berret, Giorgio Metta and Francesco Nori

Abstract Inspired by the biomechanical and passive properties of human muscles, we present a novel actuator named passive noise rejecting Variable Stiffness Actuator (pnrVSA). For a single actuated joint, the proposed design adopts two motor-gear groups in an agonist-antagonist configuration coupled to the joint via serial non-linear springs. From a mechanical standpoint, the introduced novelty resides in two parallel non-linear springs connecting the internal motor-gear groups to the actuator frame. These additional elastic elements create a closed force path that mechanically attenuates the effects of external noise. We further explore the properties of this novel actuator by modeling the effect of gears static frictions on the output joint equilibrium position during the co-contraction of the agonist and antagonist side of the actuator. As a result, we found an analytical condition on the spring potential energies to guarantee that co-activation reduces the effect of friction on the joint equilibrium position. The design of an optimized set of springs respecting this condition leads to the construction of a prototype of our actuator. To conclude the work, we also present two control solutions that exploit the mechanical design of the actuator allowing to control both the joint stiffness and the joint equilibrium position.

L. Fiorio (✉) · F. Romano · A. Parmiggiani · G. Metta · F. Nori
iCub Facility Department, Istituto Italiano di Tecnologia, 16163 Genova, Italy
e-mail: Luca.Fiorio@iit.it

F. Romano
e-mail: Francesco.Romano@iit.it

A. Parmiggiani
e-mail: Alberto.Parmiggiani@iit.it

G. Metta
e-mail: Giorgio.Metta@iit.it

F. Nori
e-mail: Francesco.Nori@iit.it

B. Berret
CIAMS, University of Paris-Sud, Universit Paris-Saclay, 91405 Orsay, France
e-mail: Bastien.Berret@u-psud.fr

B. Berret
Institut Universitaire de France (IUF), Paris, France

© Springer International Publishing AG, part of Springer Nature 2019
G. Venture et al. (eds.), *Biomechanics of Anthropomorphic Systems*, Springer Tracts in Advanced Robotics 124, https://doi.org/10.1007/978-3-319-93870-7_11

1 Introduction

In the past decades industrial robotics has been the dominating sector in the robotic sales worldwide. However, there are indications that today's robotics market is about to undergo substantial changes. Recently, promising advances have been obtained in the emerging sector of "compliant robots", mostly from research laboratories and universities. Compliant robots are intended to interact with unstructured and dynamic environments. Possible application scenarios include human-robot interaction and collaboration, walking and running robots, prosthetic devices and exoskeletons, and more applications will probably be conceived as the sector grows.

The development of compliant robots has been possible thanks to the increment of processing power of digital controllers and the design of novel actuators.

The first compliant robots have been conceived by exploiting force sensors and classical stiff actuators, composed only by electric motors and gears. As an example, by exploiting accurate force measurements and fast control loops it has been possible to perform challenging interaction tasks, as reported by Albu-Schaffer and Hirzinger [1].

However, there are intrinsic limitations to what the controller can do to modify the behavior of the system because inertia and friction play a dominant role in defining its bandwidth. To overcome this limitation, roboticists have developed a new set of systems endowed with an intrinsic (i.e. passive) compliance. From a mechanical standpoint, advances were closely tied to the development of new actuators that try to introduce at the mechanical level the advantages of compliance. Series Elastic Actuators (SEA), firstly introduced by Pratt and Williamson [28], nowadays represent an established technology to drive robotic joints. To overcome the constant stiffness limitation of the SEA, several Variable Stiffness Actuators (VSA) have also been introduced more recently (see [35, 36]). Conceptually, as detailed in Table 1, the novel actuators differ from their stiff counterpart in the way joint compliance is achieved. However, joint position control is still achieved through active feedback (i.e. software feedback).

Relying on active feedback in artificial agents (such as humanoid robots) might not be a practical strategy to deal with external perturbations, specifically considering the growing amount of sensors (e.g., distributed force/torque sensors (see [17]),whole-body distributed tactile sensors (see [12]), gyros and accelerometers (see [34]) which are currently available and have to be acquired and centrally processed to perform complex actions. Furthermore, closed-loop stability with respect to model inaccura-

Table 1 Comparison of position and compliance control method for different robotic actuators

Actuator	Joint Position	Joint Compliance
Stiff Actuator	ACTIVE	ACTIVE
SEA/VSA	ACTIVE	PASSIVE
pnrVSA	PASSIVE	PASSIVE

cies or friction and backlash in transmissions, can be maintained under small delays in the feedback loop, while most closed loop systems become unstable for large delays.

In humans, the Central Nervous System (CNS) does not always rely on feedback loops to achieve an efficient trajectory control of limbs subject to environmental noise. Indeed, as described by Paillard [24], our sensory feedback (visual or proprioceptive) is too slow. In this context, as described by Hogan [19], one of the most interesting characteristics of biological actuators is the ability to passively compensate for external perturbations, without explicitly relying on active feedback. We name this feature *passive noise rejection (pnr)*.

Starting from these premises, we propose a definition of passive noise rejecting VSA (pnrVSA) as the set of actuators that combine intrinsic compliance with the ability to passively compensate external disturbances (see Table 1). Furthermore we propose the design of a novel single-joint pnrVSA, based on four non-linear springs and two electric motors in agonist-antagonist configuration. The distinguishing characteristic of the proposed mechanism is a closed force path that connects the actuator output joint to the actuator frame, allowing for the implementation of a passive mechanical feedback.

The work is organized as follows. Section 2 introduces the key mechanical features of a *pnr* design. Section 3 presents the "case study" prototype focusing on its main features, introduces the dynamical model of the actuator and the analytical description of the sensitivity of its internal states. We highlight the issue caused by the gear friction providing the mathematical representation of friction propagation among the actuator. Furthermore, we deduce which conditions the non-linear springs should satisfy to reduce the effects of gear friction. Section 4 introduces two different control strategies which exploit the peculiarities of the proposed device. Section 5 presents the results of the simulations together with the experimental tests to prove the effectiveness of the non-linear springs in reducing the effects of the gear friction.

2 Background

The design of a robotic manipulator that can reliably interact with an unstructured environment has to consider also impact loads and manipulation tasks in unstable force fields.[1] In such situations, the actuator has to provide the necessary force or torque to maintain the manipulator as close as possible to the desired trajectory. We define the activity of compensating for the errors in joint position due to external perturbations as "noise rejection". If we consider non-*pnr* actuators, this compensation task is completely entrusted to the digital controller (i.e. active feedback), while in *pnr* actuators part of this compensation is achieved thanks to the passive properties of the system.

---

[1] An example of unstable force field manipulation is represented by the task of keeping a screwdriver in the slot of a screw, as reported by Burdet et al. [8].

2.1 Passive Noise Rejection Designs

To easily understand the difference between a *pnr* and a non-*pnr* design we can compare the mechanical structure of the two systems. Figure 1a represents a simplified model of a non-*pnr* actuator using a linear motor. The motor (ϑ) is linked to the joint (q) through a transmission and a variable stiffness spring (c). This elastic element in between the transmission and the joint is the key component of a typical VSA.[2] As observed before, a disturbance that affects the position of the joint can be compensated only by relying on the motor through active feedback. Depending on the feedback delay and on the frequency and amplitude of the external perturbation, Berret et al. [3] have shown that the system can become unstable.

Figure 1b represents instead a simplified model of a *pnr* actuator. The spring elements connecting the joint to the environment, *a* and *b*, are typically not present in non-*pnr* designs. Nevertheless, they play a crucial role in determining the overall system passive noise rejection. In practice when $a = b = 0$ the system is free-floating with respect to the environment, and noise can drive the system arbitrarily far from the target configuration. The advantage given by the spring elements *a* and *b* resides in the closed loop paths that connect the joint directly to a fixed frame (i.e. the actuator frame). Thanks to one of these paths, either *frame − a − c − joint* or *frame − b − joint*, the joint has a unique equilibrium position. In this case, when an external perturbation deviates the joint from its equilibrium position the elastic elements generate a passive restoring force.

In a work by Berret et al. [4], we studied the effect of disturbances acting on a *pnr* actuator. Disturbance has been represented by stochastic variables acting as forces on the joint. Computations showed that the passive noise rejection monotonically increases with the stiffness of the elastic elements *a*, *b* and *c*. In a sense, passive noise rejection is increased by augmenting the stiffness of the path that connects the joint to the fixed frame. Indeed, the mechanical bandwidth of the connection is closely related to the passive noise rejection: the higher the bandwidth the faster the passive response (meant as the restoring force) of the system. In this sense, the most important advantage of the mechanical feedback is that feedback happens physically without the typical delays of active control loops.

In this work, we focus on actuators like the ones in Fig. 1b, i.e. not having *a* and *b* simultaneously zero.

2.2 Agonist-Antagonist Design

The agonist-antagonist arrangement is a design suitable to construct a *pnr* actuator. As seen in the introduction, a nice example of agonist-antagonist system with *pnr* is given by the human actuation model. In particular, muscles arranged in antagonistic

---

[2]The same model can represent a classical SEA by using a constant stiffness spring, or a stiff actuator by removing the spring and connecting the joint directly to the transmission.

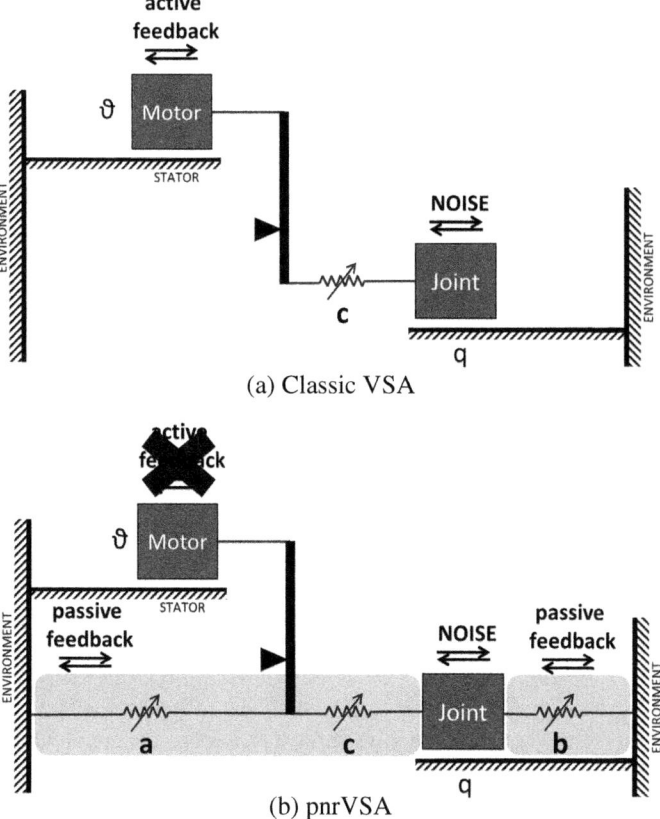

Fig. 1 Conceptual design of an actuation system using a linear motor. **a** Schematic representation of a classical VSA. **b** Generic representation of a pnrVSA. The components a, b and c represent the elastic elements, while ϑ and q represent the motor and the joint respectively

pairs create two closed paths that connect the joint (the limb) to a fixed frame (the previous limb).

The main properties of the biological muscles have been reported by Hill and Gasser [18] using the model represented in Fig. 2a, used to model the tension dynamics of various isolated frog muscles. It can be proven (see [21], p. 23) that its mechanical model is equivalent to the one shown in Fig. 2b and therefore the overall muscle force can be written as:

$$F = F_{SE}(K_{SE}, l_2) = F_{PE}(K_{PE}, l_1) + P(L_j, f(t)),$$

where K_{SE} is the series nonlinear elastic element, K_{PE} is the parallel nonlinear elastic element, which in series with K_{SE} account for the passive tension properties of the muscle, and P is the active force generated by the contractile element depending

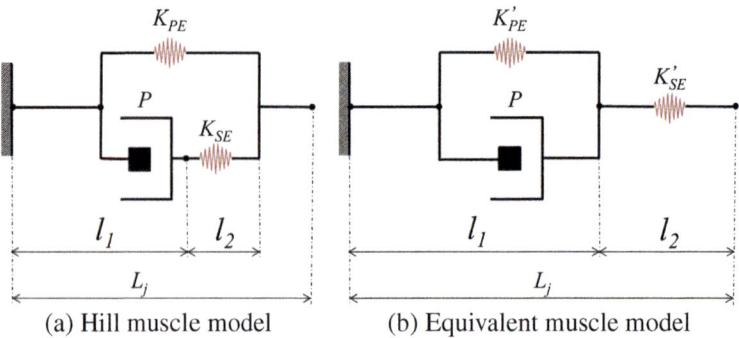

(a) Hill muscle model   (b) Equivalent muscle model

Fig. 2 Mechanical circuit diagrams showing two equivalent muscle models. *Source* [23]

on the muscle history activation $f(t)$ and the overall length L_j. Indeed, biological muscles behave both as unidirectional force sources and non-linear springs, and, consequently, thanks to the agonist-antagonist arrangement, the joint has a unique equilibrium position that depends on the "co-contraction" level of the agonist and antagonist muscles. When a disturbance deviates the joint from its equilibrium position, the muscles thanks to their passive elastic properties generate a restoring force that compensates for the external disturbance. Nevertheless, the limited stiffness of tendons and muscles puts some bounds on the joint stiffness that can be achieved in humans. In support to this analysis, different research studies analyzed the relationship between the joint passive properties and human motor control (see [9, 11, 27]).

When we consider robotic actuators, different agonist-antagonist actuator designs have been presented in literature. In general, as described by Migliore et al. [22], two internal motors act on the same joint through non-linear springs. Concordant actuation of the internal motors causes only the displacement of the output joint, while opposite actuation determines a pure joint stiffness variation. Other examples, include the quasi-antagonistic design described by Eiberger et al. [13], the bidirectional design described by Petit et al. [26] and the cross-coupled design described by Tonietti et al. [33] and Schiavi et al. [30]. Nevertheless, the majority of these designs belong to the category of classical VSA, while only few possess also the *pnr* property. As an example, Bicchi et al. [6] propose a one degree of freedom joint driven by means of two artificial muscles which, similarly to the biological counterparts, create two closed paths connecting the joint to the actuator frame. The NeurArm described by Vitiello et al. [37] instead, is a two degrees of freedom planar robotic arm specifically designed and developed for investigating models of human motor control principles and learning strategies.

To the best of our knowledge, however, a design based on rotary actuators which are more commonly employed when designing robotics arms, has not been proposed yet.

3 Actuator Design

Conventional agonist-antagonist systems are constituted by two actuation modules arranged symmetrically around the output joint. In our case, the design problem to be solved was to conceive a module possessing the *pnr* property, and that could have been arranged in an antagonist configuration. With reference to Fig. 1 we can identify the key mechanical components of this module considering a, ϑ and c. The complete design is described in the following sections, where details on a specific pnrVSA implementation are given.

3.1 Conceptual Design

As shown in Fig. 3 we followed a design based on non-linear springs (red coiled elements) which can vary their equilibrium configuration thanks to the agonist-antagonist arrangement. The springs are connected through wires (red lines) to the actuator output joint q, the actuator frame and to the motor capstans ϑ and ϑ^a. Within a module, each spring has a different behavior when the motor capstan rotates. The spring connecting the capstan to the frame, named K_{PE}, works in parallel with respect to the capstan (spring elongation is proportional to capstan angular displacement); the spring K_{SE} connecting the capstan to the output joint, instead, behaves as series elastic element.

   Considering the whole system, we can describe the behavior of the output joint with respect to angular displacements of both capstans. Intuitively, a clockwise rotation of the capstan ϑ coupled with the same counterclockwise rotation of ϑ^a stretches all springs causing no movement of the joint q. Conversely, rotating the two capstans in the same clockwise direction results into a pure movement of the joint without affecting its stiffness.

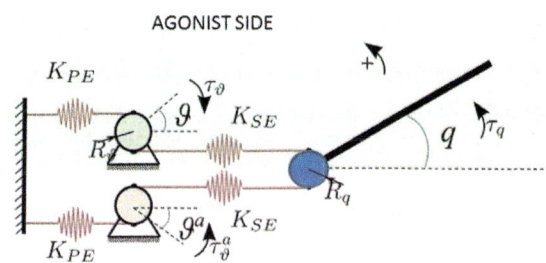

Fig. 3 pnrVSA prototype diagram: the output joint q is controlled by means of internal actuation torques τ_ϑ and τ_ϑ^a, while ϑ and ϑ^a represent capstan angular positions. τ_q is an external torque acting on the output joint

3.2 Analytical Model and Analysis

In the following section, we describe the analytical model of the proposed *pnr*VSA. Subsequently, the model is used to describe the main properties of the system.

3.2.1 Notation

With reference to the actuator schema in Fig. 3, we define the following quantities:

- ϑ and ϑ^a represent the angular position of the agonist and antagonist capstans, τ_ϑ and τ_ϑ^a are the associated torques.
- q is the joint position, τ_q the associated torque.
- I_ϑ is the inertia of the capstan and R_ϑ its radius.
- I_q is the joint inertia and R_q is the radius of the joint pulley.
- U_{PE} represents the potential energy of the parallel springs.[3] We named U_1 the agonist parallel spring and U_4 the antagonist one.
- U_{SE} represents the potential energy of the series springs. We named U_2 the agonist serial spring and U_3 the antagonist one.
- $l_i, i = 1, \ldots, 4$ is the spring elongation for the ith spring, i.e.:

$$
\begin{aligned}
l_1 &= -R_\vartheta \vartheta \\
l_2 &= -R_\vartheta \vartheta - R_q q \\
l_3 &= R_\vartheta \vartheta^a + R_q q \\
l_4 &= R_\vartheta \vartheta^a
\end{aligned}
$$

We also use the following notation throughout the rest of the paper:

- Given a time function $f(t) \in \mathbb{R}^n$, the first and second order time derivatives are denoted by $\dot{f}(t)$ and $\ddot{f}(t)$, respectively.
- Given a function $f(x)$, we denote the first, second and third order derivatives with respect to its argument by $f'(x) = \frac{\partial f}{\partial x}, f''(x) = \frac{\partial^2 f}{\partial x^2}$ and $f'''(x) = \frac{\partial^3 f}{\partial x^3}$, respectively. When the function argument is clear, we will use the "prime" notation, for sake of conciseness.
- Given the angular position variables and the associated torques, for the sake of notational simplicity, we define:

$$
\hat{\vartheta} := -R_\vartheta \vartheta
$$

$$
\hat{\vartheta}^a := R_{\vartheta^a} \vartheta^a
$$

---

[3]To make the analysis as general as possible, the non-linear spring potential energies are kept unspecified in the theoretical analysis.

$$\hat{q} := -R_q q$$

$$\hat{\tau}_\vartheta := \frac{\tau_\vartheta}{R_\vartheta}$$

$$\hat{\tau}_{\vartheta^a} := \frac{\tau_{\vartheta^a}}{R_{\vartheta^a}}$$

$$\hat{\tau}_q := \frac{\tau_q}{R_q}$$

3.2.2   System Modeling

The dynamical model of the mechanical system can be written by using the Euler-Lagrange formulation:

$$\begin{cases} I_\vartheta \ddot{\vartheta} - R_\vartheta \frac{\partial U_1}{\partial l_1}(-R_\vartheta \vartheta) - R_\vartheta \frac{\partial U_2}{\partial l_2}(-R_\vartheta \vartheta - R_q q) = \tau_\vartheta \\ I_{\vartheta^a} \ddot{\vartheta}^a + R_{\vartheta^a} \frac{\partial U_4}{\partial l_4}(R_{\vartheta^a} \vartheta^a) + R_{\vartheta^a} \frac{\partial U_3}{\partial l_3}(R_{\vartheta^a} \vartheta^a + R_q q) = \tau_{\vartheta^a} \\ I_q \ddot{q} - R_q \frac{\partial U_2}{\partial l_2}(-R_\vartheta \vartheta - R_q q) + R_q \frac{\partial U_3}{\partial l_3}(R_{\vartheta^a} \vartheta^a + R_q q) = \tau_q. \end{cases} \quad (1)$$

The state of the mechanism is thus $x := \begin{bmatrix} \vartheta & \vartheta^a & q & \dot{\vartheta} & \dot{\vartheta}^a & \dot{q} \end{bmatrix}^\top$, and we are interested in controlling the joint position variable q. It is worth noting that the torque applied at the joint side τ_q, for which no direct control is available, has to be interpreted as an **external disturbing torque**. On the other hand, the torques applied at the agonist τ_ϑ and antagonist τ_{ϑ^a} sides of the actuator represent the **internal actuation torques**.

Given an equilibrium configuration $x_{\text{eq}} = \begin{bmatrix} \vartheta_{\text{eq}} & \vartheta_{\text{eq}}^a & q_{\text{eq}} & 0 & 0 & 0 \end{bmatrix}^\top$, the torques must satisfy the following equation:

$$\begin{cases} -U_1'(\hat{\vartheta}) - U_2'(\hat{\vartheta} + \hat{q}) = \hat{\tau}_\vartheta \\ U_4'(\hat{\vartheta}^a) + U_3'(-\hat{q} + \hat{\vartheta}^a) = \hat{\tau}_{\vartheta^a} \\ -U_2'(\hat{\vartheta} + \hat{q}) + U_3'(-\hat{q} + \hat{\vartheta}^a) = \hat{\tau}_q, \end{cases} \quad (2)$$

where we adopted the "hat" and prime notation for the sake of simplicity.

Definition 1 Given the system (1), we define **co-contraction** as the control action that satisfies:

$$\dot{\hat{\vartheta}} > 0,$$

$$\dot{\hat{\vartheta}}^a > 0.$$

As seen in Sect. 3.1, this control action stretches all the springs.

Definition 2 Considering the vector of the actuator internal/external torques $\tau = \left[\hat{\tau}_{\vartheta}, \hat{\tau}_{\vartheta^a}, \hat{\tau}_q\right]^{\top}$ and the vector of the capstan/joint position $\alpha = \left[\hat{\vartheta}, \hat{\vartheta}^a, \hat{q}\right]^{\top}$, by exploiting the implicit function theorem, we define as **sensitivity matrix** the Jacobian matrix (see the Appendix 6 for the complete matrix):

$$\frac{\partial \alpha}{\partial \tau} = \begin{bmatrix} \frac{\partial \hat{\vartheta}}{\partial \hat{\tau}_{\vartheta}} & \frac{\partial \hat{\vartheta}}{\partial \hat{\tau}_{\vartheta^a}} & \frac{\partial \hat{\vartheta}}{\partial \hat{\tau}_q} \\ \frac{\partial \hat{\vartheta}^a}{\partial \hat{\tau}_{\vartheta}} & \frac{\partial \hat{\vartheta}^a}{\partial \hat{\tau}_{\vartheta^a}} & \frac{\partial \hat{\vartheta}^a}{\partial \hat{\tau}_q} \\ \frac{\partial \hat{q}}{\partial \hat{\tau}_{\vartheta}} & \frac{\partial \hat{q}}{\partial \hat{\tau}_{\vartheta^a}} & \frac{\partial \hat{q}}{\partial \hat{\tau}_q} \end{bmatrix}. \tag{3}$$

The analytical expression of $\partial \alpha / \partial \tau$ will play a crucial role in Sect. 3.3 during the modeling of the effects of stiction on the output joint q. In particular, we will focus on the quantities referred to the sensitivity of the output joint to the internal actuation torques $\frac{\partial \hat{q}}{\partial \hat{\tau}_{\vartheta}}$ and $\frac{\partial \hat{q}}{\partial \hat{\tau}_{\vartheta^a}}$, aiming at characterizing how the equilibrium configuration for q is affected by static friction acting on ϑ and ϑ^a.

The analytical expression for $\partial \alpha / \partial \tau$ will also play a major role in Sect. 4. In particular, it will be used to design a control policy that maintains the joint equilibrium configuration and regulates the actuator intrinsic stiffness.

3.2.3 System Properties

We now state two important properties of the considered system, together with the sufficient conditions to hold, when co-contracting control actions are used.

Proposition 1 *Suppose a control action satisfies Definition 1, that is, it is a co-contracting control action. If the springs are designed such that holds:*

$$U_i'' > 0$$
$$U_i''' > 0$$

with

$$i = 1, \ldots, 4,$$

then, the control action is increasing the individual stiffness of each spring. As a consequence, $\frac{\partial \hat{\tau}_q}{\partial \hat{q}}$, i.e. the joint level stiffness, is monotonically increasing.

We compute the joint level stiffness, defined as the sensitivity of the equilibrium configuration $\hat{q}$ with respect to variations of the external torque $\hat{\tau}q$. Analytically, this quantity coincides with $\partial \hat{q} / \partial \hat{\tau}_q$ and therefore it can be extracted as the element $(3, 3)$ in the matrix $\partial \alpha / \partial \tau$:

$$\frac{\partial \hat{q}}{\partial \hat{\tau}_q} = -\frac{U_1'' U_2'' U_3'' + U_1'' U_2'' U_4'' + U_3'' U_4'' U_1'' + U_1'' U_3'' U_4''}{(U_1'' + U_2'')(U_3'' + U_4'')}.$$

Easy calculations show that we have:

$$\frac{\partial \hat{q}}{\partial \hat{\tau}_q} = -\frac{1}{\frac{1}{\frac{1}{U_1''}+\frac{1}{U_2''}} + \frac{1}{\frac{1}{U_3''}+\frac{1}{U_4''}}}.$$

This last equation represents the intuitive result that the joint stiffness is the series of parallel of springs, nominally the series of U_1, U_2 in parallel with the series of U_3, U_4. It is indeed intuitive to conclude that global stiffness will increase if all individual stiffnesses (U_1, U_2, U_3, U_4) are increased.

At the control level, this is a desirable property because increasing the joint stiffness increases also the actuator *pnr*.

We now focus our attention on the quantity $\partial \hat{q}/\partial \hat{\tau}_\vartheta$ which represents the sensitivity of the joint position $\hat{q}$ with respect to the internal torque $\hat{\tau}_\vartheta$. Thanks to the symmetry of the system, the properties hereafter discussed will hold for the analogous quantity $\partial \hat{q}/\partial \hat{\tau}_{\vartheta^a}$.

Proposition 2 *Suppose a control action satisfies Definition 1, that is, it is a co-contracting control action. If the parallel and series springs are selected such that*

$$\frac{U_1''}{U_1'''} < \frac{U_2''}{U_2'''}$$

holds, then, the control action leads to decreasing value of $\partial \hat{q}/\partial \hat{\tau}_\vartheta$.

From the expression of $\partial \alpha/\partial \tau$, we have:

$$\frac{\partial \hat{q}}{\partial \hat{\tau}_\vartheta} = \frac{U_2''}{U_1'' + U_2''}\frac{\partial \hat{q}}{\partial \hat{\tau}_q},$$

and inverting the expression above we have:

$$\left(\frac{\partial \hat{q}}{\partial \hat{\tau}_\vartheta}\right)^{-1} = \underbrace{\frac{U_1'' + U_2''}{U_2''}}_{g(\hat{\vartheta},\hat{q})} \underbrace{\left(\frac{\partial \hat{q}}{\partial \hat{\tau}_q}\right)^{-1}}_{\text{joint stiffness}}. \tag{4}$$

From Eq. (4), we notice that this property is guaranteed if both the joint stiffness and the function $g(\vartheta, q)$ are increasing. From *Proposition* 1, we already know that the joint stiffness is increasing under the selected control action. Therefore we are left with guaranteeing that $g(\hat{\vartheta}, \hat{q})$ is non-decreasing in $\hat{\vartheta}$.

Easy computations show that:

$$\frac{\partial g}{\partial \hat{\vartheta}} > 0 \quad \Longleftrightarrow \quad \frac{U_1''}{U_1'''} < \frac{U_2''}{U_2'''}.$$

At the control level, this is a desirable property because it allows to have finer control over the variable $\hat{q}$ (since identical variations in the control variables τ_ϑ, τ_ϑ^a will correspond to smaller variations in the equilibrium configuration for $\hat{q}$).

3.3 Stiction Compensation

Static friction, due to its discontinuous nature, can produce undesired behaviors that are rather difficult to compensate. In particular, with reference to agonist-antagonist VSA, friction is often identified as one of the most evident and most adverse drawbacks. All antagonistic actuators are based on the primary idea of co-contracting both agonist and antagonist motor sides of the system to increase joint stiffness. As a consequence, internal forces increase. Among internal forces we have also friction and stiction components that, due to their non-linear nature, greatly degrade the performance of the antagonistic actuators. This decay in performance is, together with inertia issues, a limit for controlling the actuator mechanical bandwidth while performing torque control or dynamic tasks involving human or environment interaction. Different approaches have been proposed to reduce these drawbacks. As an example, we have classical integrator, action and disturbance observer described by de Wit et al. [10], adaptive controllers described by Tomei [32], sliding mode control described by Parra-Vega and Arimoto [25] or model-based friction compensation described by Armstrong [2] and Bona and Indri [7]. All of these approaches have the advantage of being fast and accurate, but most of them reduce the benefit of inherent passive compliance to a certain extent. Controllers indeed introduce an additional compliance which acts upon the series, variable or fixed, elasticity and that could be difficult to tune.

As we have seen in Sect. 2.1, one of the key features of the pnrVSA is the unique equilibrium position of its output joint. However the presence of static friction[4] gives rise to a set of "indifferent" equilibrium configurations. This range of equilibrium positions, named "dead-band" $\triangle q$, rapidly increases together with actuator co-contraction. To understand this behavior we can imagine that when the output joint q is displaced from its unique equilibrium position, the synergistic action of the two closed path ($K_{PE} + K_{SE}$ for both the agonist and the antagonist modules) should generate a restoring force. In our system this behavior is abated because the gear stiction prevents the rotation of the capstans.

Starting from the observation that both friction and spring restoring forces can be represented as a function of the actuator internal forces, we explored the possibility of mechanically compensating stiction's adverse effects by exploiting the actuator elastic elements. We derived one analytical condition over the potential energy of the elastic elements to ensure that during co-contraction the increase of the spring restoring forces is "greater" than the increase of the friction forces.

---

[4]In our actuator the main source of static friction are the gearboxes that have been used to connect the electric motors to the capstans (see Fig. 4).

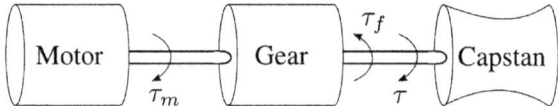

Fig. 4 The picture shows the motor subsystem: the motor is connected to the capstan by the gear. Friction effects between gear teeth greatly influence the torque transmission from motor to gear and vice-versa. *Source* [15]

As described by Fiorio et al. [16], to perform our qualitative analysis we assumed that the torque due to static friction is a function of the torque acting on the capstan[5]: $\tau_f = \tau_f(\tau)$, with τ being either τ_ϑ or τ_{ϑ^a} depending on the considered capstan. The analytical description of the dead-band can then be found considering the sensitivity matrix defined in Eq. 3. In particular, we can think of the dead-band as the consequence of an uncertainty that affects the torques that are applied at the capstans, where the amplitude of this uncertainty can be represented by the stiction torque. In this perspective, the total actuator dead-band Δq can be seen as the sum of the contributions coming from the agonist and antagonist side, yielding:

$$\Delta q = \frac{\partial \hat{q}}{\partial \hat{\tau}_\vartheta} \hat{\tau}_f(\hat{\tau}_\vartheta) + \frac{\partial \hat{q}}{\partial \hat{\tau}_{\vartheta^a}} \hat{\tau}_f^a\left(\hat{\tau}_{\vartheta^a}\right). \tag{5}$$

Equation 5 analytically defines the global actuator dead-band as a function of the internal system torques (and thus the co-contraction level), and describes how the static friction on the gearbox is reflected on the output joint.

To guarantee that co-contraction reduces the effect of gearbox friction on the joint equilibrium position, we formulated a differential condition on the spring potential energies. To make the analysis as general as possible, the conditions were expressed by assuming a generic functional dependence between stiction τ_f and applied torque τ. Similarly the non-linear spring potential energies were kept unspecified. We then studied how Eq. (5) varies with variations of $\hat{\tau}_\vartheta$ and $\hat{\tau}_{\vartheta^a}$. The major outcome of this analysis is the condition in order for the dead-band Δq to decrease with co-contraction:

$$\frac{\partial}{\partial \hat{\tau}_\vartheta} \left(\frac{U_{SE}''}{U_{SE}'' + U_{PE}''} \hat{\tau}_f \right) < 0. \tag{6}$$

If the potential energies of the parallel and series spring are selected such that condition (6) is satisfied, we are guaranteed that during co-contraction the deadband decreases instead of increasing. This is a significant result because it shows that it is possible to passively compensate the increase of static friction. In other words, this improvement in the actuator design adds a passive property that accomplishes a control objective mechanically.

---

[5]This assumption derives from the fact that co-activation increases internal forces. Certain friction forces, such as stiction, increase with gear teeth normal forces and therefore an increased stiction should be expected in response to an increased level of internal forces.

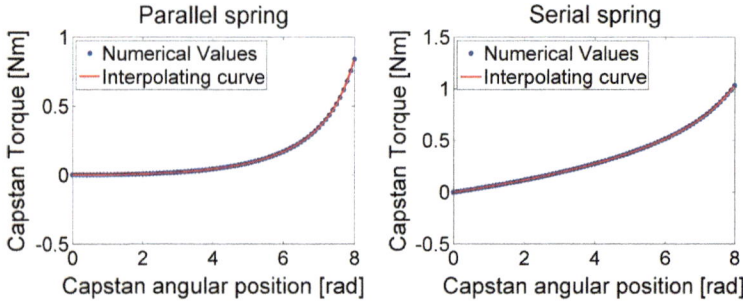

Fig. 5 The plots show the non-linear force-displacement characteristics of the parallel and series springs. The characteristics have been computed through a numerical optimization. *Source* [16]

3.4 Embodiment Design

In this section we present an overview of the actuator mechanical design focusing on the non-linear springs. The design process and the actuator performances have been extensively reported by Fiorio et al. [14].

3.4.1 Non-linear Spring Design

The final actuator design includes the four non-linear springs whose force-displacement characteristic has been optimized and customized in order to have light and compact solutions for both the parallel elastic element K_{PE} and the series elastic element K_{SE}. The differential inequality of Eq. 6 has been solved analytically. For this purpose we considered the condition as a single functional of $\hat{\tau}_\vartheta$ relating the potential energy of both springs to each other. With this approach we solved the differential inequality and made explicit the relation between serial and parallel elastic elements. Eventually, the analytical solution has been optimized for our setup by relying on a numerical optimization. The construction of an optimized set of springs, respecting these conditions, led to the construction of a version of our actuator which showed that for increasing levels of co-contraction the effect of stiction, and thus the dead-band effect, decreases (see tests in Sect. 5). Figure 5 shows the force-displacement functions of the non-linear springs that have been designed complying with condition (6).

To construct the non-linear springs, we exploited the idea of "non-circular spool", where the change in stiffness is achieved through a cam of varying radius, specialized into two different custom solutions. Regarding the K_{SE} spring a steel cable is wound around a non-circular spool in parallel to a linear torsional spring. Figure 6 shows a model of the non-linear spring. For the parallel elastic element K_{PE} we had to consider that due to the wide range of motion of the output joint, its non-circular spool performs more than one rotation and has been therefore realized with a three-

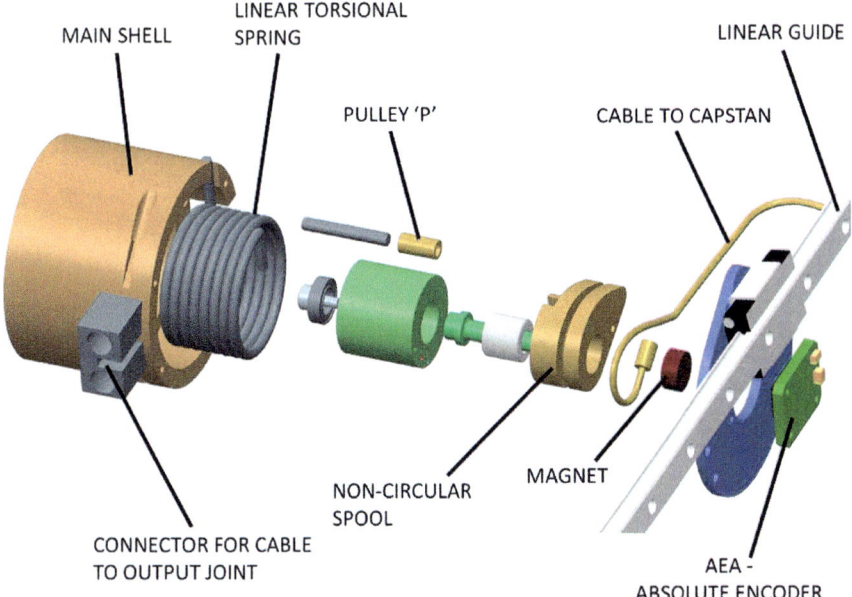

Fig. 6 The exploded view of the K_{SE} elastic element, which can be seen as a non-linear extensional spring

dimensional cut of the non-circular profile. Furthermore, in order to reduce the total spring size, the design has been optimized connecting directly the motor shaft to the non-circular spool. A steel cable connected to a linear compression spring is then wound on the non-circular spool (see Fig. 7b). The complete actuator CAD is shown in Fig. 7a, while in Fig. 8 the shape of both the non-circular spools is depicted.

3.4.2 Actuator Construction

The actuator has been constructed using aluminum for most of its components. In particular we used "ERGAL 7075-T6 temper" alloy. We selected this material because it has a low density (2810 [kg/m^3]), good mechanical properties (ultimate tensile strength of 510–540 [MPa]) and can be easily machined using Computer Numerical Control (CNC) machines. Only for some critical components we had to use a more robust material. In particular, all the non-circular spools and some other components of the non-linear springs were made with the stainless steel "17-4 PH". Another interesting peculiarity of the non-circular spools is that, due to their complex shape, some of them have been manufactured using Selective Laser Sintering (SLS). This technique is an additive manufacturing technology that uses the laser as a power source to melt powdered material to create a solid structure. All

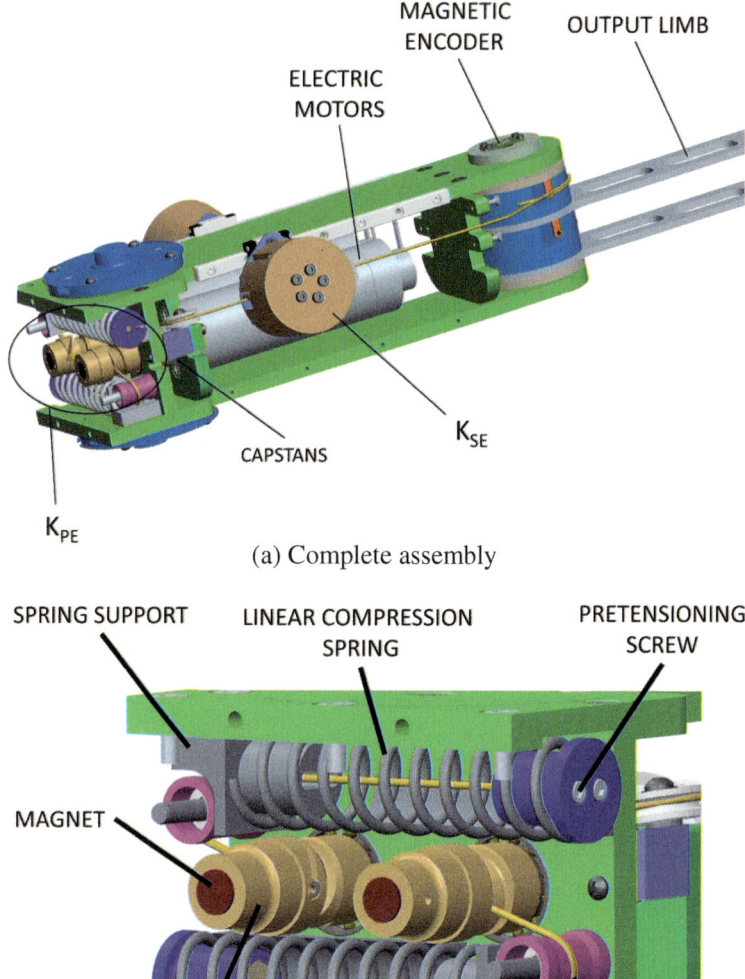

(a) Complete assembly

(b) K_{PE} design

Fig. 7 In **a**, the complete assembly: the two electric motors are connected, through their relative gears, to the capstans. Each capstan winds the cable (yellow) that stretches the K_{SE} springs. In **b**, a detailed view of how the parallel elastic elements have been integrated directly on the motor shaft: the extruded non-circular spools, that realize the parallel elastic elements K_{PE}, are fastened together with the capstans on the motor shafts. The pretensioning screws are used to align the capstans with the zero configuration (i.e. all springs elongations are zero). *Source* [14]

Fig. 8 The plots show the profiles of the non-circular cams together with the CAD models of the spools. The non-circular profiles have been used to design the cable grooves. In particular due to the overlap in the profile of the parallel spring we had to design the spool extruding the cable groove in 3D

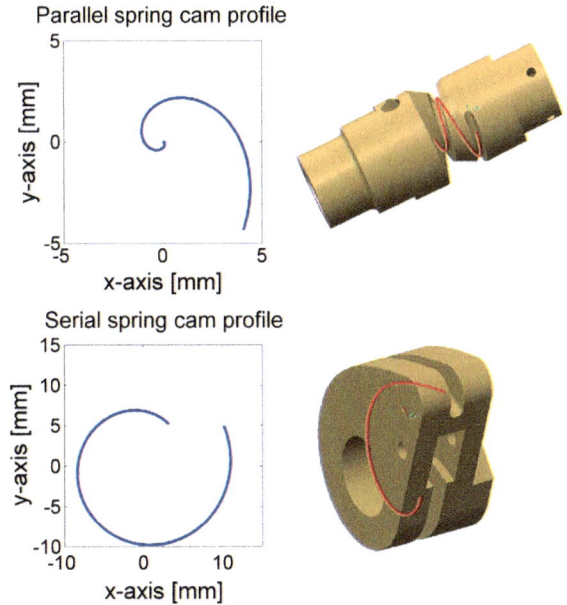

the mechanical connections between springs, frame, capstans and pulleys have been realized through steel cables. Figure 9 from *a* to *d* shows the 3D printed components and the assembled series and parallel springs. At the bottom (*e*) the complete actuator is shown.

4 Controller Design

In this section we describe two different control solutions that exploit the properties of the proposed passive noise rejecting variable stiffness actuator. In particular, we first design a control law capable of decoupling the regulation of the joint-stiffness from the joint equilibrium configuration (see also [23]). To test the passive noise rejecting property of the actuator, we also design a second type of control action. In particular, we formulate a Stochastic Optimal Control (SOC) problem (see also [5, 29]), and we apply the obtained control policy as a purely open-loop action, leaving to the actuator the responsibility of rejecting eventual disturbances. The main advantage of the SOC [20] strategy is the possibility to generate a complete motor plan characterized by a "global" perspective. In particular, this motor plan, is the optimal solution with respect to some optimization criteria, while taking into account stochastic information due to noise. End-effector positioning accuracy and energy consumption are just an example of possible optimization criteria that can be specified to the control problem. The latter, is of paramount importance due to

(a) Series spring - spool   (b) Parallel spring - spool

(c) Series spring assembled   (d) Parallel springs assembled

(e) Complete pnrVSA assembled

Fig. 9 From **a** to **d** some of the components of the actuator are shown, while at the bottom the complete actuator is depicted. In the pictures **a** and **b** it is possible to see a comparison between the coarse metal printed component and the finished one

the fact that co-contraction is energetically expensive, i.e. it is important to find the minimum level of co-contraction necessary to achieve the task.

Nevertheless, it is important to notice that both controllers generate "open-loop" solutions which fully exploit the native noise-rejection property of the actuator, without the need to rely on state estimations and feedback gains during control.

4.1 Stiffness and Position Control

The control action is computed considering the time derivative of the output joint position. Given that we do not have direct control on $\hat{\tau}_q$, assume that $\dot{\hat{\tau}}_q = 0$ so that the time derivative of $\hat{q}$ results to be:

$$\dot{\hat{q}} = \underbrace{\left[\frac{\partial \hat{q}}{\partial \hat{\tau}_\vartheta} \quad \frac{\partial \hat{q}}{\partial \hat{\tau}_{\vartheta a}} \quad \frac{\partial \hat{q}}{\partial \hat{\tau}_q} \right]}_{N} \underbrace{\left[\begin{array}{c} \dot{\hat{\tau}}_\vartheta \\ \dot{\hat{\tau}}_{\vartheta a} \\ 0 \end{array} \right]}_{\delta\tau}, \tag{7}$$

where $\delta\tau$ is the time derivative of the variable τ introduced by *Definition 2*, i.e. $\delta\tau = \dot{\tau}$, while an analytical expression for N is given by the last row of the sensitivity matrix (3). The generic solution of this control problem (imposing, as usual, direct control only on the internal torques) is given by the following:

$$\delta\tau = \left[\begin{array}{c} U_2''(U_3'' + U_4'') \\ U_3''(U_1'' + U_2'') \\ 0 \end{array} \right] \cdot k_1 \cdot v + \left[\begin{array}{c} -U_3''(U_1'' + U_2'') \\ U_2''(U_3'' + U_4'') \\ 0 \end{array} \right] \cdot k_2 \cdot u, \tag{8}$$

with:

$$k_2 = U_1'' U_2'' U_3'' + U_1'' U_2'' U_4'' + U_1'' U_3'' U_4'' + U_2'' U_3'' U_4'',$$
$$k_1 = k_2^2 \left(\left(U_2'' \left(U_3'' + U_4'' \right) \right)^2 + \left(U_3'' \left(U_1'' + U_2'' \right) \right)^2 \right).$$

The control action $\delta\tau$ in Eq. (8) is composed of two terms. The first term, parametrized by the free variable $v \in \mathbb{R}$, acts directly on the time derivative of $\hat{q}$, i.e. is responsible of changing the equilibrium configuration of the output joint. The second variable $u \in \mathbb{R}$, instead, monotonically changes the joint stiffness while maintaining a constant value for $\hat{q}$, i.e.

$$\dot{\hat{q}} = N \left. \delta\tau \right|_{v=0} = 0, \quad \forall u \in \mathbb{R}.$$

4.2 Stochastic Optimal Control (SOC)

In the second control strategy we consider a stochastic non-linear control-affine system:

$$\mathrm{d}\mathbf{x} = \mathbf{a}(\mathbf{x}(t), t)\mathrm{d}t + B(\mathbf{x}(t), t)\mathbf{u}(\mathbf{x}(t), t)\mathrm{d}t + C(\mathbf{x}, t)\mathrm{d}\mathbf{w}, \qquad (9)$$

where $\mathbf{x} \in \mathbb{R}^n$ is the state of the system, $\mathbf{u} \in \mathbb{R}^m$ is the control input, $\mathbf{w} \in \mathbb{R}^m$ is brownian noise, $\mathbf{a}(\cdot)$ is the drift term, $B(\cdot)$ is the control matrix and $C(\cdot)$ is the diffusion matrix.

We also define the cost-to-go at time t_0, state $\mathbf{x}_0$ as:

$$J(\mathbf{x}_0, t_0) = \mathbb{E}[\phi(\mathbf{x}(t_f)) + \int_{t_0}^{t_f} \mathscr{L}(\mathbf{x}, t) + \frac{1}{2}\mathbf{u}^\top R\mathbf{u}\mathrm{d}t], \qquad (10)$$

where $\phi(\cdot)$ is a final state-dependent cost and $\mathscr{L}(\cdot)$ is the running cost term and $\mathbb{E}$ is the expected value operation.

The optimal cost-to-go J^* must satisfy the stochastic version of the Hamilton-Jacobi-Bellman (HJB) equation, which is a second order non-linear partial differential equation. The resulting optimal control $\mathbf{u}^*$ can then be expressed as

$$\mathbf{u}^* = -R^{-1}B^\top \frac{\partial J^*(\mathbf{x}, t)}{\partial \mathbf{x}}. \qquad (11)$$

The HJB equation can be transformed into a linear second order partial differential equation by performing a logarithmic transformation, $\psi = exp(\frac{1}{\lambda}J)$ and by assuming that $C = B\sqrt{\lambda R^{-1}}$, see [31]. It can then be shown that the optimal control can be expressed at each state/time as a path integral which can be approximated via importance sampling methods.

It is worth noting that the control action in Eq. (11) is a state feedback action. In order to test the passive noise rejecting capabilities of the proposed actuator, we deliberately modify the control action to result in a pure open-loop action. It is thus responsibility of the noise rejecting property of the mechanism to ensure robustness against external disturbances.

5 Simulations and Experimental Tests

5.1 Spring Design Validation

To validate the non-linear spring design of Sect. 3.3 we tested the actuator with two different sets of series and parallel springs. The first set, named "optimized", has been designed complying with the condition (6), while the second set, named "quadratic",

Table 2 Behavior of the dead-band "DB" for the pnrVSA prototype with two different set of parallel and series non-linear springs

Torque (Nm)	Optimized set DB (deg)	Quadratic set DB (deg)
0.12	104.61	3.50
0.25	100.81	27.90
0.50	87.73	69.90
1.00	61.62	85.50
1.50	49.81	Entire RoM

comprises two non-linear springs whose force-displacement relationship is quadratic. Both the spring sets satisfy the requirements derived in Sect. 3.2.3.

During the experiments the dead-band was measured for different levels of co-contraction by moving by hand the output joint within the actuator Range of Motion (RoM). The co-contraction level was controlled by setting equal absolute values for the internal torques of the agonist τ_ϑ and antagonist side τ_{ϑ^a}.

Table 2 shows the behavior of the dead-band for increasing level of co-contraction for the two sets. The quadratic set shows an increasing dead-band, which reaches the output joint RoM when the absolute value of the internal torques τ_ϑ and τ_{ϑ^a} is about 1.5 Nm. On the contrary the optimized set of springs is effective in inverting the dead-band mechanical behavior. In particular, due to the numerical procedure adopted to compute the non-linear springs potential energy functional (see also [16]), we have a large dead-band for low levels of co-contraction.

5.2 Control Simulations

We first tested the control law (8) on a simulation of the actuator represented in Fig. 3. Springs were designed to comply to the conditions derived in Sect. 3.3 and therefore satisfy all the required conditions outlined in Sect. 3.2.3.

Figure 10 shows the results of applying (8) to the system. In particular, the plots (a), (b) and (c) shows the stiffness, the joint position and the sensitivity $\partial \hat{q}/\partial \hat{\tau}_{\vartheta^a}$ when we applied (8) with $v = 0$ and $u \neq 0$. In the plot (b) we can notice both the transient response (in blue) and the constant equilibrium configuration of the output joint (in red). The plots (d), (e) and (f), instead, show the results of applying (8) with the opposite choice of control parameters, i.e. $u = 0$ and $v \neq 0$. It is possible to notice that while the equilibrium configuration of the output joint changes, the stiffness remains constant.

Secondly, we tested the stochastic optimal control in Sect. 4.2. As depicted in Fig. 11, in this case we simulated a two-DoF arm equipped with a pnrVSA at each joint. The arm was required to push against a wall with a constant force while subject to an external disturbance chosen to be a divergent force field at the contact point.

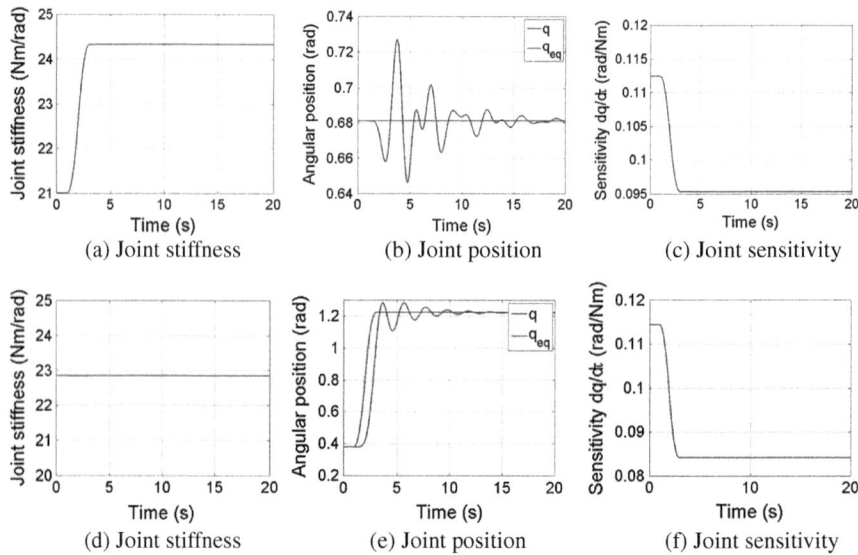

Fig. 10 **a, b** and **c** show the system response to the control law in Eq. (8) with $v = 0$ and $u \neq 0$. **d, e** and **f** show the system response to the same control action, but with the choice $u = 0$ and $v \neq 0$

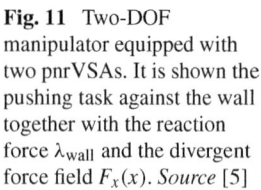

Fig. 11 Two-DOF manipulator equipped with two pnrVSAs. It is shown the pushing task against the wall together with the reaction force λ_{wall} and the divergent force field $F_x(x)$. *Source* [5]

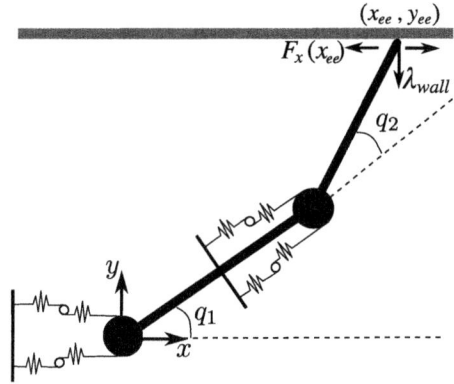

The control policy has been pre-computed with two different stiffness configuration: an high stiffness and a low stiffness configuration. The plots (a) and (b) of Fig. 12 shows the x-coordinate of the end-effector during the application of the two control policies. The equilibrium configuration was required to be 0.3 m. The stiffer solution, on the right, exhibits a lower error.

Finally, we tested the stochastic optimal control algorithm with a two-DOF manipulator equipped with classical VSAs (without *pnr*). As shown in Fig. 12c, the manipulator in this case is not able to maintain the desired position, neither for the high stiffness configuration.

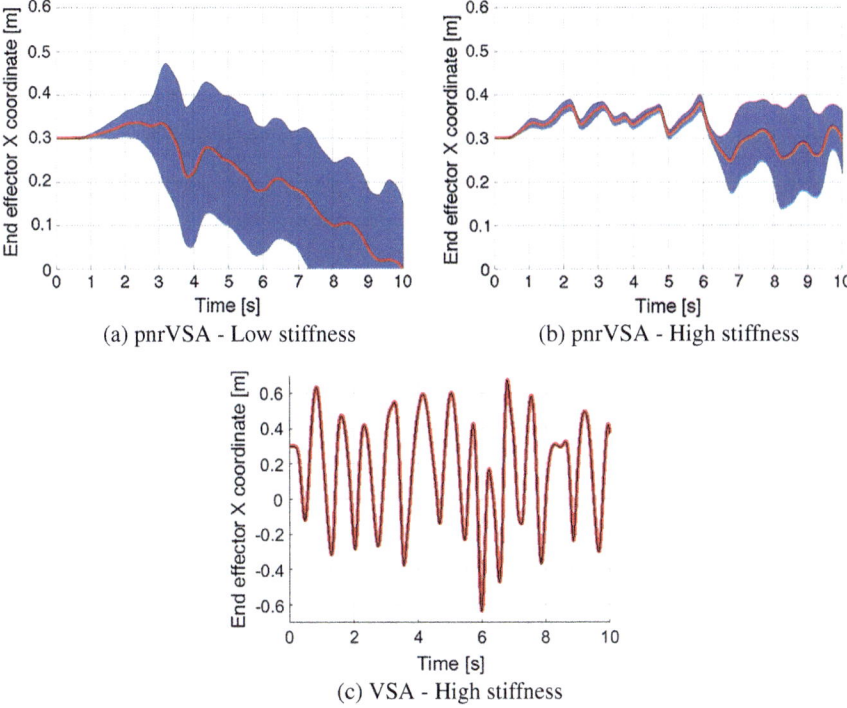

(a) pnrVSA - Low stiffness

(b) pnrVSA - High stiffness

(c) VSA - High stiffness

Fig. 12 a and **b** show the X-coordinate of the two-DoF arm end-effector when controlled with the SOC algorithm of Sect. 4.2. Plots show the average (red) and standard deviation (shaded blue) over 20 simulation trials. **c** shows the behavior of the same two-DoF manipulator equipped with classical VSAs. *Source* [29]

The main result of the simulations is that the proposed actuator makes it possible to mimic some effects of muscle co-contraction in humans, which is useful to cope with noise and sensorimotor delays affecting physical/biological systems. The price to pay to achieve this behavior is a waste of energy associated with the need for stretching additional springs. This property is nevertheless similar to the large energy expenditure of muscle co-contraction and this moreover justifies the use of optimal control techniques to reduce this energy consumption to a minimum. The fact that we stick to open-loop control laws is just to illustrate the properties of pnrVSA for such an extreme case, but this does not prevent the use of feedback laws in real applications.

6 Conclusions and Future Work

This work presents the mechanical design of a novel actuator capable of actively changing its passive noise rejection characteristic. The proposed actuator is composed of two independent motors in an agonist-antagonist configuration and its design takes inspiration from the muscles configuration in biological systems. Crucial elements in the proposed system are four non-linear springs whose force-displacement characteristic has been customized for our specific needs.

The problem of quantifying analytically the effects and the propagation of internal static friction has been addressed and solved. Eventually, some important control properties of the novel variable stiffness actuator have been also characterized and tested in simulation. After a characterization of the system stiffness, we proposed a control action that guarantees a monotonically increasing joint stiffness, a desirable property for augmenting the system disturbance rejection. This property was guaranteed with minimal requirements on the spring potential energies (basically positiveness of the derivatives). We also showed how adding a planning element to the control (e.g. using SOC) can help attenuating external disturbances without explicitly relying on feedback.

The proposed actuator mimics a functional property of human antagonist muscle apparatus, but to make it more human-like, more work would be required. We aim in the next future at designing a new actuator to fully exploit the potential of our methodology. Future works include also the realization of a two degrees of freedom robotic arm actuated by three of these actuators: two acting on a single joint, one spanning two joints in a polyarticular-like configuration.

Appendix

Sensitivity Matrix Computation

Let's represent (2) in a compact way, with the following definition:

$$
\begin{cases}
-U_1'(\hat{\vartheta}) - U_2'(\hat{\vartheta} + \hat{q}) = \hat{\tau}_\vartheta \\
U_4'(\hat{\vartheta}^a) + U_3'(\hat{\vartheta}^a - \hat{q}) = \hat{\tau}_{\vartheta^a} \\
-U_2'(\hat{\vartheta} + \hat{q}) + U_3'(\hat{\vartheta}^a - \hat{q}) = \hat{\tau}_q
\end{cases}
\quad \Longleftrightarrow \quad f(\alpha,\tau) = 0.
$$

By resourcing to the implicit function theorem, the equation $f(\alpha,\tau) = 0$ locally defines a function $\alpha(\tau)$ (equilibrium configuration) with sensitivity:

$$
\frac{\partial \alpha}{\partial \tau} = -\left[\frac{\partial f}{\partial \alpha}\right]^{-1} \frac{\partial f}{\partial \tau}.
$$

as easily follows by numerical derivation of the constrain equation $f(\alpha(\tau),\tau) = 0$:

$$\frac{\partial f}{\partial \alpha}\frac{\partial \alpha}{\partial \tau} + \frac{\partial f}{\partial \tau} = 0 \rightarrow \frac{\partial \alpha}{\partial \tau} = -\left[\frac{\partial f}{\partial \alpha}\right]^{-1}\frac{\partial f}{\partial \tau}.$$

Using the analytical expression of f given by (2), we obtain:

$$\frac{\partial f}{\partial \alpha} = \left[\begin{array}{ccc}\frac{\partial f}{\partial \hat{\vartheta}} & \frac{\partial f}{\partial \hat{\vartheta}^a} & \frac{\partial f}{\partial \hat{q}}\end{array}\right] = \begin{bmatrix} -U_1'' - U_2'' & 0 & -U_2'' \\ 0 & U_4'' + U_3'' & -U_3'' \\ -U_2'' & U_3'' & -U_2'' - U_3'' \end{bmatrix},$$

and:

$$\frac{\partial f}{\partial \tau} = \begin{bmatrix} -1 & 0 & 0 \\ 0 & -1 & 0 \\ 0 & 0 & -1 \end{bmatrix}$$

which eventually results in the following expression:

$$\frac{\partial \alpha}{\partial \tau} = \begin{bmatrix} -U_1'' - U_2'' & 0 & -U_2'' \\ 0 & U_4'' + U_3'' & -U_3'' \\ -U_2'' & U_3'' & -U_2'' - U_3'' \end{bmatrix}^{-1}$$

$$\frac{\partial \alpha}{\partial \tau} = \begin{bmatrix} \frac{\partial \hat{\vartheta}}{\partial \hat{t}_\vartheta} & \frac{\partial \hat{\vartheta}}{\partial \hat{t}_\vartheta^a} & \frac{\partial \hat{\vartheta}}{\partial \hat{t}_q} \\ \frac{\partial \hat{\vartheta}^a}{\partial \hat{t}_\vartheta} & \frac{\partial \hat{\vartheta}^a}{\partial \hat{t}_\vartheta^a} & \frac{\partial \hat{\vartheta}^a}{\partial \hat{t}_q} \\ \frac{\partial \hat{q}}{\partial \hat{t}_\vartheta} & \frac{\partial \hat{q}}{\partial \hat{t}_\vartheta^a} & \frac{\partial \hat{q}}{\partial \hat{t}_q} \end{bmatrix} =$$

$$\begin{bmatrix} -(U_2''U_3'' + U_2''U_4'' + U_3''U_4'') & -U_2''U_3'' & U_2''(U_3'' + U_4'') \\ U_2''U_3'' & U_1''U_2'' + U_1''U_3'' + U_2''U_3'' & -U_3''(U_1'' + U_2'') \\ U_2''(U_3'' + U_4'') & U_3''(U_1'' + U_2'') & -(U_1'' + U_2'')(U_3'' + U_4'') \end{bmatrix}$$

$$\cdot \frac{1}{U_1''U_2''U_3'' + U_1''U_2''U_4'' + U_1''U_3''U_4'' + U_2''U_3''U_4''}$$

Actuator Specifications

The specifications of the actuator are recapped in Fig. 13.

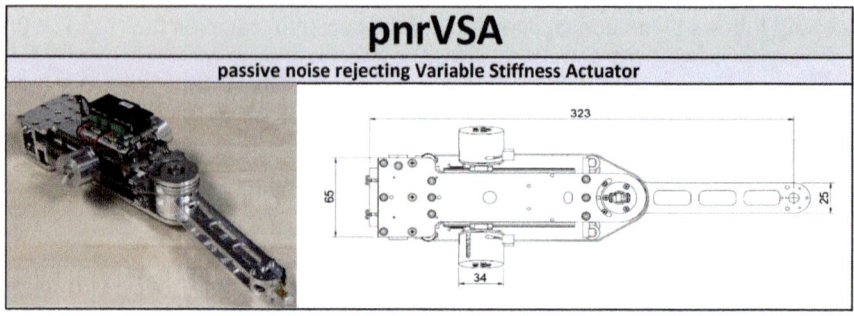

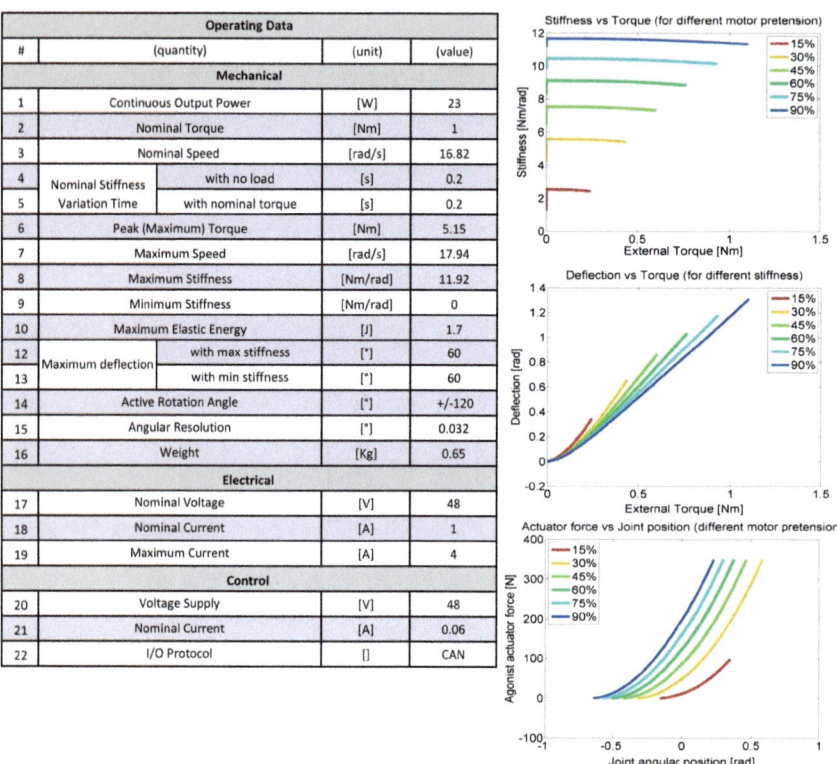

Fig. 13 The VIACTORS VSA datasheet of the pnrVSA. In the plots on the right hand side we report the pnrVSA characteristic curves for different internal motor pretensions. This pretension has to be interpreted as the applied torque at motor capstan, ranging from 15 to 90% of the stall torque. The VIACTORS Variable Stiffness Joint Datasheet was developed within the VIACTORS project, which is a part of the EU 7th Framework Programme. *Source* [14]

References

1. Albu-Schaffer, A., Hirzinger, G.: Cartesian impedance control techniques for torque controlled light-weight robots. In: IEEE International Conference on Robotics and Automation, 2002. Proceedings. ICRA '02. vol. 1, pp. 657–663 (2002). https://doi.org/10.1109/ROBOT.2002. 1013433

2. Armstrong, B.: Dynamics for robot control: Friction modelling and ensuring excitation during parameter identification. Dissertation, Stanford University (1988)

3. Berret, B., Ivaldi, S., Nori, F., Sandini, G.: Stochastic optimal control with variable impedance manipulators in presence of uncertainties and delayed feedback. In: International Conference on Intelligent Robots and Systems (IROS2011), pp. 4354–4359 . IEEE (2011)

4. Berret, B., Sandini, G., Nori, F.: Design principles for muscle-like variable impedance actuators with noise rejection property via co-contraction. In: 2012 12th IEEE-RAS International Conference on Humanoid Robots (Humanoids), pp. 222–227 (2012). https://doi.org/10.1109/ HUMANOIDS.2012.6651524

5. Berret, B., Yung, I., Nori, F.: Open-loop stochastic optimal control of a passive noise-rejection variable stiffness actuator: application to unstable tasks. In: 2013 IEEE/RSJ International Conference on Intelligent Robots and Systems, pp. 3029–3034 (2013). https://doi.org/10.1109/ IROS.2013.6696785

6. Bicchi, A., Tonietti, G., Piaggio, E.: Design, realization and control of soft robot arms for intrinsically safe interaction with humans. In: Proceedings of the IARP/RAS Workshop on Technical Challenges for Dependable Robots in Human Environments, pp. 79–87 (2002)

7. Bona, B., Indri, M.: Friction compensation in robotics: an overview. In: 44th IEEE Conference on Decision and Control, 2005 and 2005 European Control Conference. CDC-ECC '05, pp. 4360–4367 (2005). https://doi.org/10.1109/CDC.2005.1582848

8. Burdet, E., Osu, R., Franklin, D.W., Milner, T.E., Kawato, M.: The central nervous system stabilizes unstable dynamics by learning optimal impedance. Nature **414**(6862), 446–9 (2001a). https://doi.org/10.1038/35106566

9. Burdet, E., Osu, R., Franklin, D.W., Milner, T.E., Kawato, M.: The central nervous system stabilizes unstable dynamics by learning optimal impedance. Nature **414**(6862), 446–449 (2001b). https://doi.org/10.1038/35106566

10. de Wit, C.C., Olsson, H., Astrom, K., Lischinsky, P.: A new model for control of systems with friction. IEEE Trans. Autom. Control **40**, 419–425 (1994)

11. De Luca, C.J., Mambrito, B.: Voluntary control of motor units in human antagonist muscles: coactivation and reciprocal activation. J. Neurophysiol. **58**(3), 525–542 (1987). http://jn. physiology.org/content/58/3/525, http://jn.physiology.org/content/58/3/525.full.pdf

12. Del Prete, A., Nori, F., Metta, G,. Natale, L.: Control of contact forces: the role of tactile feedback for contact localization. In: 2012 IEEE/RSJ International Conference on Intelligent Robots and Systems (IROS) (2012)

13. Eiberger, O., Haddadin, S., Weis, M., Albu-Sch äffer, A., Hirzinger, G.: On joint design with intrinsic variable compliance: derivation of the DLR QA-joint, pp. 1687–1694 (2010)

14. Fiorio, L., Parmiggiani, A., Berret, B., Sandini, G., Nori, F.: pnrVSA: human-like actuator with non-linear springs in agonist-antagonist configuration (2012)

15. Fiorio, L., Romano, F., Parmiggiani, A., Sandini, G., Nori, F.: On the effects of internal stiction in pnrVIA actuators. In: 2013 13th IEEE-RAS International Conference on Humanoid Robots (Humanoids), pp. 362–367 (2013) https://doi.org/10.1109/HUMANOIDS.2013.7030000

16. Fiorio, L., Romano, F., Parmiggiani, A., Sandini, G., Nori, F.: Stiction compensation in agonist-antagonist variable stiffness actuators. In: Proceedings of Robotics: Science and Systems, Berkeley, USA (2014)

17. Fumagalli, M., Ivaldi, S., Randazzo, M., Natale, L., Metta, G., Sandini, G., Nori, F.: Force feedback exploiting tactile and proximal force/torque sensing. Theory and implementation on the humanoid robot iCub. Autonom. Robots **33**(4), 381–398 (2012)

18. Hill, A., Gasser, H.: The dynamics of muscular contraction (1924)

19. Hogan, N.: Adaptive control of mechanical impedance by coactivation of antagonist muscles. IEEE Trans. Autom. Control **29**(8), 681–690 (1984). https://doi.org/10.1109/TAC.1984.1103644
20. Kappen, H.J.: Optimal Control Theory and the Linear Bellman Equation, p. 363387. Cambridge University Press, Cambridge. https://doi.org/10.1017/CBO9780511984679.018
21. McMahon, T.: Muscle, Reflexes, and Locomotion (1984)
22. Migliore, S. A., Brown, E. A., DeWeerth, S. P.: Biologically inspired joint stiffness control. In: Proceedings of the 2005 IEEE International Conference on Robotics and Automation, ICRA 2005, April 18–22, 2005, Barcelona, Spain, pp. 4508–4513 (2005). https://doi.org/10.1109/ROBOT.2005.1570814
23. Nori, F., Berret, B., Fiorio, L., Parmiggiani, A., Sandini, G.: Control of a single degree of freedom noise rejecting-variable impedance. In: Proceedings of the 10th international IFAC symposium on Robot Control (SYROCO2012) (2012)
24. Paillard, J.: Fast and slow feedback loops for the visual correction of spatial errors in a pointing task: a reappraisal. Can. J. Physiol. Pharmacol. **74**, 401–417 (1996)
25. Parra-Vega, V., Arimoto, S.: A passivity based adaptive sliding mode position-force control for robot manipulators. Int. J. Adapt. Control Signal Process. **10**, 365–377 (1996)
26. Petit, F., Chalon, M., Friedl, W., Grebenstein, M., Albu-Schäffer, A., Hirzinger, G.: Bidirectional antagonistic variable stiffness actuation: analysis, design & implementation. In: ICRA, pp. 4189–4196 (2010)
27. Polit, A., Bizzi, E.: Characteristics of motor programs underlying arm movements in monkeys. J. Neurophysiol. **42**(1), 183–194 (1979)
28. Pratt, G., Williamson, M.: Series elastic actuators. In: 1995 IEEE/RSJ International Conference on Intelligent Robots and Systems 95. 'Human Robot Interaction and Cooperative Robots', Proceedings, vol. 1, pp. 399–406 (1995)
29. Romano, F., Fiorio, L., Sandini, G., Nori, F.: Control of a two-DOF manipulator equipped with a PNR-variable stiffness actuator. In: 2014 IEEE International Symposium on Intelligent Control (ISIC), pp 1354–1359 (2014).https://doi.org/10.1109/ISIC.2014.6967620
30. Schiavi, R., Grioli, G., Sen, S., Bicchi, A.: VSA-II: a novel prototype of variable stiffness actuator for safe and performing robots interacting with humans. In: IEEE International Conference on Robotics and Automation, 2008. ICRA 2008, pp. 2171–2176 (2008). https://doi.org/10.1109/ROBOT.2008.4543528
31. Theodorou, E., Buchli, J., Schaal, S.: A generalized path integral control approach to reinforcement learning. J. Mach. Learn. Res. **11**, 3137–3181 (2010)
32. Tomei, P.: Robust adaptive friction compensation for tracking control of robot manipulators. IEEE Trans. Autom. Control **45**(11), 2164–2169 (2000)
33. Tonietti, G., Schiavi, R., Bicchi, A.: Design and control of a variable stiffness actuator for safe and fast physical human/robot interaction. In: ICRA, pp 526–531. IEEE (2005)
34. Traversaro S, Pucci D, Nori F (2015) In situ calibration of six-axis force-torque sensors using accelerometer measurements. In: 2015 IEEE International Conference on Robotics and Automation (ICRA), IEEE, pp. 2111–2116
35. Van Ham, R., Sugar, T.G., Vanderborght, B., Hollander, K.W., Lefeber, D.: Compliant actuator designs. IEEE Robot. Autom. Mag. **16**, 81–94 (2009)
36. Vanderborght, B., Albu-Sch äffer, A., Bicchi, A., Burdet, E., Caldwell, D.G., Carloni, R., Catalano, M.G., Eiberger, O., Friedl, W., Ganesh, G., Garabini, M., Grebenstein, M., Grioli, G., Haddadin, S., Hoppner, H., Jafari, A., Laffranchi, M., Lefeber, D., Petit, F., Stramigioli, S., Tsagarakis, N.G., Damme, M.V., Ham, R.V., Visser, L.C., Wolf, S.: Variable impedance actuators: a review. Robot. Autonom. Syst. **61**(12), 1601–1614 (2013)
37. Vitiello, N., Cattin, E., Roccella, S., Giovacchini, F., Vecchi, F., Carrozza, M.C., Dario, P.: The neurarm: towards a platform for joint neuroscience experiments on human motion control theories. In: 2007 IEEE/RSJ International Conference on Intelligent Robots and Systems, pp. 1852–1857. IEEE (2007)

Bipedal Locomotion: A Continuous Tradeoff Between Robustness and Energy-Efficiency

Mehdi Benallegue and Jean-Paul Laumond

Abstract Walking is a mechanical process involving all the limbs the body and subject to balance constraints. The structure of the body can allow the emergence of a stable walking using few to no energy, but this motion is sensitive to external perturbations and prone to fall easily. On the other hand, with strong actuation and anticipation, much more disturbances can be overcome, but at the cost of high energy consumption. The choice between these two modes of locomotion is more than a context-dependent binary selection. It is a continuous tradeoff, not only providing a span of different walking control policies but building the framework of a precise classification of these controllers. In this chapter, we discuss, for the humans and the robots, the two extreme walking paradigms and how to sort the solutions ranging between them. We analyze also the incentives behind the decision to change to a more robust or more efficient control policy. Finally, we show that the versatility of a walker depends deeply on the relationships lying between its controller dynamics and its mechanical design, both being ideally built together in a process of *codesign*.

1 Two Competing Humanoid Walking Paradigms: Energy-Efficiency Versus Robustness

Let us observe the walking gaits of the two women in Fig. 1a. Both are walking on the paved ground in a very regular, quiet and synchronous way. The area is empty of obstacle. They are not taking care where placing their feet. The man and the woman in Fig. 1b walk from the same area towards a ground paved with larger stones. At some stage the woman stumbles. She temporarily loses her balance as her left foot enters a cavity between two stones. She did not watch her step. These two examples are extracted from [14]. They illustrate the influence of the ground texture on the

M. Benallegue (✉)
AIST, 1-1-1 Umezono, Central 1, Tsukuba, Ibaraki 305-8560, Japan
e-mail: mehdi.benallegue@aist.go.jp

J.-P. Laumond
LAAS-CNRS, Univ. de Toulouse, 7 Ave du Colonel Roche, 31400 Toulouse, France
e-mail: jpl@laas.fr

© Springer International Publishing AG, part of Springer Nature 2019
G. Venture et al. (eds.), *Biomechanics of Anthropomorphic Systems*, Springer Tracts in Advanced Robotics 124, https://doi.org/10.1007/978-3-319-93870-7_12

(a) Walking without thinking. Reflex-based walking is stereotyped and robust enough to absorb slightly textured pavements.

(b) Transiting from slightly to significantly textured pavements requires attention.

Fig. 1 Pavements in Roma. On the bottom part of the pavement, the walker has to anticipate which stones will be used for the next steps. On the other hand, walking on the pavement at the top part of the figure does not require any anticipation of the foot placements: one can "walk without thinking", i.e. without requiring any vision modality to plan where to place the feet

way to walk and they reveal the existence of different gait control strategies. Walking on a rough pavement requires more attention than walking on a flat floor. A purely reflex-based walking allows us to wander safely on the thinly textured pavement. The same reflex-based gait fails when walking on the other pavement. The woman stumbles because she did not change her walking control strategy: she maintains a reflex-based gait while approaching the rough pavement would have required foot placement anticipation.

Let us now imagine a humanoid robot walking in the same conditions. The question of footstep anticipation echoes two opposite paradigms addressing the locomotion control of humanoids.

The first paradigm is based on clever mechanical designs that take advantage of gravity. In the 1980s, the seminal work by McGeer [16] showed how a simple compass-like mechanical system can walk down a slope without requiring any actuator. Walking stability derives from the natural swing of the legs. It does not require any energetic provision other than the gravity. This work opened the development of the so-called passives walkers (see [4] for an introductory survey). By adding actuation to simple compass-like mechanisms it is possible to devise humanoid walkers walking efficiently on a flat ground. At each step, the robot falls and catches itself in a controlled manner. One of the best performance has been reached on May 1–2, 2011, when the 4-legged bipedal robot Cornell Ranger walked non-stop 40.5 mile ultra-Marathon (about 65 km in 186,076 steps) while only using 5 cents worth of electricity [13]. By considering simultaneously clever a mechanical design together with a control architecture, passive walkers are very energy-efficient. However, as they are not anticipating their footsteps, they are very fragile with respect to the ground perturbations. Cornell Ranger requires walking on the flat surface of a sport stadium. It would probably fall on the low textured pavement in Fig. 1a. It would definitely fall when facing the rough pavement in Fig. 1b.

A second control paradigm steers also the development of humanoid robotics. With respect to passive walkers, which are mainly centered on the design of clever leg design, the paradigm applies to whole-body mechanical structures including legs, torso, arms and head, for a total of about 30 articulated joints. Most joints (if not all) are actuated. These robots are built not only to walk but also to perform manipulation tasks with their arms. Doing so, locomotion appears as a critical issue, but an issue among a large variety of tasks, such as climbing stairs, grasping, manipulating or carrying objects. All these tasks require a whole anthropomorphic body. As bipedal walking is concerned, the approaches have been influenced by Kajita's seminal work at the beginning of the 2000s [12]. The idea is to guarantee the ability of the robot to control *actively* the center of mass during all the trajectory by ensuring the dynamic feasibility of the desired motion. The method has been extended in various manners, including the introduction of the so-called Capture Point that defines a condition to guarantee the ability to stop in a short enough time [19]. Most of them require a preview control managing an anticipation of foot placements [30]. We describe the feasibility constraints and the different kinds of control in the rest of the chapter.

In fact, despite the apparent opposition of these paradigms, the objective of this chapter is to give a unified view based on mechanics and control theory to better understand how energy-efficiency and robustness are competing criteria in the design of humanoid robot architectures. We argue that this competition between two criteria may be viewed as a continuous tradeoff better than as a binary opposition.

2 Human and Humanoid Walking: Complexity, Underactuation and Redundancy

Human bodies are geometrical shapes composed of multiple bodies attached together. Therefore if we neglect the deformations of soft tissues of humans, the shape of their bodies depend only on the angle at every joint. The value of the set of joint angles is called a *configuration*. Every joint of a human being is attached to some of the 600 muscles of the body allowing to actuate and control this configuration. On the other hand, the position of the human on the ground can be represented by two linear coordinates and the orientation by a heading angle. Unlike joint angles there is no muscle dedicated to control these degrees of freedom, the human is *underactuated*.

The generation of translations and most rotations relies exclusively on the production of external forces through contacts with the environment using careful combination of muscle activations. The high number of muscles involved makes this generation a *high dimensional* problem. This dimensionality allows to achieve these forces in infinite ways: the human body is *redundant* regarding this task. Nevertheless, the relative velocity of a moving person and the ground requires the human to constantly create new contact points, allowing for the well-known cyclic motion of bipedal walking.

However, the generation of motion using contact forces triggers a dynamics of the multibody in terms of forces and kinematics. This dynamics is nonlinear and non-holonomic, making it unsuitable to be controlled purely in the geometrical space but requires higher order time-derivatives of the kinematics. It may even be necessary to predict the outcome of every motion beforehand to ensure that the motion is not leading to violating a constraint.

Indeed this dynamics is *constrained*. Some of the constraints are related to the body itself, for example respecting joint position, velocity and torque limits. Some are related to the safety of the motion, and include avoiding self-collisions and collisions with the environment. But there are other constraints that characterize specifically locomotion. The most important one is related to fall avoidance. Most falls during locomotion are due to the unilaterality of contacts, i.e. the unability to pull on the ground instead of pushing. On flat grounds, this amounts to say that the center of pressure (CoP) also called Zero Moment Point (ZMP) lies inside the convex hull of the contact surface [27]. This leaves a narrow space for the motion of the center of mass (CoM) in order to avoid falling, especially because the CoM is high-positioned compared to the contact area. Finally, humans usually avoid slipping while walking and must limit the tangential components of the contact forces.

The combination of these constraints with the mechanics of the multibody provides a natural *instability* of the overall dynamics, not in the sense of balance, but in terms of pure dynamics. In other words, even if we respect the constraints at each instant, we may easily be drifted to a state for which the constraints have no other choice than being violated. This can be easily understood by considering the inverted pendulum, a classic model used to describe the dynamics of walking. With the slightest perturbation, a point-contact inverted pendulum leaves the stable vertical position and follows a fast falling trajectory. Fortunately for humans, they have wider contact area and another leg to catch-up, but this adds another constraint to the motion: the walker needs to be sure that they are able to catch-up at every instant. We see in the following sections how fuzzy this concept can be, especially in the context of stochastic perturbations.

The statements provided above may seem relatively straightforward, especially for people familiar with walking dynamics and control, but we wanted to summarize in few words the core of the problem of walking motion generation: a high-dimensional problem for a redundant and underactuated systems subject to a nonlinear, non-holonomic, constrained and unstable dynamics. To this we have to add that walking should consume the least possible resources while being robust to perturbations, to be versatile to allow the achievement of concurrent tasks, and to have variability to allow the expression of emotions [6]. Finally, larger scale locomotor trajectories on the plane carry their own load of geometric properties and dynamical characteristic [24].

Humans share all these properties with machines designed to imitate them: humanoid robots. And even if the number of muscle and degrees of freedom are much lower for these systems, the problem is still tough. Hundreds of researchers work hard to be able to achieve reliable and efficient walking motions with various kinds of bipedal systems, and all agree that the reached performances on this regard are far from being comparable to any average human.

Indeed, most humans are capable to overcome all these difficulties and even produce one of the most energy efficient locomotor control of the animal kingdom [20]. How humans are able to accomplish that prodigy is part of the mystery surrounding the most common and vital action of a human life. Nevertheless, we have several hypotheses which may shed light on this phenomenon. This should profit both to humans through relevant diagnosis and therapy for disorders and to robots through a wiser design and effective control of the systems.

Hereinafter, we present some of the ideas discussing the mechanisms of walking motion generation. When viewing this topic through the spectrum of controllability, in the sense of control theory, we have a new understanding on the trade-off at the core of this chapter.

3 Accessibility and Controllability for Walking Systems

The control of walking motion has a twofold objective: (i) allow locomotion, i.e. displacements and rotations of the body in the world and (ii) ensure balance, i.e. avoid falling. Both purposes have to relate the kinematics of the body to the contact forces that must be applied to modify them. Here the center of mass pops as the most convenient point to consider since it can be seen both as a kinematic object depending only on the limb positions, and as a relevant link with the dynamics . In other words, allowing the center of mass to reach a given position while keeping above a minimum height constitutes the successful locomotion control.[1]

As stated in the previous section, the underactuation of walking systems is the lack for actuator to control the position of the CoM. The second order time-derivative of CoM position is proportional to the vector of the resulting force applied to the robot, namely the sum of weight and contact forces. These contact forces depend on the position of the contact and joint torques, usually actuated. Therefore this classic problem of walking motion generation is simply a dynamical system with a state variable to command using a control input variable.

To tackle this kind of problems in classic control theory, we must ensure the *accessibility* or attainability of the desired state starting from the current one. A state is accessible if there is a control input allowing to make the system reach this state in finite time [17]. Despite the simplicity of the definition, the accessibility condition is not easy to verify for any state due to the complexity of the dynamics. Furthermore, even if the state is accessible, we still need to design a control policy to guarantee the stability of the trajectory regarding perturbations.

---

[1]From this sentence until the end of the chapter, body heading angle is excluded from the objectives of locomotion. Nevertheless, adding orientations does not change drastically the core of the discussion.

Instead of attempting to solve this difficult problem, roboticists resort to two solutions. The first solution is to restrain the target space to trajectories which are known to be stable. These trajectories can be the outcome of a given set of controls or to use the natural passive dynamics of the mechanical structure. This offers usually energy efficient and dynamic motions but reduces the variability of the gait maintaining it on a unique trajectory, which makes it also sensitive to external perturbations. Indeed, this solution includes the case of fully passive dynamic walking systems [4].

The second solution is to ensure the more strict condition of *controllability*. The controllability of a system is the accessibility to any state from any other state. The local controllability of the system at a given initial state is the the property that, for any neighborhood of this point, the set of states accessible without leaving this neighborhood is also a neighborhood containing the initial state. This means that we can reach neighboring states without having to go arbitrarily far. This property is also called local-local controllability [10]. or small-time controllabilty [25].

If this property applies to all the states lying in an open set, then the system is said to be locally controllable in this set [8]. A simplified illustration of these properties is shown in Fig. 2. We will show later how this condition applies to a walking controller. However, constraining the system to respect this condition may reduce the accessible state space to narrow areas and generate inefficient walking motions because of the stiff actuation and the slow dynamics that it requires.

In fact, the dichotomy between these solutions describes extreme cases for gait control, and the most prominent current solution to achieve bipedal locomotion are located between these extremes, either by allowing to have weaker controllability

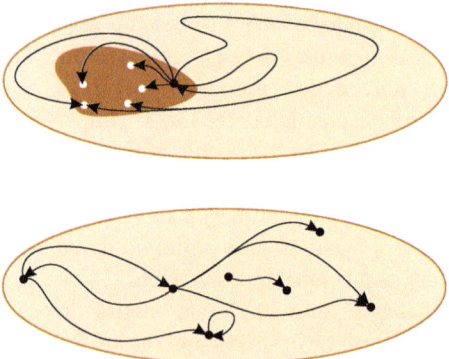

Fig. 2 A sketch of two definitions of controllability. In the upper part a representation of the local controllability at a point. The black point represents the initial state. Depending on the control, there are many possible trajectories starting from this point, represented by the arrows. The set of points accessible from the black point without leaving the light elliptic area is a neighborhood of the black point represented by the dark area. If this holds for any neighborhood of the black point, the system is locally controllable at this point. At the bottom is a representation of the local controllability at an open set represented by the light ellipse, which means that the system is locally controllable at any point of this set

conditions or by resorting to piece-wise solutions switching between stable uncontrolled trajectories and more controllable segments. We have then an axis of controllability in which each controller is located according to the tradeoff between them, and this allows to make a classification.

4   The Controllability Classification for Gait Controllers

The study of the controllability constraining walking controllers allows to classify them. This arrangement, together with some schematic representation of their properties, are illustrated in Fig. 3. From the left to the right of the figure we place the systems constrained to the strongest controllability conditions to the weakest ones.

To understand the controllability constraints, we need to ensure the feasibility of the desired trajectories. The dynamics of a usual walking systems is constrained by physics. The most critical constraint regarding balance is the unilaterality of the contacts: an unbalanced set of forces applied on the foot would lead it to roll on an edge or to simply lose contact with the ground. If this happens unexpectedly, we usually consider that the body is falling. Since CoM kinematics is related to contact forces, constraints on forces turn into constraints on the control of CoM position.[2]

The strongest controllability guarantee is reached in the case of quasi-static motions [9], shown on the left of Fig. 3. These motions are slow enough to neglect the effects of dynamics. In that case, as long as the CoM projects vertically strictly inside the convex hull of the surface area, the forces are feasible and the walker is considered to be balanced. The CoM position is locally controllable in all the space defined by this constraint, and may be controlled using simple inverse kinematics. These motions are very slow and require stiff actuation and high gain stabilization, but they can be made very robust against ground unevenness and many other kinds of perturbations.

However, most advanced robots do not use this constraining techniques and many trade weaker condition of controllability for more energy efficiency and dynamic motions. This new constraint is usually expressed in terms of center of pressure (CoP). This point, also referred to as zero moment point (ZMP) always lies in the support polygon [29]. In addition, these methods require its position to be *strictly* inside. This guarantees that the foot will not rotate around an edge and and gives the controller a neighborhood of feasible center of pressure positions. This provides us with the *local-local controllability* of CoM accelerations. However, respecting this condition every moment is not a sufficient guarantee of balance, and not even of the control stability. The controller must not diverge in any future. One way is to ensure that the null CoM velocity is accessible: if the robot is able to safely stop in finite time in the future, then it may be considered balanced now, this is ensured for example

---

[2]Of course other constraints also have to be enforced, such as avoiding to slide and all kinds of joint limitations, but they are easier to guarantee in this case.

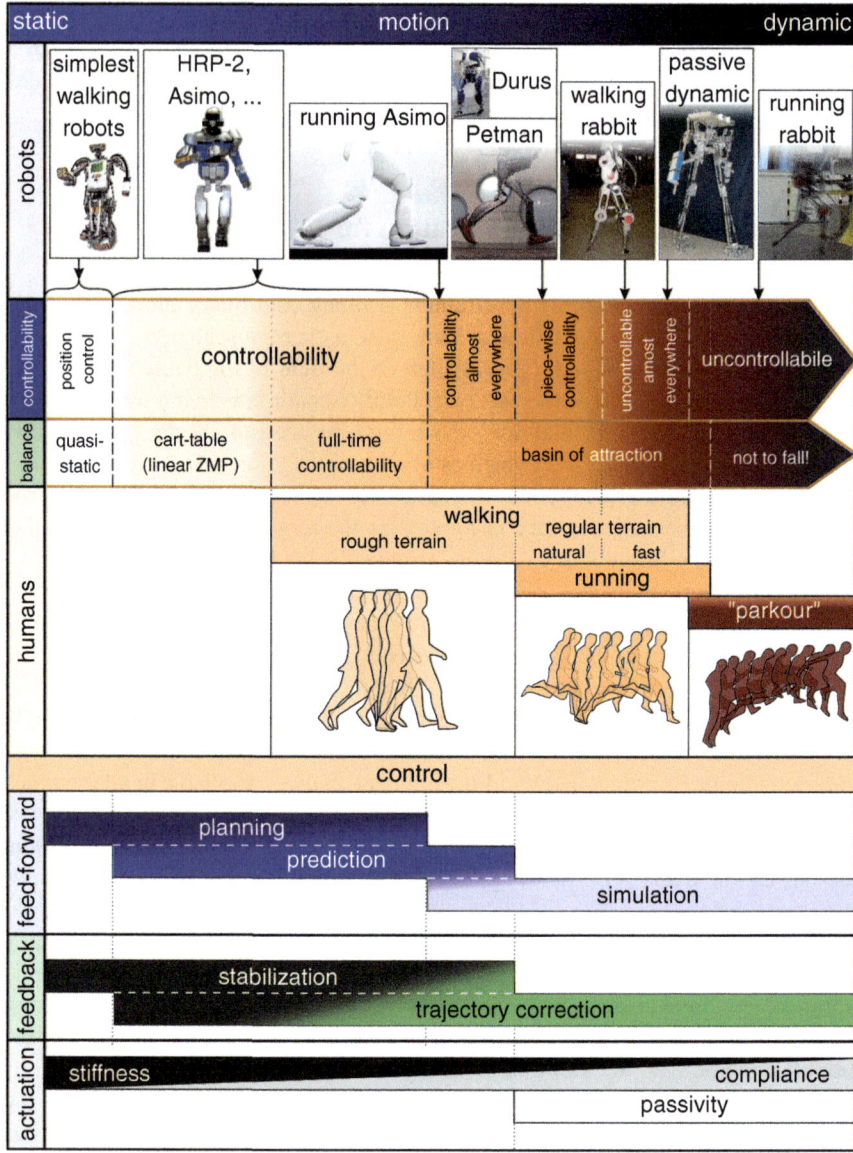

Fig. 3 Walking motion classification. The double arrow represents two parallel axes, they classify controllability properties and balance criteria. To every class of controller corresponds a class of balance criteria. On this double axis, we place most prominent representative robots at the top, and we show the span of performance of human locomotion. On the bottom part of the figure, we show some proper characteristics of these controllers in terms of feed-forward, feedback and actuation

by the so-called capture point methods [19]. Anther way is to be sure that a stable periodic gait is accessible, this is done usually by preview control techniques [30].

There are two main models used to achieve this control, either by maintaining a constant height of the CoM and neglecting gesticulation effects, which gives a linear control of the CoM [12], or by solving the nonlinear problem through various ways, these approaches are represented on Fig. 3 at the right of quasi-static approaches. These methods allow to generate relatively fast walking, more efficient and to overcome many kinds of obstacle. However, the controllability requires the full actuation of the legs chain and stiff stabilization to guarantee the constraints. There is few if any consideration for the natural dynamics of the structure and thus energy efficiency is far from optimality.

Nevertheless, once the controllers reached these performances, we may target more dynamic or efficient motion and allow the robot to go through a period of controllability. This uncontrollability happens usually when the area of the support surface reaches zero, i.e. when it reduces to an edge, a point or when there is no contact. Uncontrollability may happen also when a joint is not or no longer actuated. In both cases the robot is subject to its passive dynamics, in what can be seen as a ballistic-like trajectory, and retrieving controllability requires the robot to actuate all the joints or to create a new contact. The duration of this uncontrollable phase can go from very short instants for the case of running Asimo [26] to almost all the cycle, keeping only short instants of controllability. A representation of these kinds of walking control are on Fig. 4. Since the controllability is not guaranteed during all the cycle, the balance criterion is the stability of the system, or more precisely its *stabilizability* around the performed trajectory. A system is said to be stabilizable around a trajectory when all its non-controllable dynamics are stable. In other words, all the perturbations that cannot be actively compensated have spontaneously vanishing effects. The set of states which verify this property is generally called the *basin of attraction* of the system. All these cases of piece-wise controllability are represented in Fig. 3.

The last kind of walking systems are always underactuated, their motion relies all the time, partly or fully, on the passive dynamics of the system. The generated trajectories are then usually very efficient, look "natural" and can be really dynamic. The motions generated by such systems can be either periodic, non periodic or even chaotic [7].

A stable periodic trajectory is called *limit cycle* and the idea of the basin of attraction of this cycle is valid as a balance criterion. however, in the case where the motion becomes more unstructured and more chaotic, this concept no longer applies and the only criterion asked to the system is to avoid falling.

All this class of systems usually suffers from a high sensitivity to external perturbations and they may fall or stop stepping. This is mainly because of the narrow volume of the basin of attraction. This class of systems is represented in Fig. 3 at the right.

In the conclusion of the robotics perspective to this classification, we observe that when we lose controllability, the balance criterion that can apply becomes more general, and includes the more controllable ones.

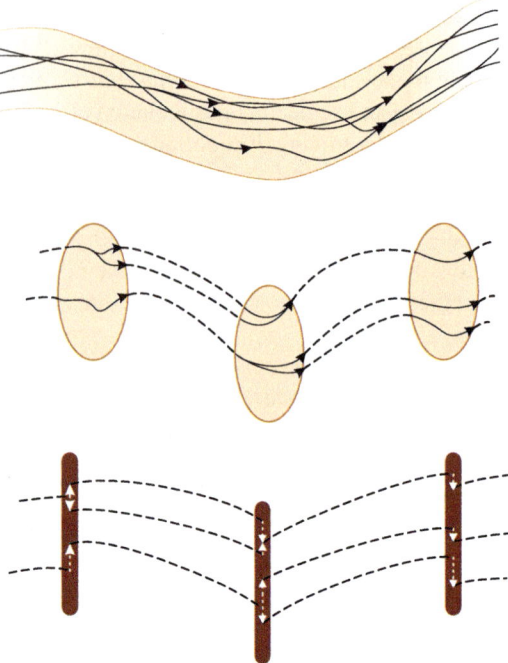

Fig. 4 Examples of controllability. In the upper part a representation of the controllability-constrained walking controllers, the system remains in the controllable colored area. At the middle a representation of piece-wise controllable systems where the controller uses the controllable area to modify the trajectory before going to uncontrollable trajectories represented by dashed lines. At the bottom a representation of motions uncontrollable almost everywhere, where there are only instants which allow the system to modify the state quasi-discontinuously using impacts and impulsions, represented by the dotted white arrows, before going for a ballistics-like uncontrollable trajectory

We may wonder where humans are placed in this categorization. Fig. 3 gives an answer to this question. As we explained in the beginning of this chapter, the control of walking motion depends on the context. When subject to perturbations, though textured terrains or external forces, humans tend to have a stiff actuation, every step is achieved cautiously and the foot does not leave the ground before the next one has firm contact. This is very similar to a controllability constraint, the dynamics should be as actuated as possible. When the environment is more structured, a more confident walk is generated, where humans allow a weaker muscle activation and rely on the efficient passive dynamics. The control mainly corrects for the perturbations and ensures they are not strong enough to drive them out of the basin of attraction. Other locomotion modes exist for humans. Running involves an uncontrollable flying phase and fast and dynamic single support phases. The motion remains periodic with a basin of attraction. Finally the very dynamic *parkour* locomotion is very unstructured and usually admits no period and no basin of attraction [15].

When studying humans versatility, we may notice that we are missing a robot able to span this kind of spectrum. More precisely we have to state that only few robots are able to achieve piece-wise controllability, and these examples are far from reaching satisfactory performances allowing to be used in order to perform other tasks than simple locomotion. We believe that a more important development of this part of the classification should be the target to aim to reach efficient and robust walking control. However, the ability to use the passive dynamics of the walker deeply relies on the suitability of the mechanical structure in terms of geometry and mass distribution. A careful design of both the mechanical structure and the control is a key factor in the success of this quest.

Other properties are associated with this classification, and some of them are displayed in Fig. 3. However, the balance criteria described here only consider regular surfaces and no perturbations. Indeed, even if the criteria are respected, there can be a disturbance leading the system to fall. In the next section we discuss how we may generalize these criteria to stochastic perturbations cases.

5   Dynamic Balance, Definitions and Criteria

In the previous section, we have intentionally spoken about balance without providing a precise definition, not only counting on the intuitive understanding of this concept, but to leave it for a deeper discussion. In fact there is a nomenclature debate on the sense of balance, especially dynamic balance. One spread definition of the dynamic balance describes criteria that depend only on the current state of the walker, without considering the outcome of the motion, for example reducing it to the CoP constraint [28]. Another common definition involves the control ability of the walker and can be summarize as the capability, from the current instant to the end of the desired task, to counteract the action of gravity without falling [31], i.e. without resorting to create a contact between the environment and a part of the body which is not intended for this contact. And that is the accepted vision in this section, because it covers the case of static equilibrium as well as dynamic balance. Indeed, despite the apparent simplicity of the intuitive concept, especially in static postures, its extension to dynamical case requires to constantly take care of the future outcome of the motion. Two people with identical current states may be in different states regarding balance, simply because their control is not identical.

This can be made clearer if we formalize it using the phase space, i.e. the set of all possible values of the state of the system in terms of positions, velocities and more. In this space, there are positions where the walker is *fallen* down (when a non-contact limb comes into contact). These are undesired states that a walker has to avoid continuously. However, there are other states that are not fallen yet but will unavoidably lead to a fall, no matter the control you apply on your system.[3] These are *falling states* If we remove these two sets from the phase space, remains the set

---

[3]This is mainly due to the limitations of the walker, in terms of torques, velocities and joint positions.

called the *viability kernel* [29] A state is viable if there exists a control which enables the walker to avoid falling, at least until the end of the current task.

Using the viability kernel as a balance criterion is unconvenient as well for concluding that a state is not viable as for finding out the appropriate control to feed the system, because it possibly requires the computation of all possible controls.

Actual solutions have to decide a control policy which guarantees balance in a set of states included in the viability kernel. This set is the basin of attraction of this controller. If the state of the system is inside then the walker is safe simply using this controller. The basin of attraction does not cover all the viability kernel, and has usually no closed form and no way to compute it. But it is easy to check if a given state will lead to a fall using this policy, we just have to simulate the control until falling or reaching a known stable state. Therefore, this can be considered as a balance criterion.

Such balance test considers that the walker is lying in perfectly known environment without perturbations, and this allows to make the simulation. In the reality, the robot is subject to external disturbances. When a given controller performs safely, it is because the perturbations are small enough to keep the state in the basin of attraction. However, any controller has a range of admissible perturbation beyond which the walker is thrown to a non-viable state. We need to estimate the robustness of controllers in order to tune the parameters for better performance or simply to choose the most reliable one.

Some balance metrics provide a measure on the safety margin of a controller, either by computing the approximate volume of the basin of attraction [23] or by estimating the biggest admissible perturbation. These metrics can provide an insight on the robustness of a controller to a single disturbance, but they don't take into account how fast the systems comes back to the steady dynamics, which means that a system which is robust to a large perturbation may be sensitive in case of repetitive small perturbations.

In fact, it is more accurate to consider that the system is subject to stochastic perturbations. For example the ground texture may be more or less rough and could be modeled as a probability distribution. In that case, other metrics try to give data about the local stability of a limit cycle by computing Floquet's multipliers [16] or the gait sensitivity norm [11]. However these metrics only describe the local behavior of the controller and don't account for the dynamics when far from the limit cycle, and in particular they don't describe the conditions within which falls happen.

Another way to see this issue is by considering that we must take into account the probability law of the disturbances. The probability to fall at the next step is never null. That means that the definition of a balanced state becomes blurry, every state has higher or lower chances to succeed. This voids the use of the basin of attraction as a balance criterion. The three layers of balance criteria : viability kernel, basin of attraction and the blurry phase space are illustrated in Fig. 5.

Since the perturbations are considered stochastic, we could possibly compute the probability for any walker to succeed in a given task without falling, for example using sampled simulations. However this probability could tend to one when motions are getting longer, for any control, making it impossible to compare between controllers.

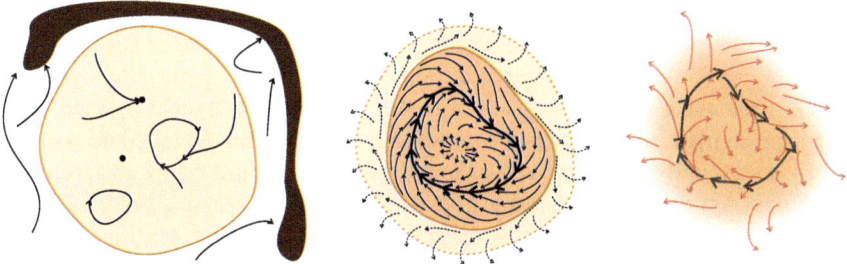

Fig. 5 **Definitions of viability**. On the left we see the failed states in the very dark area. The light color area is the viability kernel from which there exist a control input which allows to avoid to fail indefinitely. On the middle we see a flow plot when a specific controller is applied. The set of states allowing to avoid to fall is called basin of attraction. Since the controller is not perfect there is an area in the viability kernel which is not included in this basin and leads to a failure. On the right, we show the basin of attraction when the system is subject to stochastic perturbations. The boundary of this basin becomes blurry and the viability of a state can only be stated in a probabilistic manner

Fig. 6 **Metastability**. An illustration of a system (the ball) in a stable state, but subject to perturbations, it falls in another state, here with a more stable dynamics

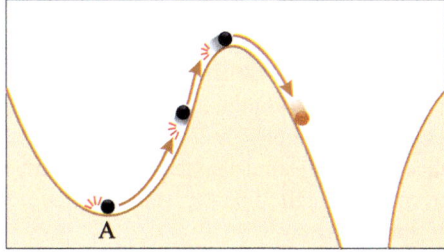

This characteristics of walking dynamics is called metastability: the property of some systems to be in a stable state, here "limit cycle", and because of perturbations to move to another more stable state, here "falling", see Fig. 6.

The study of metastable systems gave birth to a metrics called Mean First Passage Time (MFPT), which describes, for a given probability distribution of the disturbances, the average time the system stays in the initial state before switching to the more stable one. For the case of walking, time can be replaced by the distance, or the simplest, the number of steps [3].

MFPT is a metrics that gathers properties which don't meet in other ones. It is not local but covers all the reachable state space of the system, it takes into account the real perturbations that the walker meets and not causeless deviations in the phase space, and it takes into account the repetition in disturbance and their accumulation. However, their biggest drawback is that it is usually costly to compute, especially for complex systems, which made several approaches to be developed to increase the efficiency of simulations in calculating MFPT [1, 21].

This metrics or the metrics provided by the simple probability of a successful task have also the property to be easy to understand, we can clearly see if an action should be attempted or if it is too risky. Humans also have an estimation of the risk of a given action, this can be seen when they are asked to walk on a balance beam. The first steps are slow, muscles are stiff and reactive. But at a given distance from the end of the beam, they suddenly switch to a different control, faster steps, more dynamics and more passive motion which leads often, but not always, to a successful outcome. The decision triggering the switch between these locomotion modes could possibly be based on a perceived decrease in the probability of failure below an acceptable threshold. Of course, this reasobable hypothesis is difficult to assess experimentally.

We come to the conclusion that these measures are worth to be used in actual studies, not only for comparing controllers for humanoid robots, but also in the context of studying humans motion. In the next section we show how this metrics allows to reveal a balance effect of neck joint control in the dynamics of a simulated walker with human-like mass distribution.

6 Conclusion: The Example of Head Stabilization and the Perspective of Codesign

The dynamics of humans during steady walking uses passive dynamics and low actuation to an important extent. Therefore the dynamics of every limb has an influence on overall trajectory. Indeed, not only walker's geometry and mass distribution influence the motion, but also the control of any limb. This effect on the gait dynamics can be small enough to be neglected or may impact significantly the stability or balance of the walker.

For example, one interesting feature of human walking is head stabilization. Indeed, humans, among many other animals, stabilize actively the orientation of their head, especially during dynamic action such as locomotion [18]. This has several known benefits in terms of sensor inputs and the establishment of a stable egocentric frame [5]. In addition, we could study the *mechanical* effect of this feature, especially in terms of balance robustness regarding ground perturbations.

In this perspective, we simulate two models, model A and model B, operating in the sagittal frame, and having both the same human-like segment size and mass distributions. The only difference lying between these models is in terms of control. Model A has a rigid neck while Model B stabilizes its head orientation to maintain the head segment vertical. Both walkers are simulated walking on a textured ground represented by a slope inclination changing every step, according to a Gaussian distribution (see Fig. 7).

The standard deviation of the ground slope distribution determines how rough is the terrain. In particular, at zero, the ground is flat, and then both models walk indefinitely: MFPT is infinite. However, as soon as the slightest perturbation appears a difference between the controllers is visible. This comparison is shown on Fig. 8. For example in the case of 0.01 rad of standard deviation, rigid head model A makes in average 23 steps while head stabilizing model B reaches more than 3 millions. A better interpretation of these plots is that for this system the head stabilization allows

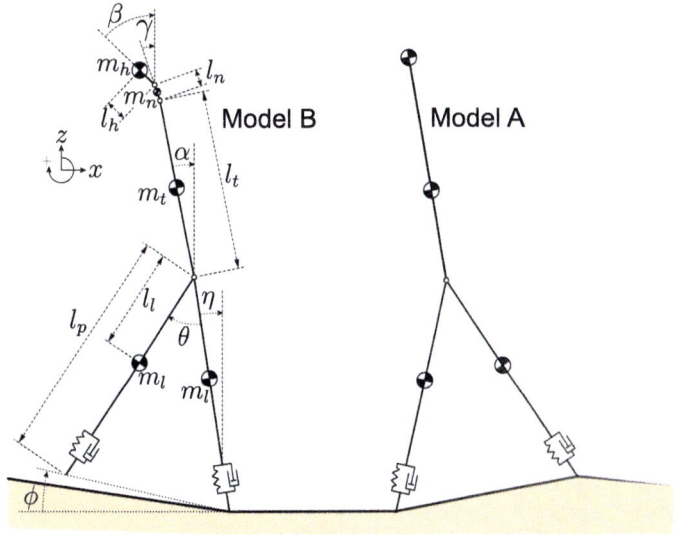

Fig. 7 **Two models on textured terrain**. An illustration the two models, model A has rigid neck and model B has an actively stabilized head. The textured terrain is simulated by modifying the slope of the terrain at every step following a probability distribution

Fig. 8 **MFPT for the two models**. For any non-null ground texture we can see that the head stabilization provides an important increase of MFPT

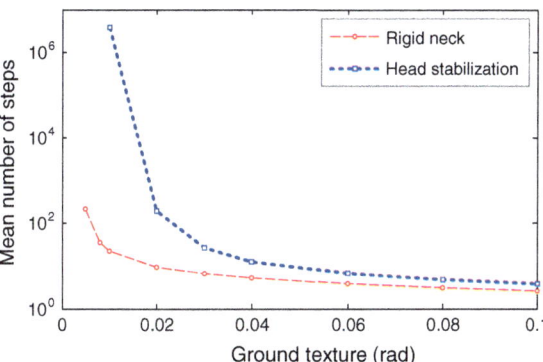

to extend the range of texture on which the walker can walk safely. The details on these models, their control, these results and their interpretations can be found in [14].

Beyond this example, we see how the relation between the mechanical structure and the control is important in the generation of efficient walking motions. In fact if we want to reach efficient walking with robust capable robots, we need to design the mechanical structure and the control at the same time. This approach is called Codesign. Indeed, the mechanical structure should not be considered as an abstract structure allowing to perform any walking controller, and an efficient walking controller cannot be a robot-agnostic black box. That is because the energy required to perform the same motion depends on mass distribution of the structure performing it.

The objective of Codesign is to produce versatile robot able to find at each moment the appropriate trade-off between robustness and energy efficiency. They should be able to be fully actuated in the presence of disturbances and to release the passivity for economic walking on known environment. But only few works are tackling this issue, with no instance of actual robot produced today [2, 22].

Acknowledgements This work was partially funded by the European Research Council grant Actanthrope (ERC-ADG 340050).

References

1. Benallegue, M., Laumond, J.-P.: Metastability for high-dimensional walking systems on stochastically rough terrain. In: International Conference on Robotics, Science and Systems (2013)
2. Buondonno, G., Carpentier, J., Saurel, G., Mansard, N., De Luca, A., Laumond, J.-P.: Optimal Design of Compliant Walkers. To appear in IEEE/IROS (2017)
3. Byl, K., Tedrake, R.: Metastable walking on stochastically rough terrain. In: Proceedings of Robotics: Science and Systems IV, Zurich, Switzerland (2008)
4. Collins, S., Ruina, A., Tedrake, R., Wisse, M.: Efficient bipedal robots based on passive-dynamic walkers. Science **307**(5712), 1082–1085 (2005)
5. Farkhatdinov, I., Michalska, H., Berthoz, A., Hayward, V.: Modeling verticality estimation during locomotion. In: Romansy 19 - Robot Design, Dynamics and Control CISM International Centre for Mechanical Sciences, vol. 544, pp. 359–366 (2013)
6. Felis, M.L.: Modeling Emotional Aspects in Human Locomotion. Ph.D. thesis, Heidelberg University (September 2015)
7. Garcia, M.S.: Stability, Scaling, and Chaos in Passive-Dynamic Gait Models. Ph.D. thesis, Cornell University Ithaca, NY (1999)
8. Gonçalves, J.B.: Local controllability of nonlinear systems. Syst. Control Lett. **6**(3), 213–217 (1985)
9. Goodwine, B., Burdick, J.: Gait controllability for legged robots. In: 1998 IEEE International Conference on Robotics and Automation, 1998. Proceedings, vol. 1, pp. 484–489. IEEE, New York (1998)
10. Haynes, G.W., Hermes, H.: Nonlinear controllability via lie theory. SIAM J. Control **8**(4), 450–460 (1970)
11. Hobbelen, D.G.E., Wisse, M.: A disturbance rejection measure for limit cycle walkers: the gait sensitivity norm. IEEE Trans. Robot. **23**(6), 1213–1224 (2007)
12. Kajita, S., Kanehiro, F., Kaneko, K., Fujiwara, K., Harada, K., Yokoi, K., Hirukawa, H.: Biped walking pattern generation by using preview control of zero-moment point. In: IEEE International Conference on Robotics and Automation, vol. 2, pp. 1620–1626 (September 2003)
13. Kelly, M., Sheen, M., Ruina, A.: Off-line controller design for reliable walking of ranger. In: 2016 IEEE International Conference on Robotics and Automation (ICRA), pp. 1567–1572. IEEE, New York (2016)
14. Laumond, J.-P., Benallegue, M., Carpentier, J., Berthoz, A.: The yoyo-man. Int. J. Robot. Res. p. 0278364917693292 (2017)
15. Liu, L., Yin, K., van de Panne, M., Guo, B.: Terrain runner: control, parameterization, composition, and planning for highly dynamic motions. ACM Trans. Graph. **31**(6), 154–1 (2012)
16. McGeer, T.: Passive dynamic walking. Int. J. Robot. Res. **9**(2), 62–82 (1990)
17. Murray, J.D.: Mathematical Biology: I. An Introduction, vol. 17. Springer Science & Business Media, Berlin (2007)

18. Pozzo, T., Berthoz, A., Lefort, L.: Head stabilization during various locomotor tasks in humans. Exp. Brain Res. **82**, 97–106 (1990)
19. Pratt, J., Carff, J., Drakunov, S., Goswami, A.: Capture point: a step toward humanoid push recovery. In: 2006 6th IEEE-RAS International Conference on Humanoid Robots, pp. 200–207. IEEE, New York (2006)
20. Rubenson, J., Heliams, D.B., Maloney, S.K., Withers, P.C., Lloyd, D.G., Fournier, P.A.: Reappraisal of the comparative cost of human locomotion using gait-specific allometric analyses. J. Exp. Biol. **210**(20), 3513–3524 (2007)
21. Saglam, C.O., Byl, K.: Quantifying the trade-offs between stability versus energy use for underactuated biped walking. In: 2014 IEEE/RSJ International Conference on Intelligent Robots and Systems (IROS 2014), pp. 2550–2557. IEEE, New York (2014)
22. Saurel, G., Carpentier, J., Mansard, N., Laumond, J.-P.: A simulation framework for simultaneous design and control of passivity based walkers. In: 2016 IEEE International Conference on Simulation, Modeling, and Programming for Autonomous Robots SIMPAR, San Francisco, United States (December 2016)
23. Schwab, A.L., Wisse, M.: Basin of attraction of the simplest walking model. In: Proceedings of ASME Design Engineering Technical Conferences and Computers and Information in Engineering Conference, 21363 (2001)
24. Sreenivasa, M., Laumond, J.-P., Mombaur, K., Berthoz, A.: Principles Underlying Locomotor Trajectory Formation, pp. 1–17. Springer Netherlands, Dordrecht (2016)
25. Sussmann, H.J.: Lie brackets, real analyticity and geometric control. Differ. Geom. Control Theory **27**, 1–116 (1983)
26. Takenaka, T., Matsumoto, T., Yoshiike, T.: Real time motion generation and control for biped robot-1st report: walking gait pattern generation. In: IEEE/RSJ International Conference on Intelligent Robots and Systems, 2009. IROS 2009, pp. 1084–1091. IEEE, New York (2009)
27. Vukobratović, M.: On the stability of anthropomorphic systems. Math. Biosci. **15**(1–2), 1–37 (1972)
28. Vukobratović, M., Borovac, B., Potkonjak, V., Jovanović, M.: Dynamic balance of humanoid systems in regular and irregular gaits: an expanded interpretation. Int. J. Humanoid Rob. **6**(01), 117–145 (2009)
29. Wieber, P.-B.: On the stability of walking systems. In: Proceedings of the International Workshop on Humanoid and Human Friendly Robotics, Tsukuba, Japan (2002)
30. Wieber, P.-B., Tedrake, R., Kuindersma, S.: Modeling and control of legged robots. In: Springer Handbook of Robotics, pp. 1203–1234. Springer, Berlin (2016)
31. Wieber, P.-B.: Viability and predictive control for safe locomotion. In: IEEE-RSJ International Conference on Intelligent Robots & Systems, Nice, France (2008)

An Overview of Humanoid Robots Technologies

O. Stasse and T. Flayols

Abstract Humanoid robots are challenging mechatronics structures with several interesting features. Choosing a humanoid robot to develop applications or pursue research in a given direction might be difficult due to the strong interdependence of the technical aspects. This paper aims at giving a general description of this interdependence and highlight the lessons learned from the impressive works conducted in the past decade. The reader will find in the annex a table synthesizing the characteristics of the most relevant humanoid robots. Without focusing on a specific application we consider two main classes of humanoid robots: the ones dedicated to industrial application and the ones dedicated to human-robot interaction. The technical aspects are described in a way which illustrates the humanoid robots bridging the gap between these two classes. Finally this paper tries to make a synthesis on recent technological developments.

1 Mechanical Structure

1.1 General Design Principal

Humanoid robots are complex mechatronic systems. As such, it is necessary to consider the mechanical structure, the computational system and the algorithms as a whole and for a given application. The robot's size, weight and strength are important factors when designing its structure. Let us consider two general classes of applications: physical performances while doing motion generation and validation of biological and/or cognitive models. The ATLAS robot from Boston Dynamics is

---

This contribution is a translation and a revision of [58].

O. Stasse (✉) · T. Flayols
CNRS - LAAS, Toulouse, France
e-mail: o.stasse@laas.fr

T. Flayols
e-mail: t.flayols@laas.fr

© Springer International Publishing AG, part of Springer Nature 2019
G. Venture et al. (eds.), *Biomechanics of Anthropomorphic Systems*, Springer Tracts in Advanced Robotics 124, https://doi.org/10.1007/978-3-319-93870-7_13

an example of the first category, while the Kenshiro robot [45] from Tokyo University is an example of the second category.

When the goal is to have a robot with walking speed performances around 2–3 km/h, the knowledge from walking robots such as HRPs robots from Kawada Industries or the LOLA robot from the Technological University of Munich, shows that there are two mechanical points to take into account [41]: the mass distribution on one side, and the undesirable mechanical resonances on the other. The last point implies to suppress compliance at the level of the joints and the links. For this reason, most of the humanoid robots are very rigid in order to achieve a high precision control. When the human-robot interaction is a major constraint during the design phase, the control precision is not the main objective. The security level necessary to allow a physical interaction with a human is then obtain by introducing actuators with low power and flexible mechanisms in the transmissions, such as the one described in Sect. 3. There exists robot designs which try to synthesize several constraints with more specific objectives. For instance, the HRP-4 humanoid robot is lighter (39 kg) for a size of 1.5 m with 34 degrees of freedom (DoFs). This is the result of a compact power electronics and a skeleton made of carbon fiber. The drawback is that the robot segments are more flexible and the low power actuators limits the load that the robot can hold.

1.1.1 Mass Distribution

Balance is an example of relation between control, computing capabilities and the mechanical structure. For walking robots evolving on flat floors, the main criteria for balancing is the point on the ground where there is no angular momentum. This point is called the Center-Of-Pressure (CoP). To maintain the balance of a humanoid robot evolving on flat ground, it is necessary to find in real time a control allowing to keep the CoP inside the convex hull of the contacts points on the ground. To solve the associated control problem in an efficient way, it is usually assumed that the robot is behaving like a point mass model. This assumption is valid when the robot limbs are light, and when the mass distribution is concentrated on the waist. The Center-Of-Mass (CoM) is then globally fixed with respect to the waist. When this assumption is not valid, then the control needs to use more complex models which are more difficult to solve. More precisely, the controllers then need to consider a three mass model when the legs masses are not negligible, or a five mass model if the arms are too heavy as well. Such models need more complex control techniques. For instance the team from the Technological University of Munich had to redo a design phase with its walking robot LOLA to improve the mass distribution [41].

The mass distribution depends mainly on the actuators. To limit their inertial effects on the robot dynamic, it is mandatory to bring the actuator as close as possible to the root of the link on which it is fixed. To transmit the motion to the joints, various mechanisms can be used: lever, ball-screw, pulleys with driving bells. For instance LOLA is using levers [41], while HRP-4 is using ball-screw for some joints and pulleys for others [32].

1.1.2 Human Robot Interaction

Human robot interaction implies to control the robot forces for safety. This can be done in an active manner (using a feedback loop) or passively (by mechanical design). Robots with high gains controlled actuators are not able to estimate correctly the forces applied by a human. If performance is still the main objective, it is better to integrate at an early stage supplementary sensors (force sensors or artificial skin cf. Sect. 4) at high frequency (1 kHz) as it is done for the Kuka LWR robot. Another approach consists in using servomotors with which it is possible to lower the gains and hence to allow the actuator to be compliant. This strategy is used for instance by the Poppy robot [39].

1.1.3 Mechanical Resonance

Mechanical resonance consists in vibrations of the robot mechanical structure related to the link deformation (for instance the legs) or the actuators control. As specified before, this passivity might be desired to perform human robot interaction. It can be mechanically integrated by the actuators to reject impacts as it is done in the COMAN humanoid robot at the IIT [65]. It might be undesired and be the consequences of design constraints such as for HRP-4 [32]. In all these cases, it is strongly advised to perform a frequency analysis of the system when it is possible. The team of the LOLA robot [41] realized analysis for each link to detect the weakness of some pieces such as the hip and the knee to avoid undesirable deformation. It is possible to use beam theory on the femur and the tibia to evaluate the leg deformation when it is exposed to strong torques and forces. For tall robots (1.6–2 m), not handling properly the passive dynamic might induce strong impacts during the landing phase [26].

1.1.4 Estimating the Leg Deformation

It is possible to use beam theory to evaluate the link deformations and the system natural frequencies. A link submitted to a load q has the following dynamics:

$$\frac{\partial^2}{\partial x^2}\left(EI\frac{\partial^2 u}{\partial x^2}\right) + \rho A\frac{\partial^2 u}{\partial t^2} = q \tag{1}$$

with E the material Young modulus, I the second order moment of area of the link's cross-section, ρ the material density, A the section surface, u and x respectively the deflection and its position along the static axis. In the case of an homogeneous beam of size L, parameters E, I, ρ and A are constants. If we consider the natural frequencies when there is no load, we solve the previous differential equations by using the Fourier Transform. The frequency is therefore:

$$\omega = \lambda^2 \sqrt{\frac{EI}{\rho AL^4}} \qquad (2)$$

and the solution can be written as:

$$v(x, t) = \bar{v}(x) sin(\omega t) \quad \text{with}$$
$$\bar{v}(x, t) = A_1 sin(\beta x) + A_2 cos(\beta x)$$
$$+ A_3 sinh(\beta x) + A_4 cosh(\beta x)$$
$$\beta = n\pi, \ \lambda = L\beta$$

where A_1, A_2, A_3, A_4 are constants. The objective is to maximize this frequency to avoid the one occurring during the application. However only the section and the moments surface can be changed. The coefficient E/ρ suggests materials with very high rigidity such as carbon fiber. The beam geometry is represented by I/A. This ratio is maximal for a geometry with high second order moments with respect to the small transverse sections.

1.2 Robot Skeleton

The kinematic structure of humanoid robots has long been inspired by the human structure, and more precisely based on the study of Saunders [54]. For this reason, numerous humanoid robots have the structure depicted in Fig. 1. The legs are made of 3 rotation joints at the hip level (to simulate a spherical joint), one joint for the knee flexion-extension and two joints for the ankle (flexion-extension and pronosupination). This structure has one advantage: it has an analytical solution to the problem of finding a configuration for the legs with respect to a given position of the waist and the feet. For all these reasons, it is find in numerous robots such as the HRPs series, HUBO [40], ASIMO and REEM-C. The counterparts of this structure are the performance limits in the kinematic chains. Adding a passive toe joint allows to increase the robot walking speed [56]. To limit singularities and kinematic constraints, recent humanoid robots such as S-One from Schaft or ATLAS have more joints. For instance, Schaft as one more DoF on its legs on the sagittal plane. More generally in the context of the DARPA Robotics Challenge (DRC) [11], ROBOSIMIAN from JPL [36] and CHIMP [60] from CMU are ape-like robots which allow to perform more extended locomotion modes than bipedal ones. S-One has highly redundant arms which permit to avoid singularities and kinematic limits. In a general manner, if dexterity is a primary objective, it is strongly advised to use an arm with 7 DoFs to avoid singularities. This has however a direct impact on the number of motor and therefore on the arm mass, its electronic complexity and the its weakness.

Fig. 1 Classical structure of
a humanoid robot

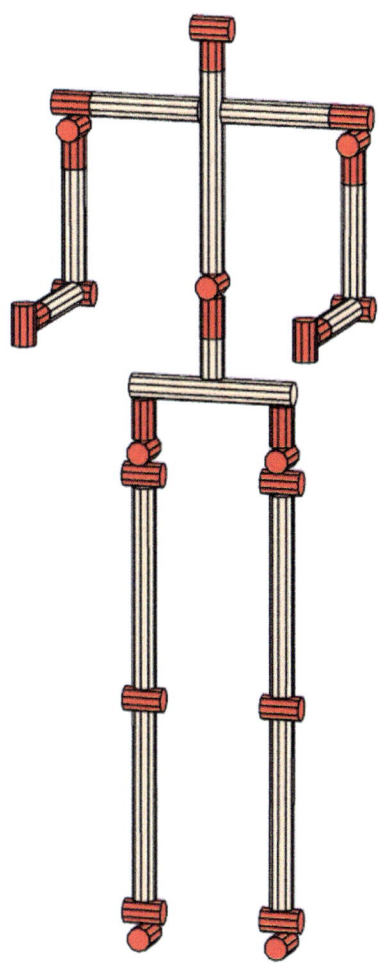

1.3 Grippers

Grippers for humanoid robots should be chosen according to the application. It is necessary to find a compromise between dexterity and the load that the hand can carry. A dexterous hand allowing fine manipulation such as the one of the DLR [5] involves a structure with the same numbers of DoFs than the whole robot skeleton. For instance, the complex hands developed by the Japanese National Institute of Advanced Industrial Science and Technology (AIST) [30] for the HRP-3 have 3 fingers and 3 DoFs by fingers, with a thumb of 4 DoFs. This hand is equipped with force sensors at each finger. The fingers are allowing a maximal push of 15 N thanks to a mechanical structure using motors and Harmonic Drive (HD). If aesthetic criteria are important, one has to take into account the proportion of the hand size with the rest of the body [33]. For instance, the DLR hand is too long and too big to be integrated in a robot such as HRP-2. The challenge is to integrate motors, cables and power electronics in the forearms. For this reason, the AIST hand is provided with a box at the wrist. It contains an embedded computer for the control which limits the cabling to be added to the initial structure. The box is also used to protect the hand (it is the same protection that is used for the Q-RIO robot). This, however, increases the weight at the robot arm extremity and needs to be taken into account in the robot dynamics. Complex hands with several DoFs are heavier, take more space but are also more fragile, and, at the same time, can not lift up an heavy load. ASIMO, in its initial version, could not take a load heavier than 500 g [30]. Another example of complex hand is the one developed for iCub [55]. It has 19 DoFs, touch sensors and can perform complex manipulation. For all these reasons, simpler grippers can be chosen. For instance, the gripper used by the humanoid robot HRP2-Kai [34] can be used to climb a ladder. The humanoid robot TALOS [59] is able to handle 6 kg while stretching its arm. The humanoid robot HRP-2 is able to sustain 15 kg forces during multi-contacts walking [8]. Under-actuated hands are a frequent choice which simplifies the control and the actuation, with a slightly more complex transmission. For instance, the HRP-2 from Tokyo University and AIST are equipped with a parallel mechanism using one DoF. It is less dexterous but it enables the manipulation of thirty different classes of objects with masses up to 5 kg. Robots such as NAO are using a single DoF by hand, all the fingers are folding or unfolding together at the same time. This kind of grasp provides an automatic adaptation to the object shape. The last version of the humanoid robot ASIMO is integrating hands with hydraulic actuators of small size directly mounted in the forearm. With such hands, the robot can pereform complex motions such as opening a thermos and pour water in a glass.

1.4 Mobility

There exists a certain number of robots qualified as "humanoids" for which the lower part does not use two legs but a mobile platform or a higher number of legs. A famous robot in this category is the PR-2 robot from the former company Willow Garage. Other famous robots with the same structure are the ARMAR series, and the REEM-B from PAL-Robotics. The former is able to recognize objects, to manipulate them, and for instance to put them in a dishwasher. Such robots are very efficient for mobile manipulation. When the mass to manipulate is too big with respect to the part of the robot in contact with the ground, the dynamical effects need to be taken into account. The balance problems handled by humanoid robots and for instance the angular momentum regulation are the same in this context. However recent advances on multi-contacts generation show the possibility to generate motions which are outside the scope of the ones using a mobile platform (Fig. 2). HRP-2 is also able to

Fig. 2 Non-planar contacts
to extend the reachable space
(left) [37] and to perform
generalized locomotion
(right) [8]

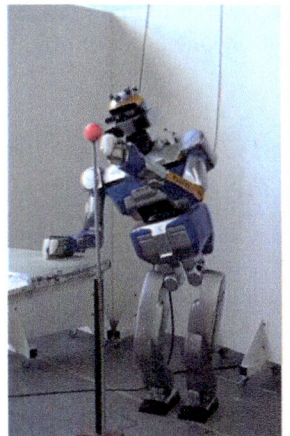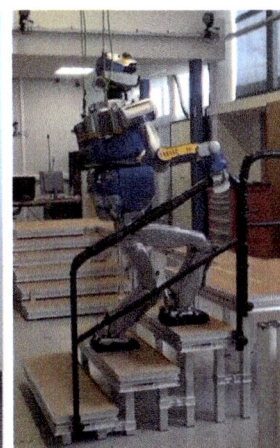

make contact with a desk to expand its workspace and reach another target. If such motion can be planed, only recently they can be generated on line. The multi-contact problem has a very high complexity in its general formulation. It exists some efficient approximations but which have not yet been implemented on a physical robot. If the DRC has shown the possibility for the robots such as S-One to handle complex environments, most of the contacts are planar (but not necessary in the same plane). This is possible thanks to a robust control of the CoP able to handle slope variations of 10°.

1.5 Foot

Feet are very important for humanoid robots: they must insure a contact with sufficient friction to avoid slippage; they must handle impacts during landing; they must handle contact transition during the double support phase; finally their mechanical structure is often constrained by the size. In addition as most of the humanoids are walking by controlling their CoP, it is necessary to have 6 axis force sensors at the ankles, or pressure sensors under its feet to measure the CoP Position.

To avoid robot slippage, humanoid robots have soils with shapes and materials which provide a sufficient friction (typically with a friction coefficient >1).

To dissipate impacts, the LOLA robot [41] for instance is using Sylomer with an elastic modulus higher at the heel and smaller for the rest of the foot. Some feet include dampers [41, 46].

Toe joints do have numerous interests. They enable the transfer of the CoP quickly to the front, and thus make the robot walk faster than a structure without toe-joint [56]. They are allowing the robot to kneel without being in a singular position at the foot level. Finally if we add a spring to the toe-joint, it is possible to store energy to use it during the take-off phase or to make the robot step on the spot [27]. This implies a more complex control. One needs to insure that the mobile part is able to handle the impacts.

1.6 Constraints from the Environment

Using humanoid robot in industrial environments may expose them to dust, humidity or even rain. The HRP-3 humanoid robot has been designed to face such constraints [31]. It is compatible with norm IEC-IP52 (International Electrotechnical Commission, Ingress Protection). The first number means that the system is protected against dust which would keep the robot from working properly. The second number expresses a protection against vertical rain. This protection level has been obtained thanks to tests on the HRP-2 actuators. To track the actuator weakness, dust resistance has been tested with talc, while water resistance was tested with fluorescent liquid. The robot has been reinforced through sealed ball bearings, liquid, torques and silicon joints (Table 1).

Table 1 TB: Timing Belt, HD: Harmonic Drive, IS: Infinite screw, G: Gear, PG: Planetary Gearset. The data comes from the robot H7 of Tokyo University [21]. Motors M150, M90, M20, M5 are respectively the Maxon motors 148877, 118778, 118754, 118736. The motor M9 is a HD system RH8C-6006

Joints	Motor	Actuator	Reduction	Joint limit	Max velocity (s)	Torque (Nm)
Toe	M20	PG+EDD	177	−60 15	441	56
Ankle - R	M90	TB+HD	121	−35 30	433	91
Ankle - P	M90n	HD	121	−90 90	433	91
Knee - P	M150	HD	161	0 150	283	144
Hip P	M90	TB+HD	181.5	−135 40	289	143
Hip R	M90	TB+HD	180	−35 30	291	142
Hip Y	M20	HD	120	−30 90	650	38
Shoulder Y	M90	HD	120	−90 0	437	94
Shoulder P	M90	HD	120	−180 90	437	94
Shoulder - R	M90	HD	120	−30 100	437	94
Elbow - P	M90	HD	120	−140 0	437	94
Wrist Y	M20	PG+TB	118	−150 120	661	37
Wrist R	M20	TB+HD	137.5	−90 90	567	44
Finger R	M5	VSF+EDD	125	−65 30	696	48
Neck P	M20	TB+HD	150	−60 60	520	48
Neck - Y	M9	HD	50	−90 90	1198	N/A

2 Computational Structure

2.1 Embedded Computer

Currently, control laws used to maintain a humanoid robot balance needs power-ful computational capabilities which are not available on CPU without acceleration mechanisms. For instance, the HRP-4 robot from Kawada Robotics is using a Pentium M at 1.6 Ghz which is not equipped with the Intel Turbo Boost technology in contrast with processor $i7 - 2710QE$ which can be found in the REEM-C humanoid robot. This technology makes it possible to increase the CPU internal clock according to the current kind of operations (mathematical for instance) and according to the CPU temperature. The embedded motherboard has been chosen according to temperature tests made on the robot. Bigger robots are able to embed bigger mother-boards equipped with powerfull CPU. Some care is however taken to avoid high heat dissipation. It is recommended to have enough space to have powerful CPUs which can be updated easily. This is for instance the case for the TALOS robot which has all his computers on one box which can be easily changed.

2.2 Communication Buses

Communication buses are a very important component of humanoid robots. They can limit the control bandwidth available on the robot. They need to be extremely robust to electromagnetic perturbances. For this reason, it is strongly advised to have a bus with a high bandwidth to comply with hard real-time constraints. However, depending on the data type, such constraints might be hard to fulfill or have a different nature. For instance, in control, deterministic dead-line are of primary importance. However in vision, the throughput is of utter importance. For all these reasons, the communication structure may have multiple layers. For instance we may have PCI/PCI104/PCIe for the connection between the motherboards, the network cards and the analog/digital converter boards. To have a strong connection between the motors and the encoders, it is necessary to have very robust buses. Typically the ones from the automotive industry are used such as CAN (HRP3, iCub, HUBO-2) or EtherCAT (Pyrene, ATLAS, WALKMAN). However, CAN has a limited bandwidth (1 Mbit/s) and does not allow to handle multiple point to point connections. Therefore humanoid robots often use several buses (HPR3, iCub). In order to solve this problem, the SERCOS-III protocol has been chosen by the robot LOLA [41]. This protocol has the following advantages:

- transmission by Quality of Service which reserve bandwidth to prevent collisions;
- higher efficiency during the transmission, obtained by using ASICS to control the buses and by using optic fiber as the physical support. SERCOS III extend this approach to Ethernet but it implies to have a switch for each node;

- Data transfer in real-time and in none real-time;
- Connection possible between slaves;
- Connection to proprietary devices by adding fields to the data packets (force sensors, inertia measurement system).

2.3 Wireless Communication

Wireless communications are generally handled by 802.11a/b/g/n switches. The last letter specifies the protocol version. These protocols match various frequencies. The latest (g/n) correspond to higher frequencies and are allowing higher bandwidths. For instance with protocol 802.11 it is possible to transmit uncompressed VGA images at 10 Hz on WIFI using a 5 GHz frequency. However wireless communications main weakness is their sensitivity to environment where frequency overloading is possible. This is typically the case in events with a large public. In this case, people with their smartphone are overloading the network. This can be handled by carefully analyzing the frequency which can be used in the context of the robotic application.

2.4 Middleware

Middleware: A middleware is a software framework which enables distributed processes to exchange data on heterogeneous platforms. This "software bus" uses an object map to offer a simple and coherent interface to access objects and to guarantee data transmission.

In humanoid robotics, there are three well known middlewares: ROS, YARP and OpenRTM (Corba). There exists other middleware such as NAOqi from Softbank Robotics. However, nowadays, ROS has the leading position. It is due to a large support from the robotics community which permits to quickly have access to innovative algorithms for 3D reconstruction (KinFu, Octomap) or motion planning (OpenRave, OPMPL) and to system simulator such as Gazebo.

Middlewares propose a client-server architecture communicating by messages. Data types are usually specified through an Interface Description Language.

2.4.1 ROS

ROS (Robot Operating System) exchanges data by subscription and publication at frequencies up to 100/200 Hz. For static or rarely changing data, another system called parameter server is used. An application is a graph of nodes communicating through topics. Each node provides or consumes data (or both) and can also provide services. The interface to these services is also specified by an interface description language. From a functional point of view, ROS provides a package system which simplifies the creation and handling of packages. A build farm system provided by the Open Source Robotics Foundation (OSRF) creates binary packages from the source codes provided by contributors. This results in a better diffusion of the software created by the community.

ROS is supported in a stable manner on Ubuntu, but there exists also experimental versions on Mac OS X, Windows and other distributions on Linux. Moreover, the sources of core components are available to adapt to other operating system. The main inconvenient of ROS is its lack of real time capabilities. It is often necessary to use an additional middleware. The most used open source solution is OROCOS with Xenomai. This solution is implemented on the robot REEM-C. ROS is a very active project and is supported by the Open Source Robotics Foundation. The most common software license used in ROS is BSD.[1] However, to provide a solution which can be deployed in factories, a group of industrial are working together to propose a certified solution.

2.4.2 OpenRTM

OpenRTM(Open source Robotic Technology) is a middleware designed and developed by the AIST at the same time than the humanoid robot series HRPs [3]. OpenRTM is an implementation of the Robot Technology Components standard validated by the Object Management Group (OMG). OpenRTM is more oriented towards real-time application. The standard formalizes the messages sent between the components as data-flow. It is possible to specify the scheduling class of one software component. For instance, a component providing sensor data is specified as being periodic at the reading frequency of the sensors. This specification can be used to prove Real Time Quality of Service (if this is feasible on the platform). The data used in the control part of the application is specified as ports. Output ports correspond to data provided by the component, input ports are data that are needed by the components. It is also possible to access components in an asynchronous manner through services (like ROS). The data and the service calls are formalized by an Interface Description Language. As OpenRTM is based on CORBA, it relies on its tools. The OpenRTM community is mostly Japanese. The ecosystem includes the OpenHRP simulator, the human-robot interaction package OpenHRPI and a set of software support named OpenINVENT. However this project does not seem to be very active anymore. Most of the OpenRTM real-time components are implemented directly using the real time operating system. We can notice however that in the case of the robots HRPs, OpenRTM is delivered with a balancing component, a walking pattern generator and a simulator quite realistic. For instance the DLR biped robot TORO, uses OpenHRP for simulation.

---

[1]BSD: Berkeley Software Distribution License is a very permissive free license used quite a lot in software distribution.

2.4.3   YARP

The goal of YARP (Yet Another Robot Platform) [17] is to provide a message transmission system through ports by using different protocols and between machines with various operating systems. This is achieved by abstracting the communications as do CORBA and ROS, and by using libraries such as ACE to abstract the operating system functionalities (process, threads, file access, timers). It is worth noting that ACE, developed by the group implementing TAO, a real-time middleware, is also used by OpenRTM. The interface with different languages is performed thanks to SWIG which enables to generate the various wrappers. Hence YARP is compatible with Linux, MAC OS, Windows and can be interfaced with numerous languages (http://www.swig.org). The data port concept exists in YARP as well as the object directory to localize object over a distributed architecture. Interacting with the hardware is possible through the specialization of an abstract DeviceDriver. With this mechanism, YARP supports various hardware without propagating their specificities over the software infrastructure. This integration is handled through the components in other robotics middlewares.

To conclude this part, OROCOS is a middleware used mostly to encapsulate the real-time interface of robots and is often found in conjunction with ROS.

3   Actuators

Actuators for humanoid robots need to fulfill the following criteria: a high ratio between power and mass, the capability to produce high torques at low speed, a relatively small size and back drivability. This part describes various technologies developed for humanoid robots.

3.1   Actuators with Electric DC Motors

Humanoid robots of human size, such as the HRP-2 humanoid robot, Johnny or HUBO-DRC are using DC electric motors with Harmonic Drives (HD) to transform speed into torque. Recent work on finding appropriate motors and reduction ratio can be found in [68] using the an upper and a lower bound on the inertia of the robot. The highest angular momentum take place at the level of the hip and the knee. For this reason, in the case of the Walkman robot, the most powerful motors are located at these two places, as well as for the H7, a very impressive humanoid robot developed in the 90s by Tokyo University 1. The main advantages of such motors are their size, and the compromise between speed and provided torque. To put the motors as close as possible to the rotation axis of the link on which they are fixed, various connecting systems are used: connecting rod, parallel link, but the timing belt is the most frequent. The main inconvenient regarding approaches using HD is the difficulty to model the force coming from the interaction with the world by tracking the motor current. For this reason, the control that has been used for many years is based on high gains position control.

3.2 Actuators with Electric AC Motors

Robots using electric DC motors are limited in speed. Foot step correction on non smooth or partially known terrain, as well as the generation of high speed motion, implies to have motors able to deliver more torque at a higher speed. Electric AC motors can deliver such high performance, but with a more complex control and a power supply of higher capacity. The TORO humanoid robot developed at DLR uses such motors together with a technology able to identify the system parameters and a torque sensor on the joint side. This enables to reject rejecting perturbations through torque control [51]. When considering S-One, the robot build by Schaft, the key is the cooling system of the motors. The thermal power P dissipated by the motor can computed as follows:

$$P = R_e(T_c)I_a^2 \qquad (3)$$
$$I_a^2 = I_q^2 + I_d^2 \qquad (4)$$

where R_e is the electric resistance of the wire in the motor field coil, I_q and I_d are the currents with respect to the axis q (quadrature) and d (direct) of the power separately excited direct-current motor. The resistance depends upon the motor core internal temperature T_c and can be approximated with the following linear model:

$$R_e = K_{re1}T_c + K_{re0} \qquad (5)$$

where K_{re1} and K_{re0} are the coefficients of the linear model. This model is identified by randomly choosing currents and by measuring the motor core temperature together with the potential applied to the two motor axis. The motor is cooled down by the enveloping structure in which a refrigerant liquid is circulating. For Schaft, it is a Maxon motor at 200 W which has been modified as depicted in Fig. 3.

For powerful motions it is necessary to provide high current. As it is not possible for batteries without damaging them, a solution is to couple batteries with super capacitors to provide the peak currents. However the size (see for instance [66], 390 × 194 × 112 mm) might be a strong disadvantage. Another consequence is the need for a communication bus which is resilient to the electro-magnetic field generated by the high intensity (150 A [66]). Finally, the speed reached by this kind of motors is stressful for the reductors, thus this significantly reduces the HarmonicDrive^{TM} lifespan.

Fig. 3 Principle of the
cooling system for the AC
motor used in the S-One
robot

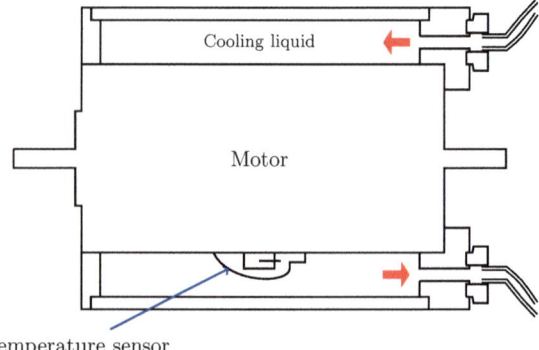

3.3 Hydraulic and Pneumatic Actuators

3.3.1 Hydraulic Actuators

It exits several humanoid robots using hydraulic systems such as DB and CB developed by SARCOS [9]. The most well known is the ATLAS robot from Boston Dynamics. The main advantages with hydraulic actuators are their power and the possibility to perform force control. The price to pay is usually the size of the pump which might be a problem for autonomy [1]. Moreover the necessity to have a servo-valve for each actuator increases the robot weight. Finally the tubes and the connectors which are transporting the liquid to the piston can led to leakage problems. Recently such limitations have been avoided by reducing the pump size and their integration in the robot itself as it is the case for ATLAS. However the noise produced by the pump is extremely annoying and involves wearing a hearing protective helmet. Another approach to avoid such problems are the electro-hydrostatic actuators (EHA) designed for robotics. The EHA are an answer to both performances and size issues. In [1] a system has been developed including micropump, micro-valves, a tank and a passive distributor in a size of $8 \times 4 \times 4$ cm. This actuator allows to pull weights of 25 kg with a speed of 2 cm/s. If there is no reference on this subject, it is probable that such technology is used for the ASIMO hands presented in 2011. Similar works have been realized for the hands of the humanoid robots ARMAR (using air) [35]. More recently a reversible EHA including a torque sensor has been proposed at Tokyo University [29].

3.3.2 Pneumatic Actuators

The most common pneumatic actuator is the Mac Kibben muscle [62]. It consists in an air chamber inside a highly resistant fabric sheath crimped at both extremities of the chamber. When this chamber inflates, the highly resistant sheath is in contact, and is contracting the muscle. This is producing a traction force related to the air pressure put in the air chamber. It is possible to use models developed in the pneumatic industry to model the relationship between the pressure and the generated force. However friction resulting from the interaction of the sheath and the air chamber is introduces non linear phenomenon which are making the control of pneumatic actuator difficult. From the practical viewpoint, air dissipation is easier to handle than hydraulic one. However, inflating such actuators imposes two practical limits: a problem of embeddability, crucial for humanoids, and the inflating speed. This last point does not provide the necessary reactivity to handle strong perturbations [48, 67].

3.4   Cable-Driven Actuators

Cable-driven actuators can be reversible and are controlled using torques. It is also possible to have very integrated actuators, as for instance in the case of iCub which has hands with 9 degrees of freedom each. Such technology is also used in the flexible hands of the DLR humanoid robot (TORO) [13]. An important practical default is the weakness of the cables which do not handle well over tension, therefore finding the proper size for the cable is important when the actuator is submitted to strong forces. This is particularly crucial for walking. One possible solution is to introduce compliance in the actuator. Tokyo University used cable driven actuators with variable impedance by introducing non linear springs in their robot Kojiro [38]. Practically, controlling a robot with compliance is more complex than a purely rigid robot. A team from CEA demonstrated the feasibility of an exoskeleton powered with cable screw [19].

3.5   Variable Stiffness Actuators

Variable stiffness actuators (VSA) are based on a control paradigm where the deformable part of a mechanism is desired or controlled [18, 63]. This deformable part is generally introduced through a spring (Fig. 4) which is absorbing undesired forces, and stores/releases energy at the right time. When the spring parameters are

Fig. 4 Serial elastic actuator

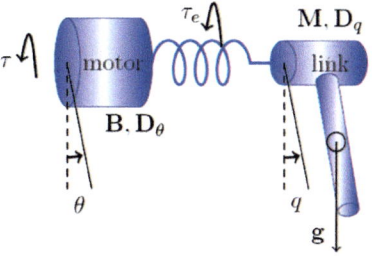

fixed the actuator is called a Serial-Elastic Actuator (SEA). The COMAN robot from
IIT is built on this principle [65]. It can absorb external impacts. The problem lies
in the structure resonance and the stiffness which must be adapted to various phases
such as walking/manipulation. Recent mechanisms (CompAct-VSA) have been proposed to modify the actuator stiffness such that the impedance varies along a motion,
which is particularly useful for explosive motions such as hammering a nail. Stiffness is low at the start of the motion, and it is strong at the impact to maximize the
force transmitted from the hammer to the nail. An equivalent principle is used on a
robot, but in a reverse way, where using a passive toe-joint, the robot stores energy
at landing and reuses it for taking off and performing hopping [28]. In general, in
addition to the SEA, there are two classes of VSA: the antagonistic VSA (Fig. 5),
and VSA with a serial spring (Fig. 6). For the SEA, the actuator dynamics can be
written as follows:

$$\mathbf{M}\ddot{q} + \mathbf{C}(q, \dot{q})\dot{q} + \mathbf{D}_q\dot{q} + \tau_e(\phi) + \mathbf{g}(q) = \tau_{ext}$$
$$\mathbf{B}\ddot{\theta} + \mathbf{D}_\theta\dot{\theta} - \tau_e(\phi) = \tau \tag{6}$$

where $\phi = q - \theta$, B and M are respectively the inertia matrices of the robot bodies
and motors, q and θ are the motor angular positions and the body general coordinates,
D_q and D_θ are respectively the damping of the motor and the body, $\mathbf{C}(q, \dot{q})\dot{q}$ are the

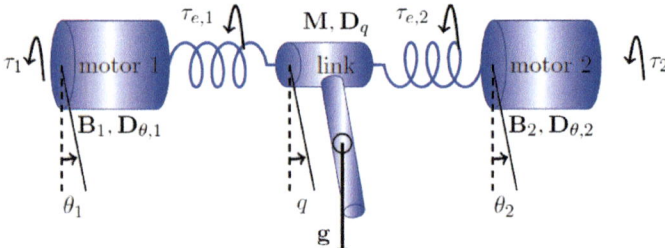

Fig. 5 Antagonist actuator

Fig. 6 Variable stiffness
actuator

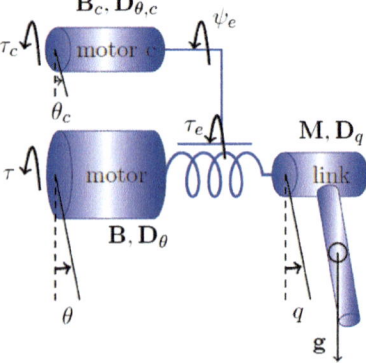

Centrifugal and Coriolis terms, τ_e and τ_{ext} are respectively the torque on the spring and the torque on the actuator while τ is the motor torque.

In the case of the Antagonist VSA, the actuator dynamics is written in the following manner:

$$\mathbf{M}\ddot{q} + \mathbf{C}(q, \dot{q})\dot{q} + \mathbf{D}_q\dot{q} + =$$
$$(\tau_{e,1}(\phi_1) + \tau_{e,2}(\phi_2)) + \mathbf{g}(q) = \tau_{ext}$$
$$\mathbf{B}_1\ddot{\theta}_1 + \mathbf{D}_{\theta,1}\dot{\theta}_1 - \tau_{e,1}(\phi) = \tau_1 \tag{7}$$
$$\mathbf{B}_2\ddot{\theta}_2 + \mathbf{D}_{\theta,2}\dot{\theta}_2 - \tau_{e,2}(\phi) = \tau_2$$

where $\phi_1 = q - \theta_1$, $\phi_2 = q - \theta_2$, θ_i is the motor angular position i, $\tau_{e,i}$ is the torque on spring i and τ_i is the torque of motor i. Finally, it exists a certain number of actuators which have a specific motor to modify the stiffness. This is particularly useful for solving integration problems. Such actuators are called serial variable impedance actuators. However instead of having a simple relationship between two motors acting on the same body, we need to consider the following non linear relationship:

$$\mathbf{M}\ddot{q} + \mathbf{C}(q, \dot{q})\dot{q} + \mathbf{D}_q\dot{q} + \tau_e(\theta_c, \phi) + \mathbf{g}(q) = \tau_{ext}$$
$$\mathbf{B}\ddot{\theta} + \mathbf{D}_\theta\dot{\theta} - \tau_e(\theta_c, \phi) = \tau \tag{8}$$
$$\mathbf{B}_c\ddot{\theta}_c + \mathbf{D}_{\theta,c}\dot{\theta}_c - \psi_e(\theta_c, \phi) = \tau_c$$

where $\phi = q - \theta$, $\mathbf{B}_c$ and $\mathbf{D}_{\theta,c}$ are respectively the auxiliary motor inertia and the the auxiliary motor damping and ψ_e is the torque applied to the mechanism which modifies the spring stiffness. Several humanoid robots have been designed with tendons which can be considered as SEA (robots Ecce [42], Kotaro [45]), with an antagonist [23], or a serial variable impedance actuator (robot Kenshiro [45]).

For robots with variable stiffness, two points need to be tackled: how to find the desired stiffness? and how to control this stiffness? The stiffness at a point on a structure is usually defined mathematically as the ratio between the force and the deflection at the point considered. For actuators with variable stiffness, it is defined as:

$$\kappa = \frac{\partial \tau}{\partial q} \tag{9}$$

However, the stiffness is not a quantity which is directly measurable. Thus it must be computed through an observer. The observation quality depends on the knowledge of a certain number of physical parameters which can be difficult to obtain from a robot provider. This may imply complex identification processes which do not always provide a sufficient precision. For this reason, numerous approaches try to create approximations or probabilistic representations of this quantity. Figure 7 illustrates the VSA developed by the IIT for the humanoid robot COMAN. It has a motor changing the stiffness of the actuator by changing the position of the pivot P around which a motion is realized. External load applied at point A is creating a torque τ_{ext} which is moving the axis by an angle ϕ. The springs (blue spirals) of stiffness k_s generates a force along their axis, perpendicular to PB. The projection of this force

Fig. 7 The actuator called
CompactVSA

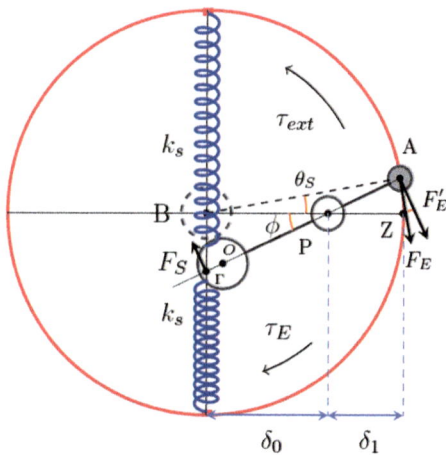

along the axis PO is compensated by the constraint which prohibits any motion of
P. The orthogonal projection F_S is applied at Γ and is expressed in the following
manner:

$$F_s = 2k_s \overline{PO} sin\phi cos\phi \tag{10}$$

The term $sin\phi$ corresponds to the spring deflection and $cos\phi$ to its projection on the
axis (PO). Due to the lever mechanism, a force F'_E is applied on (A) such that:

$$F'_E = F_s \frac{\overline{P\Gamma}}{\overline{AP}} \tag{11}$$

By making a series of derivations and simplifying assumptions described in details
in [64], it is possible to obtain the following dynamical model:

$$\begin{aligned}
I\ddot{q} + N\dot{q} + \tau_E &= \tau_{ext} \\
B_1\ddot{\theta}_1 + \phi_1\dot{\theta}_1 - \tau_E &= u_1 \\
B_2\ddot{\theta}_2 + \phi_2\dot{\theta}_2 + \tau_R &= u_2
\end{aligned} \tag{12}$$

with I the segment inertia, N the Centrifugal and Coriolis effects, B_1 and B_2 the
two matrices inertia matrices, ϕ_1 and ϕ_2 the Centrifugal and Coriolis effects on the
motors. u_1 and u_2 the motor torques. τ_e, the elastic torque, is given by :

$$\tau_e = \frac{2k_s \delta_1^2 \Delta^2 \theta_s}{(\Delta - \delta_1)^2} \tag{13}$$

with $\Delta = \delta_0 + \delta_1$, and the resisting torque at motor M_2 is given by :

$$\tau_R = \frac{2k_s n^2 \theta_2 \theta_s^2 \Delta^3}{(\Delta - n\theta_2)^3} \tag{14}$$

Fig. 8 Stiffness of the
CompactVSA actuator with
$\Delta = \delta_0 + \delta_1 = 1, k_s = 1$.
The values are here
normalized. The reader
interested by the real values
of the system is invited to
read [64]

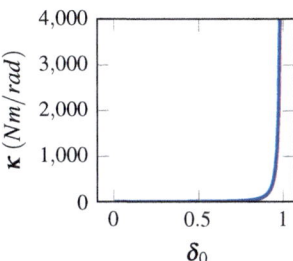

The elastic torque is the torque applied by the spring which is opposed the torque load. The resisting torque to motor M_2 is the torque resulting from the pivot point displacement. Thus, there is a non linear coupling between the two motors through the variables τ_1 and θ_s. This actuator can be controlled by using a LQR scheme [53]. VSA are generally bulky due to the varying stiffness mechanism. Here, the stiffness variation is obtained thanks to the modification of the pivot P. Contrary to other structures, this one is energetically more efficient [7]. However from the energy transmitted to the system, a part is dissipated. Indeed, the projection of force F_S orthogonal to the springs does not deform them but apply a constraint on the actuator mechanism. Having small springs does not allow a strong deflection and limit practically the stiffness variation. This limitation is compensated by the non linearity introduced by the pivot position modification which leads to the following expression of the stiffness:

$$\kappa = \frac{2k_s\delta_0^2\Delta^2}{(\Delta - \delta_0)^2} \tag{15}$$

Here the stiffness variation depends only on the distance between the spring and the pivot point δ_0. Figure 8 represents the stiffness range that can be reached by the system when δ_0 is increased with respect to Δ and for the actuator described in [64].

4 Sensors

4.1 Encoders

Encoders are an essential part to evaluate the current robot state. A very good precision is necessary, specifically for the legs, in order to prevent impact during foot landing. With an electric actuator, the encoder is generally incremental and is fixed on the motor schaft. Theoretically this enables high precision of the joint displacement thanks to the reduction ratio. Typically the theoretical articular precision of a robot such as HRP-2 is in the order of $1/1000$ of degrees. The position of the body to which the joint is connected is computed according to the relationship between the encoder and the motor. Errors are introduced by the elasticity of the Harmonic

Drive, the body flexibility and the mechanical backlash. More and more robots, such as REEM-C, ARMAR or iCub, now have a direct reading of the articular position. This permits to measure the deflection due to the actuation. Having both position before and after the actuation provides damage detection without going through a verification procedure. However, on the joint side, the encoder has to be of sufficient precision to read the values. This is important as the compliance of the actuator between the encoder on the motor side and the encoder on the joint side may enter in resonance leading to unstable oscillations. For this reason, the LOLA robot having the encoders on both side implements two types of control [41].

4.2 Force Torque Sensors

In order to maintain balance, it is very important to measure interaction forces with the environment. For this reason, humanoid robots have 6 axis force sensors below the ankles and the wrists. Using force sensors it is possible to compute the position p of the CoP in the following way:

$$p = \frac{\sum_{i=l,r} p_i f_{i,z}}{\sum_{i=l,r} f_{i,z}}$$

with p_i the position of the vertical force sensor at the ankle i (left or right). This measurement is noted $f_{i,z}$. The difficulty to integrate the sensors is the trade off between the precision and the robustness to impacts during the flying foot landing. The HRP-2 robot uses forces protected by bush rubbers. The robot H7, has gauges put at 8 support points in order to reconstruct the full external wrench [49]. The HUBO robot uses gauges put as a Maltese Cross to provide force measurements along three axis. The robot HRP-2 from DHRC [25] was equipped with a matrix of pressure of size 32×32, with each pressure sensor of dimension 4.2×7 mm under the feet. The measurements are realized at a frequency of 1 KHz. Each sensor can measure a vertical force between 0 and 20 N. The matrix can cover completely the foot of size 135×228 mm. Thanks to carbon composites, its thickness is not above 6 mm. This matrix is also able to localize more precisely the contact points with the environment and to adapt the foot pose. However the sensor submitted to strong forces (typically 600–900 N) does not resist longer than few weeks. Therefore the most common configuration is to have force sensors between the ankle and the foot soil. For instance, a sensor commercially available is the KMSi from the company IPR.

4.3 Joint Level Torque Sensors

Torque sensors measure torque applied to a joint at the output of the actuator. The goal is to implement a control law using inverse dynamics from the robot model. Several formulations exist, the most known is the Operational Space Inverse Dynamics proposed by Sentis et al. [57]. This architecture enable to implement behaviors where the robot is compliant against external contacts. This is a very important property for human-robot interaction. From a practical viewpoint, this means putting a torque sensor at each joint. This can be costly for humanoid robots (typically 26 sensors for a robot with 6 DoFs at each legs and 7 DoFs at each arm). Another solution to make the robot compliant is to add 6-axis force sensors at the hips and at the shoulders. In conjunction with a skin (cf. Sect. 4.5) this provides the contact position, while the 6-axis force sensor provides the wrench resulting from the force applied to the limbs. In case of multi-contacts, not all the amplitudes of the applied forces can be reconstructed, but this is an interesting trade-off for human-robot interaction.

4.4 Accelerometers and Gyrometers

In general a humanoid robot has at least one Inertial Measurement Unit (IMU) which consists in an accelerometer and a gyrometer. This IMU can be localized in the chest or the trunk. Together with a Kalman Filter it gives the orientation of the robot with respect to the gravity field. However without additional information, the orientation of the robot around the vertical axis can not be found. If this information is not very important for walking on flat floor, it is however very important on uneven terrain. A magnometer could provide sufficient information to recover the yaw but electric actuators are generating electromagnetic fields which are perturbating the measurement. Moreoever magnometers do not work very nicely in buildings due to their low resolution and numerous perturbations. The most common solution is to fuse inertial information with vision.

4.5 Artificial Skins

It exists several prototypes of artificial skins covering humanoid robots [2, 10, 14]. They exploit various physical modalities but the most commonly found are forces and pressure provided by capacitive touch sensors and piezo-electric. They are usually used to detect conductive materials such as human skin for human robot interaction. But, with piezo-electric sensor, it is also possible to add a cover which is transmitting current and then act as a force sensor. The goal is then to localize and estimate the forces applied on certain spots of the robot to achieve whole body manipulation. One of the best example is the control of the iCub robot interacting with humans while performing a Tai-Chi position [50, 52]. A light emitter coupled with a receptor can be used to detect approaching objects. This can be used to prevent collision [14]. Using Micro-Electro-Mechanical Systems (MEMS) gives the possibility to measure force orientation, shear forces, vibrations and their duration. Finally adding thermal sensors provides temperature. One of the major blocking point with skin is the necessity to have efficient communication buses due to the large number of sensors needed to cover the surface of the robot. The most advanced techniques are event driven and thus send information only when a contact is detected. There is also a mechanical aspect to this integration, as the cable needed for the communication buses need to be taken into account. A possible solution is to have the skin integrating directly the communication bus which simplifies the problem [6].

4.6 Vision - Lidar

Vision and by extension Lidar provide exteroceptive information, i.e. a representation of the world [15]. In the context of humanoid robotics, it is crucial to plan foot steps in unknown environments. Indeed, detecting potential contact points is necessary for model predictive control often used to make humanoid robot walk. Stereoscopic systems are relatively precises at one meter, which correspond to manipulation applications, but pure triangulation gives an error that is proportional to the square of the distance from this object. By using Simultaneous Localization and Map building (SLAM), it is possible to build more precise maps [16]. A Bayesian formulation provides the integration of various measures related together by models in an uniform mathematical background [24]. A common problem with pure vision is the lack of texture in human made environment. This is mostly important for making the link between interest points in different images. RGB-D cameras projecting a structured light in the infra-red wavelength are an excellent way of fixing this problem and are therefore very popular. When it is possible to embed a graphical processing unit (GPU) it is possible to obtain locally dense maps in real time [47]. Outdoor it is more difficult to use RGB-D because the sun disturbs infra-red readings. It is also possible to use LIDAR or time-of-flight-cameras. But they are often big and are therefore difficult to embed in a humanoid robot. For this reason, the Hokuyo laser [43] is the one most found in humanoid robot perception system. But the time taken to acquire one line is typically 25 ms. For this reason, an interesting technique is to use learning to find traversable zones. The visual description of such area is learned by coupling the images and the scan reading while stopping the robot. When the robot is in motion it uses the visual information to classify the traversable areas [43].

4.7 Audition

Using microphones is clearly needed during human robot interaction. Localization mechanisms and sources separation have been developed to recognize various locutors speaking at the same time with a robot [22] even when it is moving. The problem is complex due to the reverberation of the sounds in the environment, the shape of the robot head, and the noise created by multiple constituants (fans, motors, gears [61]). The goal is to localize the sources and separate them from each others, take into account the impact of the head on the signal propagation and to model the noise from the robot. Current approaches rely on a probabilistic formulation which are computationally involved. Using array of microphones can improve the recognition precision and quality at the cost of increased computational time. A major issue is to develop localization methods providing precise localization in real-time [44]. There already exists methods based on microphones pairs [4] and ones mixing computer vision and audition [12, 20]. Implementations for the NAO robots are available as free softwares [12, 20].

5 Conclusion

In this chapter, various technologies used in humanoid robots have been described. In the last decade several technological breakthroughs resulted in impressive robots such as ATLAS and S-One. The fact that Google bought them in late 2013, to finally sold them to Softbank Robotics in 2017 shows the potential and the limits of such robotics platform. They are now at a level where they can achieve practical task, but the existence of a viable economic application is still an open question. Besides, we can note that TORO, REEM-C or TALOS have been designed in less than 2 years, where most of the first teams in humanoid robotics took much more years to reach the same level. This gives an idea of the progresses achieved in the field recently. The reader will find in the annex a table synthesizing the characteristics of the most relevant humanoid robots in the past years. For high performance humanoid robots which are not designed to interact with humans, a very rigid structure with powerful motors, an IMU and force sensors in the feet are advised. In the case of a humanoid robot aiming at interacting with people, having compliance increases the inherent mechanical safety. It is however coming at the price of a more complex control and of resonance modes specific to the robot which need to be taken into account in the mechanical design (Table 2).

Table 2 This table describes the most relevant humanoid robots in the recent years. When data are available, it gives a glimpse of the characteristics described in this paper: robot name, builder, weight, number of degrees of freedom, actuator types, middleware, communication bus, embedded computers, kinematic structure, foot type, ankles, the proprioceptives and exteroceptives sensors are given. This non-exhaustive list gives however a rather large view of the various technologies given in the humanoid robotics field

Robots	Weight (kg)	Size (m)	Nb of DoFs/Nb of DoFs per limb	Actuators	Middleware	Embedded computer	Communication Bus	Gripper	Mobility	Feet	Force sensors	Encoders	Skin	Vision
iCub (IIT, Italy)	23	0.9	53 (6× legs, 7× arms, 3× hip, 3× neck, 3× eyes, 9× hands)	BLDC	YARP	Remote	PC104-CAN	9 dofs with cable per hands	Biped	Flat feet	ankles and hips (2), wrist and shoulders (2), 6D	Joint and motor sides	Possi-ble	3 dofs for each eye
HRP2 (Kawada, Japan)	58	1.539	30 (6× legs, 6× arms, 2× hip, 2× hands)	DC	Open-RTM	Intel Pentium M 2.8 GHz	PCI	1 parallel DoF per hand	Biped	Flat feet	Ankles (2), Wrists (2) 6D	Motor side	NA	Kinect + Point grey stereo camera
HRP3 (Kawada, Japan)	68	1.606	68 (6× legs, 7× arms, 2× hip, 6× hands)	BLDC	Open-RTM	Intel Pentium M 2.8 GHz	PC104-CAN	6 dofs per hands	Biped	Flat feet	Ankles (2), Wrists (2) 6D	Motor side	NA	CMOS camera
HRP4 (Kawada, Japan)	39	1.51	34 (6× legs, 7× arms, 6× hip, 2× head, 2× hands)	BLDC	Open-RTM	Intel Pentium M 1.6 GHz	PC104-CAN	2 dofs per hands	Biped	Flat feet	Ankles (2), Wrists (2) 6D	Motor side	NA	CMOS camera
HRP4C (Kawada, Japan)	43	1.58	42 (6× legs, 6× arms, 3× hip, 3× neck, 8× head, 2× hands)	NA	Open-RTM	Intel Pentium M 1.6 GHz	PC104-CAN	2 dofs per hands	Biped	Flat feet with toe-joint	Ankles (2) 6D	Motor side	NA	CMOS camera

(continued)

Table 2 (continued)

Robots	Weight (kg)	Size (m)	Nb of DoFs/Nb of DoFs per limb	Actuators	Middle-ware	Embedded computer	Commu-nication Bus	Gripper	Mobility	Feet	Force sensors	Enco-ders	Skin	Vision
HUBO-2 (KAIST, Korea)	56	1.25	41 (6× legs, 5× hands, 3× neck, 6× arms, 6× legs)	DC	NA	NA	CAN	5 dofs per Hands	Biped	Flat feet	Ankles (2) 3D	NA	NA	NA
Robots	Weights	Size	Nb of DOFs/Limbs	Actuators	Middle-ware	Embedded computer	Bus	Gripper	Mobility	Feets	Force sensors	Encoders	Skin	Computer Vision
LOLA (TUM, Germany)	60.88	1.8	25 (2× back, 3× hip, 2× shoulder, 1× elbow, 1× knee, 2× ankles, 1× finger)	BLDC	NA	Intel Core 2 Duo Mobile processor (T7600, 2.33 GHz)	CAN-open/SERCOS III	0 dofs	Biped	Passive absorbers, active toe-joint	Ankles (2) 6D	Motor and Joint side	NA	Two cameras
Walkman (IIT, Italy)	132	1.915	69 (6× legs, 7× arms, 3× waist, 19× hands, 2× head)	BLDC	YARP	i7 quad core x2	EtherCat	19 dofs	Biped	Flat feet	Customized 6D force sensors	Motor and joint side	NA	MultiSense 7 + Hokuyo
Poppy (INRIA, France)	3.5	0.84	25 (2× neck, 5× legs, 4× arms, 5× back)	Servo motor Dyna-mixel	Socket/HTTP access	Arduino	NA	0 dofs	Biped	Passive toe-joint	2 × 8 force sensors	Encoder inside the servomotor	NA	2 cameras

(continued)

Table 2 (continued)

Robots	Weight (kg)	Size (m)	Nb of DoFs/Nb of DoFs per limb	Actuators	Middleware	Embedded computer	Communication Bus	Gripper	Mobility	Feet	Force sensors	Encoders	Skin	Vision
REEM-C (Pal Robotics, Spain)	80	1.65	44 (6× legs, 7× arms, 7× hand, 2× torso, 2× head)	BLDC	ROS/Orocos	Intel Core i7-2710QE CPU@2.1 GHz	CAN	7 dofs (4 passive) per hand	Biped	Flat feet	Ankles (2) 6D	Joint and Motor side	NA	Stereo Camera, camera in the back, Laser
TALOS (Pal Robotics, Spain)	95	1.75	34 (6× legs, 7× arms, 2× torso, 2× head, 2× hands)	BLDC	ROS/Orocos	Intel Core i7 CPU@2.4 Ghz	EtherCat	2 dofs per hands	Biped	Flat feet	Ankles(2), Wrist(2), Torque sensors (26)	Joint and Motor side	NA	Orbec RGB-D
ATLAS (Boston Dynamics, USA)	150	1.8	30 (6× legs, 6× arms, 3× back, 3× neck)	Hydraulic	ROS/Orocos	NA	NA	Several hands can be mounted	Biped	Flat feet Plat	Ankles (2), Force sensors in the body	NA	NA	Stereo Camera, Laser, Cameras in the hands
Escher (ROMELA, USA)	77.5	1.75	(6× legs, 7× arms)	SEA	ROS/Briffs	i7/4 cores/3.2 Ghz/2.4 Ghz	CAN	4 DoFs	Biped	Flat Feet	Ankles (2) 6D ATI-Mini 58	Joint side	No	MultiSense S7, Hokuyo UTM-30LX-EW

References

1. Alfayad, S., Ouezdou, F.B., Namoun, F., Cheng, G.: High performance integrated electro-hydraulic actuator for hydraulics—Part I: principle, prototype design and first experiments. Sensors and Actuators A: Physical **169**, 115–123 (2011)
2. Alirezaei, H., Nagakubo, A., Kuniyoshi, Y.: A highly stretchable tactile distribution sensor for smooth surfaced humanoids. In: IEEE/RAS International Conference on Humanoid Robotics (ICHR) (2007)
3. Ando, M., Kurihara, S., Biggs, G., Sakamoto, T., Nakamoto, H.: Software deployment infrastructure for component based RT-systems. J. Robot. Mechatron. **23**(13), 350–359 (2011)
4. Argentieri, S., Portello, A., Bernard, M., Danés, P., Gas, B. : Binaural systems in robotics. In: The Technology of Binaural Listening. Springer, Berlin (2013)
5. Butterfass, J., Grebenstein, M., Liu, H., Hirzinger, G.: DLR-Hand II: next generation of a dextrous robot hand. In: IEEE/RAS International Confernce on Robotics and Automation (ICRA) (2001)
6. Cannata, G., Maggiali, M., Metta, G., Sandini, G.: An embedded artificial skin for humanoid robots. In: IEEE/RSJ International Conference on Intelligent Robotic Systems (IROS), pp. 434–438 (2008)
7. Carloni, R., Visser, L.C., Stramigioli, S.: Variable stiffness actuators: a port-based power-flow analysis. IEEE Trans. Robot. **28**(1), 1–11 (2012)
8. Carpentier, J., Tonneau, S., Naveau, M., Stasse, O., Mansard, N.: A versatile and efficient pattern generator for generalized legged locomotion. In: IEEE/RAS International Conference on Robotics and Automation (ICRA), pp. 3555–3561 (2016)
9. Cheng, G., Moritomo, J., Hyon, S., Ude, A., Hale, J., Colvin, G., Scroggin, W., Jacobsen, S.: CB: a humanoid research platform for exploring neuroscience. Adv. Robot. **21**(110), 1097–1114 (2007)
10. Dahiya, R., Metta, G., Valle, M., Sandini., G.: Tactile sensing-from humans to humanoids. IEEE Trans. Robot. **26**(11), 1–20 (2010)
11. DARPA: The darpa robotics challenge
12. Deleforge, A., Horaud, R.: The cocktail party robot: Sound source separation and localisation with an active binaural head. In: ACM/IEEE International Conference on Human-Robot Interaction (HRI), pp. 431–438 (2012)
13. Englsberger, J., Werner, A., Ott, C., Henze, B., Roa, M.A., Garofalo, G., Burger, R., Beyer, A., Eiberger, O., Schmid, K., Albu-Schffer, A.: Overview of the torque-controlled humanoid robot toro. In: IEEE/RAS International Conference on Humanoid Robotics (ICHR) (2014)
14. Mittendorfer, P., Cheng, G.: Integrating discrete force cells into multi-modal artificial skin. In IEEE/RAS International Conference on Humanoid Robotics (ICHR) (2012)
15. Fallon, M.F., Marion, P., Deits, R., Whelan, T., Antone, M., McDonald, J., Tedrake, R.: Continuous humanoid locomotion over uneven terrain using stereo fusion. In: IEEE/RAS International Conference on Humanoid Robotics (ICHR), pp. 881–888 (2015)
16. Filliat, D.: Cartographie et localisation simultanes en robotique mobile. In: Techniques de l'inginieur (2014) (ref. S7785)
17. Fitzpatrick, P., Metta, G., Natale, L.: Towards long-lived robot genes. Robot. Auton. Syst. **56**(11), 29–45 (2008)
18. Flacco, F.: Modeling and Control of Robots with Compliant Actuation. Ph.D. thesis, Universiá di Roma, Dipartimento di Ingegneria Informatica (2012)
19. Garrec, P.: Design of an anthropomorphic upper limb exoskeleton actuated by ball screws and cables. Univ. Polytechnique Bucarest Sci. Bull. **72**(12), 23–34 (2010)
20. Horaud, R.: HUMAVIPS project: humanoids with auditory and visual abilities in populated spaces
21. Inaba, M., Kagami, S., Nishiwaki, K.: Robot Anatomy. Iwanami Shoten, Tokyo, p. 309 (2005)
22. Ince, G., Nakadai, K., Rodemann, T., Tsujino, H., Imura, J.: Multi-talker speech recognition under ego-motion noise using missing feature theory. In: IEEE/RSJ International Conference on Intelligent Robotic Systems (IROS) (2010)

23. Jaentsch, M., Wittmeier, S., Dalamagkidis, K., Panos, A., Volkart, F., Knoll, A.: Anthrob—a printed anthropomimetic robot. In IEEE/RAS International Conference on Humanoid Robotics (ICHR) (2013)
24. Kaess, M., Johannsson, H., Roberts, R., Ila, V., Leonard, J., Dellaert, F.: iSAM2: Incremental smoothing and mapping using the bayes tree. Int. J. Robot. Res. **31**(12) (2012)
25. Kagami, S., Nishiwaki, K., Kuffner, J., Thompson, S., Chestnutt, J., Stilman, M., Michel, P.: Humanoid HRP2-DHRC for Autonomous and Interactive Behavior, pp. 103–117. Springer, Berlin (2007)
26. Kajita, S., Asano, F., Morisawa, M., Miura, K., Kaneko, K., Kanehiro, F., Yokoi, K.: Vertical vibration suppression for a position controlled biped robot. In: IEEE/RAS International Conference on Robotics and Automation (ICRA) (2013)
27. Kajita, S., Kaneko, K., Morisawa, M., Nakaoka, S., Hirukawa, H.: ZMP-based biped running enhanced by toe springs. In: IEEE/RAS International Conference on Robotics and Automation (ICRA) (2007)
28. Kajita, S., Nagasaki, T., Kaneko, K., Yokoi, K., Tanie, K.: A running controller of humanoid biped HRP-2LR. In: IEEE/RAS International Conference on Robotics and Automation (ICRA) (2005)
29. Kaminaga, H., Odanaka, K., Ando, Y., Otsuki, S., Nakamura, Y.: Evaluations on contribution of backdrivability and force measurement performance on force sensitivity of actuators. In: IEEE/RSJ International Conference on Intelligent Robotic Systems (IROS) (2013)
30. Kaneko, K., Harada, K., Kanehiro, F.: Development of a multi-fingered hand for life-size humanoid robots. In:IEEE/RAS International Conference on Robotics and Automation (ICRA) (2007)
31. Kaneko, K., Harada, K., Kanehiro, F., Miyamori, G., Akachi, K.: Humanoid robot HRP-3. In: IEEE/RSJ International Conference on Intelligent Robotic Systems (IROS) (2008)
32. Kaneko, K., Kanehiro, F., Morisawa, M., Akachi, K., Miyamori, G., Hayashi, A., Kanehira, N.: Humanoid robot HRP-4 - humanoid robotics platform with lightweight and slim body. In: IEEE/RSJ International Conference on Intelligent Robotic Systems (IROS) (2011)
33. Kaneko, K., Kanehiro, F., Morisawa, M., Tsuji, T., Miura, K., Nakaoka, S., Kajita, S., Yokoi, K.: Hardware improvement of cybernetic human HRP-4C for entertainment use. In: IEEE/RSJ International Conference on Intelligent Robotic Systems (IROS) (2011)
34. Kaneko, K., Morisawa, M., Kajita, S., Nakaoka, S., Sakaguchi, T., Cisneros, R., Kanehiro, F.: Humanoid robot HRP-2 Kai - improvement of HRP-2 towards disaster response tasks. In: IEEE/RAS International Conference on Humanoid Robotics (ICHR), pp. 132–139 (2015)
35. Kargov, A., Asfour, T., Pylatiuk, C., Oberle, R., Klosek, H., Schulz, S., Regenstein, K., Bretthauer, G., Dillmann, R.: Development of an anthropomorphic hand for a mobile assistive robot. In: International Conference on Rehabilitation Robotics: Frontiers of the Human-Machine Interface (2005)
36. Karumanchi, S., Edelberg, K., Baldwin, I., Nash, J., Reid, J., Bergh, C., Leichty, J., Carpentier, K., Shekels, M., Gildner, M., Newill-Smith, D., Carlton, J., Koehler, J., Dobreva, T., Frost, M., Hebert, P., Borders, J., Ma, J., Douillard, B., Backes, P., Kennedy, B., Satzinger, B., Lau, C., Byl, K., Shankar, K., Burdick, J.: Team robosimian: semi-autonomous mobile manipulation at the 2015 DARPA robotics challenge finals. J. Field Robot. **34**(2), 305–332 (2017)
37. Koenemann, J., Del Prete, A., Tassa, Y., Todorov, E., Stasse, O., Bennewitz, M., Mansard, N.: Whole-body model-predictive control applied to the HRP-2 humanoid. In: IEEE/RSJ International Conference on Intelligent Robotic Systems (IROS), pp. 3346–3351 (2015)
38. Kozuki, T., Motegi, Y., Shirai, T., Asano, Y., Urata, J., Nakanishi, Y., Okada, K., Inaba, M.: Design of upper limb by adhesion of muscles and bones—detail human mimetic muscoloskeletal humanoid kenshiro . In: IEEE/RSJ International Conference on Intelligent Robotic Systems (IROS) (2013)
39. Lapeyre, M., Rouanet, P., Oudeyer, P.-Y.: Poppy humanoid platform: experimental evaluation of the role of a bio-inspired thigh shape. In: IEEE/RAS International Conference on Humanoid Robotics (ICHR) (2013)

40. Lim, J., Lee, I., Shim, I., Jung, H., Joe, H.M., Bae, H., Sim, O., Oh, J., Jung, T., Shin, S., Joo, K., Kim, M., Lee, K., Bok, Y., Choi, D.G., Cho, B., Kim, S., Heo, J., Kim, I., Lee, J., Kwon, I.S., Oh, J.H.: Robot system of DRC-HUBO+ and control strategy of team KAIST in DARPA robotics challenge finals. J. Field Robot. **34**(4), 802–829 (2017)
41. Lohmeier, S.: Design and Realization of a Humanoid Robot for Fast and Autonomous Bipedal Locomotion. Ph.D. thesis, Technische Universitat Munchen (2010)
42. Schmaler, C., Jaentsch, M., Wittmeier, S., Dalamagkidis, K., Knoll, A.: A scalable joint-space controller for musculoskeletal robots with spherical joints. In: ROBIO (2011)
43. Maier, D., Stachniss, C., Bennewitz, M.: Vision-based humanoid navigation using self-supervised obstacle detection. Int. J. Humanoid Robot. **10**(12) (2013)
44. Nakamura, K., Gomez, R., Nakadai, K.: Real-time super-resolution three-dimensional sound source localization for robots. In: IEEE/RSJ International Conference on Intelligent Robotic Systems (IROS), pp. 3949–3954 (2013)
45. Nakanishi, Y., Ohta, S., Shirai, T., Asano, Y., Kozuki, T., Kakehashi, Y., Mizoguchi, H., Kuro-tobi, T., Motegi, Y., Sasabuchi, K., Urata, J., Okada, K., Mizuuchi, I., Inaba, M.: Design approach of biologically-inspired musculoskeletal humanoids. Int. J. Adv. Rob. Syst. (2013)
46. Nakaoka, S., Hattori, S., Kanehiro, F., Kajita, S., Hirukawa, H.: Constraint-based dynamics simulator for humanoid robots with shock absorbing mechanisms. In: IEEE/RSJ International Conference on Intelligent Robotic Systems (IROS) (2007)
47. Newcombe, R., Izadi, S., Hilliges, O., Molyneaux, D., Kim, D., Davison, A., Kohli, P., Shotton, J., Hodges, S., Fitzgibbon, A.: Kinectfusion: Real-time dense surface mapping and tracking. In: ISMAR (2011)
48. Niiyama, R., Nagakubo, A., Kuniyoshi, Y.: Mowgli: a bipedal jumping and landing robot with an artificial musculoskeletal system. In: IEEE/RAS International Conference on Robotics and Automation (ICRA) (2007)
49. Nishiwaki, K., Murakami, Y., Kagami, S., Kuniyoshi, Y., Inaba, M., Inoue, H.: A six-axis force sensor with parallel support mechanism to measure the ground reaction force of humanoid robot. In: IEEE/RAS International Conference on Robotics and Automation (ICRA) (2002)
50. Nori, F., Traversaro, S., Eljaik, J., Romano, F., Del Prete, A., Pucci, D.: iCub whole-body control through force regulation on rigid non-coplanar contacts. Front. Robot. AI **2**, 6 (2015)
51. Ott, C., Roa, M., Hirzinger, G.: Posture and balance control for biped robots based on contact force optimization. In: IEEE/RAS International Conference on Humanoid Robotics (ICHR) (2011)
52. Pucci, D.: iCub performing highly dynamic tai chi while interacting with humans (2016)
53. Sardellitti, I., Medrano-Cedra, G., Tsagarakis, N., Jafari, A., Caldwell, D.: Gain scheduling control for a class of variable stiffness actuators based on lever mechanisms. ITRO **29**(13), 791–798 (2013)
54. Saunders, J., Inman, V., Eberhart, H.: The major determinants in normal and pathological gait. J. Bone Joint Surg. **A**(135), 543–558 (1953)
55. Schmitz, A., Pattacini, U., Nori, F., Natale, L., Metta, G., Sandini, G.: Design, realization and sensorization of the dexterous iCub hand. In: IEEE/RAS International Conference on Humanoid Robotics (ICHR), pp. 186–191 (2010)
56. Sellaouti, R., Stasse, O., Kajita, S., Yokoi, K., Kheddar, A.: Faster and smoother walking of humanoid HRP-2 with passive toe joints. In: IEEE/RSJ International Conference on Intelligent Robotic Systems (IROS) (2006)
57. Sentis, L., Khatib, O.: A whole-body control framework for humanoids operating in human environments. In: IEEE/RAS International Conference on Robotics and Automation (ICRA) (2006)
58. Stasse, O.: Technologies des robots humanoïdes. In: Applications en robotique. Techniques de l'ingnieur (2014)
59. Stasse, O., Flayols, T., Budhiraja, R., Giraud-Esclasse, K., Carpentier, J., Del Prete, A., Souères, P., Mansard, N., Lamiraux, F., Laumond, J.-P., Marchionni, L., Tome, H., Ferro, F.: A new humanoid research platform targeted for industrial applications, TALOS (2017)

60. Stentz, A., Herman, H., Kelly, A., Meyhofer, E., Clark Haynes, G., Stager, D., Zajac, B., Bagnell, J.A., Brindza, J., Dellin, C., George, M., Gonzalez-Mora, J., Hyde, S., Jones, M., Laverne, M., Likhachev, M., Lister, L., Powers, M., Ramos, O., Ray, J., Rice, D., Scheifflee, J., Sidki, R., Srinivasa, S., Strabala, K., Tardif, J.-P., Valois, J.-S., Weghe, J.M.V., Wagner, M., Wellington, C.: Chimp, the CMU highly intelligent mobile platform. J. Field Robot. **3**(2), 209–228 (2015)
61. Takahashi, T., Nakadai, K., Komatani, K., Ogata, T., Okuno, H.G.: Improvement in listening capability for humanoid robot HRP-2. In: IEEE/RAS International Conference on Robotics and Automation (ICRA), pp. 470–475 (2010)
62. Tondu, B.: Modelling of the mackibben artifical muscle: a review. J. Intell. Mater. Syst. Struct. **23**(13), 225–253 (2012)
63. Tonietti, G., Schiavi, R., Bicchi, A.: Design and control of a variable stiffness actuator for safe and fast physical human/robot interaction. In: IEEE/RAS International Conference on Robotics and Automation (ICRA), pp. 526–531 (2005)
64. Tsagarakis, N., Sardellitti, I., Caldwell, D.: A new variable stiffness actuator (CompAct-VSA): design and modelling. In: IEEE/RSJ International Conference on Intelligent Robotic Systems (IROS) (2011)
65. Tsagarakis, N., Morfey, S., Cerda, G., Li, Z., Caldwell, D.: Compliant humanoid COMAN: optimal joint stiffness tuning for modal frequency control (2013)
66. Urata, J., Nakanishi, Y., Okada, K., Inaba, M.: Design of high torque and high speed leg module for high power humanoid. In: IEEE/RSJ International Conference on Intelligent Robotic Systems (IROS) (2010)
67. Vermeulen, J., Verrelst, B., Vanderborght, B., Lefeber, D., Guillaume, P.: Trajectory planning for the walking biped "lucy". Int. J. Robot. Res. **25**(19), 867–887 (2006)
68. Wensing, P.M., Wang, A., Seok, S., Otten, D., Lang, J., Kim, S.: Proprioceptive actuator design in the MIT cheetah: impact mitigation and high-bandwidth physical interaction for dynamic legged robots. IEEE Trans. Robot. **33**(3), 509–522 (2017)

Printed by Printforce, the Netherlands